MULTI-DIMENSIONAL IMAGING WITH SYNTHETIC APERTURE RADAR

MULTI-DIMENSIONAL IMAGING WITH SYNTHETIC APERTURE RADAR

GIANFRANCO FORNARO

GILDA SCHIRINZI

ANTONIO PAUCIULLO

VITO PASCAZIO

DIEGO REALE

ELSEVIER

ACADEMIC PRESS
An imprint of Elsevier

ISBN: 978-0-12-821655-2

For information on all Academic Press publications
visit our website at https://www.elsevier.com/books-and-journals

Publisher: Mara Conner
Acquisitions Editor: Craig Smith
Editorial Project Manager: Howi De Ramos
Production Project Manager: Erragounta Saibabu Rao
Cover Designer: Mark Rogers

Typeset by VTeX

Working together
to grow libraries in
developing countries

ELSEVIER   Book Aid International

www.elsevier.com • www.bookaid.org

Contents

Authors' biographies

Gianfranco Fornaro received the M.S. degree (summa cum laude) in electronic engineering from the University of Naples "Federico II" in 1992 and the Ph.D. in 1997. Since 1993, he has been with IREA-CNR, where he now holds the position of Research Director, working in the area of airborne and spaceborne Synthetic Aperture Radar (SAR) processing, including SAR Interferometry and SAR Tomography. In 2013, he received the "Full Professor" habilitation in Telecommunication, and in this area, he has been Adjunct Professor at several Universities in South Italy. Dr. Fornaro has been a visiting scientist at Politecnico of Milan and DLR in Oberpfaffenhofen (Germany), also during the 1996 SIR-C/X-SAR mission. He was a NATO lecturer in the Lecture Series SET 191 and SET 235, and since 2011, he has been also a lecturer at the International Summer School on Radar/SAR of the Fraunhofer Institute. He has been a convener, tutorial lecturer, chairman, and member of the program and organizing committee at the most important IEEE conferences. He has authored more than two hundred papers on SAR (peer-review journals and proceedings of international conferences). He received the Mountbatten Premium from the IEE Society in 1997, the 2011 IEEE Geoscience and Remote Sensing Letters Best Paper award, and the 2011 best Reviewers mention of the IEEE Transactions on Geoscience and Remote Sensing journal.

Gilda Schirinzi graduated summa cum laude in Electronic Engineering in 1983 at University of Naples "Federico II". In the same year, she joined the Electronic Engineering Department as a research fellow. From 1985 to 1986 she was at European Space Agency, ESTEC, The Netherlands. In 1988, she joined the Istituto di Ricerca per l'Elettromagnetismo e i Componenti Elettronici (IRECE) of the Italian National Council of Research (CNR), in Naples. In 1992, she was appointed Head of the Electromagnetics Division of IRECE, and in 1997, she became Senior Researcher. In November 1998, she joined the University of Cassino as Associate Professor of Telecommunications. In 2005, she became Full Professor. Since November 2008, she has been at the University of Naples "Parthenope" in the Telecommunication group. She has taught Signal Theory, Electrical Communications, Microwave Remote Sensing Systems, and Image Processing and Coding. Her main scientific interests are in the field of Synthetic Aperture Radar (SAR) signal processing and coding, SAR interferometry and tomography, microwave imaging techniques, and image and signal processing for remote sensing applications. Gilda Schirinzi published more than 200 technical papers, and she is Senior Member of IEEE.

Antonio Pauciullo received the Dr. Eng. with honors in 1998 and the Ph.D. in information engineering in 2003, both from the University of Naples Federico II, Naples, Italy. Since 2001, he has been with the Institute for Electromagnetic Sensing of the Environment (IREA) of the Italian National Research Council (CNR), where he holds a position of Senior Researcher. From 2004 to 2012 he was an Adjunct Professor of digital signal processing at the University of Cassino (Italy). His current research interests are in the field of statistical signal processing with emphasis on synthetic aperture radar processing.

Vito Pascazio graduated (cum laude) in 1986 at the University of Bari, Italy, in Electronic Engineering, and received in 1990 Ph.D. in Electronic Engineering and Computer Science from the Department of Electronic Engineering of the University of Napoli "Federico II", Italy. In 1990, he was first at the Research Institute on Electromagnetics and Electronic Devices (IRECE) of the Italian National Council of Research (CNR), Napoli, Italy, and then he joined the Università di Napoli "Parthenope", Italy, where he is presently Full Professor. Vito Pascazio was also the Chair of the Department of Engineering at the Università di Napoli Parthenope and Director of the National Laboratory of Multimedia Communication of National Consortium Inter-Universitary of Telecommunication (CNIT). His main research interests are in the field of Synthetic Aperture Radar (SAR) Image Processing, including SAR Interferometry, SAR Tomography, SAR Along-Track Interferometry, and Ground Based SAR; Microwave Tomographic Image Reconstruction, including Ground Penetrating Radars, Through the Wall Imaging, and De-mining radars, Body Scanning, and Microwave Biomedical Imaging; Biomedical Image processing, including Magnetic Resonance Imaging; Image Processing, including Image Compression, Speckle Filtering, Classification, Segmentation, Wavelets-based Image Processing, and Compressive Sensing; Linear and Nonlinear Statistical Signal Processing, including Markov Random Field Bayesian Image Processing. He served as a member in the Technical Committees of several international conferences. He was Co-Chair General of the IGARSS 2015 conference in Milan and holds the same role in the next IGARSS 2024 in Athens. Vito Pascazio published about 250 technical papers, and he is Senior Member of IEEE.

Diego Reale received M.S. in telecommunication engineering from the University of Cassino, Cassino, Italy, in 2007 and Ph.D. in information engineering from the University of Naples "Parthenope," Naples, Italy, in 2011. He is currently Senior Researcher at the Institute for the Electromagnetic Sensing of the Environment of the Italian National Research Council (IREA-CNR). His main research interests are framed in synthetic aperture radar (SAR) processing, with particular reference to SAR tomography, SAR interferometry, and differential SAR interferometry. His main research interests include the development and application of SAR tomography on very high-resolution SAR data for the monitoring of the built environment and critical infrastructures. Dr. Reale has been awarded at the Joint Urban Remote Sensing Event 2011 Student Competition. In 2012, his paper "Tomographic Imaging and Monitoring of Buildings with Very High-Resolution Data" was awarded as the 2011 IEEE Geoscience and Remote Sensing Letter Best Paper.

Introduction

1.1 Brief history of radar and SAR development

At the end of the 19th century the experiments conducted first by Heinrich Hertz and after by Guglielmo Marconi, highlighted the reflection of electromagnetic waves by metallic objects, an effect predicted some decades before by James Clark Maxwell in its theoretical dissertation of electromagnetism.

The idea by Nikola Tesla to exploit the reflection of radio waves for the detection of moving targets came into reality in the 20th century. It was in fact thanks to the intuition by the German inventor Christian Hülsmeyer for the application to collision avoidance, that the "telemobiloscope", i.e., the predecessor of modern radar, was patented in 1903. The concept came to the demonstration in his experiment in 1904 in Cologne, when the telemobilscope showed its capability to detect the passage of a barge some hundreds of meters away and ring a bell. However, due to the inadequacy of the high frequency electronic equipment available at that time, it took several decades for radars to make a substantial progress.

With the development related to shortwaves before, and after of microwaves, with generators as the "magnetron" capable to generate short "high-frequency" pulses, the radar experienced a huge development. Modern applications go beyond the traditional detection and tracking scopes related to surveillance, anticollision, assisted guidance, and automotive, but embrace also meteorology, altimetry, ground-penetrating and through-the-wall imaging, and finally the main topic of this book, i.e., remote sensing and Earth observation technologies. The latter have major impact for natural hazard monitoring and security as well as agriculture, forest, and urban mapping.

More specifically the early development of modern radar was carried out secretly for military uses in Germany, England, and USA, and started during and in-between the World Wars periods of the XX century.

Multi-Dimensional Imaging with Synthetic Aperture Radar
https://doi.org/10.1016/B978-0-12-821655-2.00005-X

In 1922, A. Hoyt Taylor, and Leo. C. Young at the Naval Research Laboratory (NRL) in USA demonstrated the possibility to detect the reflections of high-frequency radio waves from a wooden ships crossing a system composed by a transmitter and a portable receiver. Based on an idea suggested by Taylor, Robert Morris Page, who joined the Taylor and Young radar group at NLR, demonstrated in 1934 the possibility to operate detection of objects with a short pulse, thus giving also the possibility of target ranging, i.e., of measuring the distance of a target from the radar. Thank to this experiment, Page, Taylor, and Young are usually credited as developer and demonstrator of the world's first "true radar", i.e. radar as essentially known nowadays.

Still at NRL, in June 1936, the first radar system prototype was demonstrated to government officials: it successfully tracked an aircraft at distances up to 25 miles. Similar progress was achieved also in United Kingdom by the Royal Navy where, in 1938, it was developed by the first air-warning radar system able to operate between 30 and 50 miles. The system was immediately after installed as well as on ships. Almost in the same years, in Germany there was a significant development as well. Mainly handled by private Companies, programs for development of radar systems with detection and ranging capabilities took place. The reported operating range was slightly lower than NRL and Royal Navy ones.

The term "RADAR" was coined in 1940 by the United States Navy as an acronym for "RAdio Detection And Ranging". Radar technology experiences a sharp increase in development during World War II. It was during this period that radars were, for the first time, used as imaging systems by extending its capability to profile the backscattering along the distance (range) for bombsight systems. Ground scanning radar were built after that experiments conducted by Royal Air Force in 1941 revealed distinct returns by different land covers. H2S in 1943 was the first operative ground scanning radar system and exploiting antenna scanning capabilities to discriminate target coordinate across the range, thus producing 2D images. The development in US took place almost one year after with the AN/APQ-7, or Eagle, developed by the US Army Air Force.

Investigations to improve the Radar resolution in terms of pulse duration and sharpening of the scanning beam of the real antenna, that are the basis of modern Synthetic Aperture Radar (SAR), were however carried out in the post World War II. The resolution of a Radar is its ability to distinguish between very close targets. The smaller the distance between the targets, the better (the higher) the resolution. We could say that the resolution of a Radar is the minimum distance that two targets must have in order to be distinguished from each other. This concept will be taken up extensively in what follows.

A viable and simple solution that paved the way to the subsequent SAR development, was the "Side-Looking Radar", i.e. an imaging system with a beam directed off-nadir to the ground and moving solidly with the aircraft, thus providing an additional dimension beyond the distance. The system was thus able to illuminate a strip on the ground and provide 2D resolution capabilities. This system, later called Real Aperture Radar (RAR), was in fact capable to measure the backscattering as a function of the distance (range), across the track, and of the along track position flown by the aircraft (azimuth). Range is discriminated by the pulses, whereas the azimuth is discriminated by the movement of the beam footprint. A demonstration flight of side-looking radar imaging was carried out in 1950 over the Detroit area.

Despite the important achievement to have a simple and flexible imaging system at short/microwaves, compared to systems operating in the visible and near infrared region, radar imaging was strongly hampered by the diffraction limited angular extent, and hence by the spatial resolution, of an aperture antenna. The latter is directly proportional to wavelength and inversely proportional to aperture dimension. Therefore moving from micrometer wavelengths band, associated with the visible spectral region, to the centimeter wavelengths associated with microwaves used in radar systems, the angular aperture increases by a factor roughly of the order of 10^5. As a consequence, with the development of the electronic devices able to generate very short pulse, it was soon recognized that RAR was characterized by a rather unbalanced resolution degree, that is a high range resolution but a poor azimuth resolution.

The problem to increase the azimuth resolution was brilliantly solved by Carl A. Wiley in 1950, who was working on the idea to get high resolution by observing the Doppler shift of stationary ground targets in a moving antenna beam. What was later renamed as Synthetic Aperture Radar (SAR), that is a coherent radar able to discriminate targets along the flight direction by exploiting the Doppler effect, was reported as Doppler Beam Sharpening in a Wiley Goodyear Aircraft report in 1951. Early in 1952 Wiley got the first operational SAR. The patent application "Pulsed Radar Doppler Method and Apparatus" was filed on 13 August 1954, and classified from 1955 to 1964. Several years after, the SAR technology from satellite become of great interest both for military and civil aims. Spaceborne SAR conjugates synoptic view and cloud/sun independent observation features (Fig. 1.1); the latter allows regular and systematic observation capability, thus making

SAR a unique Earth Observation system for emergency use as well as for the observation of dynamic phenomena.

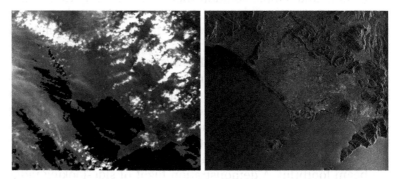

Figure 1.1. Comparison between images collected at the same instant by multispectral (MERIS) (left image) and SAR (ASAR) (right image) sensors onboard Envisat of the European Space Agency (ESA). The images show the capability of SAR to collect data relevant to the Earth surface independently of the presence of clouds.

The first civilian spaceborne imaging radar instrument (SAR) flew on Seasat in 1978 and the whole mission was handled by NASA Jet Propulsion Laboratory (JPL). Then, starting from 1981 NASA/JPL managed a sequence of four Shuttle Imaging Radar (SIR) missions: SIR-A (1981), SIR-B (1984), and two SIR-C/X-SAR missions in April and September 1994. The latter missions were carried out by NASA/JPL in cooperation with the German and Italian Space Agencies, DLR and ASI. The SIR-C/X-SAR mission was dedicated to interferometric experiments: a technology that exploits angular diversity of images acquired with two antennas, or close repeated passes of a single antenna, to reconstruct the Digital Elevation Model (DEM) of the observed scene. The efficacy of such technique for DEM reconstructions was in fact clear since the first demonstration carried out by the exploitation of the data collected by the ERS-1 and ERS-2 satellites, in the framework of the ERS program, managed by the European Space Agency (ESA) and specifically dedicated to SAR. ERS-1 satellite was launched at mid 1991, followed by ERS-2 in 1995. The two satellites were flying in a tandem configuration with a revisit time useful for SAR Interferometry (InSAR) scopes of only one-day. As explained in the book, a revisit of one day allows achieving good interferometric coherence over repeated acquisitions but does not guarantee absence of variation of the conditions of the atmosphere, which acts as a disturbance for InSAR. ERS was a successful program with ERS-2 operating till 2011, and provided data that has been exploited for many years by scientists worldwide to test new approaches and develop the SAR technology for civilian applications.

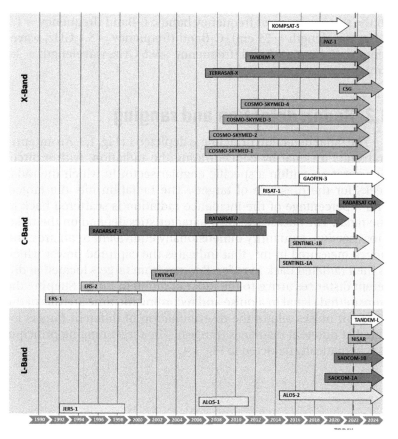

Figure 1.2. Evolution over the time of SAR sensors for the three most used frequency bands.

The NASA/JPL SAR programs exploiting the Shuttle platform culminated with the Shuttle Radar Topography Mission (SRTM). Based on the interferometric experience gained wit SIR-C/X-SAR, NASA/JPL coordinated a program with the participation of DLR and ASI aimed at collecting, at C-Band and X-Band, simultaneous interferometric data with two antennas (at each band) for the generation of what was at that time the most accurate DEM available on a quasiglobal scale. The unique characteristics of SAR operated on satellites, and its application to the natural risk monitoring and security area have been the drivers for the development of what is now considered a "new era" for spaceborne SAR characterized by increasing investments of international Space Agencies, and more recently also by the private companies. Fig. 1.2 shows the development of space programs since 1992, when the first operational sensor for interferometric use (ERS-1 by ESA) was launched, gives a clear idea of the growth of this sector. The sensors are grouped in the three most used frequency bands: L-Band (frequency $\sim$ 1.2 GHz, wavelength $\sim$ 25 cm), C-Band (frequency $\sim$ 5.4 GHz, wavelength $\sim$ 5.6 cm), X-Band (frequency $\sim$ 9.6 GHz, wavelength $\sim$ 3.1 cm).

1.2 Radar: detection and ranging

The radar detection principle is depicted in Fig. 1.3. An antenna transmits a signal by concentrating the radiation, with a directive antenna, within a specific angular sector in which the radar seeks for the presence of targets. The radiation hits the targets and a percentage of the incidence radiation is scattered back to the radar. The backscattering characteristics depend on the radar cross section, a quantity dimensionally equivalent to an area and hence measured in m^2, that indicates the captured power which is then radiated back to radar. Returns from targets located at different distances arrive to the radar at different times. Shaping the transmitted signal to a pulse and recording the time arrivals of the different pulses allows the discrimination of different targets located at different distances (ranges). The target ranging principle is schematically depicted in Fig. 1.4.

Figure 1.3. Target detection principle used in radars.

Consider now the case of two targets present at the same time in the illumination beam of a radar. The radar, assumed to be a pulsed radar, transmits a rectangular pulse that hits, at different times, the two scatterer located at different distances (ranges) r_1

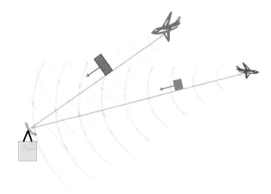

Figure 1.4. Target ranging principle used in radars.

and r_2. The echoes from the target travel with the same velocity toward the radar and they are recorded at different instants, say t_1 and t_2. The propagation velocity of an electromagnetic wave travelling in the air is known and well approximated by the speed of light in the vacuum c; measuring the time arrivals since the transmission allows therefore determining the different target ranges:

$$t_1 = 2r_1/c$$
$$t_2 = 2r_2/c \qquad (1.1)$$

where we have considered the fact that the radiation travels forth and back thus doubling the ranges. The separation in range between the scatterers (range difference) can be evaluated from the time separation $t_2 - t_1$ as follows:

$$r_2 - r_1 = \frac{c}{2}(t_2 - t_1) \qquad (1.2)$$

In order to measure the time separation, we need that the return pulses do not overlap. Therefore, the difference of the distances between the targets and the radar $(r_2 - r_1)$ must be such that the corresponding time separation $(t_2 - t_1)$ given by (1.2) is larger than the pulse duration. Therefore the lower the duration of the transmitted pulses, the lower is the distance at which the scatterers can be discriminated (resolved), that is the higher is the resolution capability of the radar. Key principles of target ranging for surveillance radars, which is the one of the basic mechanism exploited to discriminate scatterers also in radar imaging, is the subject of Chapter 2.

Another characteristic of radars is the possibility to measure the component of the target velocity along the line from the target to the radar, referred to as line of sight or briefly los. The possibility is offered by the effect described in 1842 by Christian Doppler and related to the variation of the frequency in a radiation emitted by a moving source. If the radar transmits a coherent radiation, for instance a pure sinusoidal tone at frequency f_0 then, due to the Doppler effect, the radiation backscattered by a moving target and received back by the radar will be characterized by a frequency shift which is positive or negative depending on the fact that the target is approaching or moving away from the radar. The effect is depicted in Fig. 1.5.

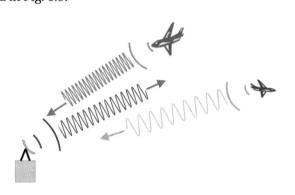

Figure 1.5. Doppler shift for moving targets.

As explained in details in Chapter 2, if the target moves with a velocity component along the range, say $v_r = -\boldsymbol{v} \cdot \hat{\boldsymbol{r}}$ with $\boldsymbol{v}$ being the velocity vector and $\hat{\boldsymbol{r}}$ the range versor which identifies the los. The frequency of the received signal will be $f_0 + f_d$ where the Doppler shift f_d is given by:

$$f_d = \frac{2 f_0}{c} v_r = \frac{2 v_r}{\lambda} \tag{1.3}$$

where λ is the wavelength associated with f_0.

The intensity of the recorded signals provides, in addition, an indication of the characteristics of the target in scattering back the impinging radiation. Such a characteristic of the target is quantified in what is referred to as radar cross section, i.e. a parameter which is discussed in the next section in order to set the radar power budget.

1.3 Radar imaging

Since the early developments of radars, it was clear the possibility to exploit the radar for imaging scopes. This can be achieved thanks to the characteristic of a radar to sense different levels of scattering from targets and the possibility to determine their ranges. Imaging systems are characterized by 2D discrimination capabilities, at least, that is for radars range and across-range. As recalled in the first section of this chapter the first developments of radar imaging systems were associated with the possibility to exploit the across range antenna beam scanning. Actually, the radar beam can be scanned across range either mechanically or electronically to provide, at different (aspect) angles, profiles along the range of the backscattering. But more can be done: a radar can be installed on a moving platform to provide range backscattering profiles at different positions along the flight direction. This is the principle of the side looking radar imaging shown in Fig. 1.6.

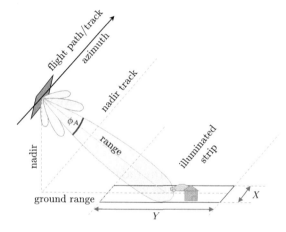

Figure 1.6. Real Aperture Radar (RAR) for side-looking ground imaging.

An antenna is installed on a platform moving along a direction referred to as azimuth and illuminates the scene on the ground at broadside direction (orthogonal to the flight path). Ground targets distributed on the scene are discriminated along the range by using the pulsed mechanism discussed for the classical surveillance radars. The antenna is pointed in the side looking direction because of the need to avoid the left and right ambiguity, related to the situation in which targets symmetrically positioned with respect to the nadir and sharing the same distance provides overlapped responses.

The discrimination of targets along the azimuth is performed by the beam. Targets located at the same range within the beam will provide contributions overlapped in time. Exploiting the relation that the 3 dB angular aperture (beamwidth) of an antenna with size L_a is given by

$$\phi_A \approx \frac{\lambda}{L_a} \tag{1.4}$$

we have that the antenna azimuth footprint is given by

$$X = \phi_A r \approx \frac{\lambda}{L_a} r \tag{1.5}$$

where r is the range of the target, which is the distance between the target and the trajectory.

As explained in Chapter 2 the use of pulse bandwidth (B_t) of the order of hundreds of MHz and the adoption of pulse compression techniques allows easily reaching achieving (slant) range resolutions equal to:

$$\rho_r = \frac{c}{2B_t} \tag{1.6}$$

and therefore of the meter or even submeter order, over a swath of width Y in ground range. The latter can be of the order of a few or tens of kilometers, depending on the exploitation of airplanes or space platforms. According to what stated above we have that the resolution along the azimuth of a RAR is:

$$\rho_A = X \tag{1.7}$$

Supposing to operate an airborne radar from a distance r of about 5 km and a rather large antenna with $L_a = 1$ m, even by operating at high microwave frequencies such as X-band ($\lambda = 3.1$ cm) would provide an azimuth footprint, and therefore an azimuth resolution, of more than 150 m. RAR systems are therefore characterized by an unbalancing of the resolution capabilities. C. Wiley had the idea to improve the azimuth resolution by using the movement of the antenna, which makes it possible to discriminate targets located forth or back with respect to the radar on the basis of the variation of their Doppler frequencies.

The capability of a coherent radar to improve the azimuth resolution by exploiting the Doppler discrimination principle can be also explained with the so-called aperture synthesis principle. The radar, during its motion along the azimuth, occupies positions corresponding to the elements of a hypothetical array of antennas. A large antenna, as large as the azimuth footprint, can be synthesized, thus providing the possibility to sharpen the beam via a data processing referred to as SAR Focusing. Fig. 1.7 shows the result of the SAR focusing operation for the ESA Envisat sensor over the Strait of Messina in South Italy: the two images before and after the focusing provide a comparison of RAR vs SAR.

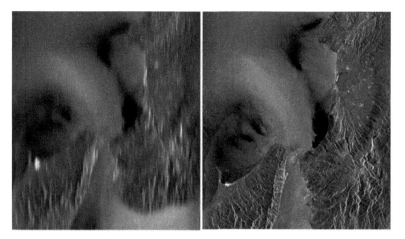

Figure 1.7. From RAR to SAR via data focusing. The image on the left shows the data received by the radar before the azimuth focusing (azimuth is vertical). The image on the right shows the result after the azimuth focusing.

The SAR model providing a mathematical description of the received signal, its spectral characteristics and the data processing operation necessary to achieve pulse compression in range and azimuth (azimuth beam-sharpening) are described in detail in Chapter 4 and 5. In particular in Chapter 4 it will be shown that the azimuth resolution is given by:

$$\rho_a = \frac{L_a}{2} \tag{1.8}$$

and therefore a meter order resolution can be achieved as well as in azimuth. To show an example of the capability of SAR to generate images with resolutions comparable to optical images, operating at frequencies much higher that microwaves, a comparison between an image collected by one of the satellite of the Very High Resolution (VHR) COSMO-SkyMed SAR constellation, and a Google Earth image of an industrial area close to the city of Matera in South Italy is provided in Fig. 1.8.

1.4 Radar equation

The radar equation describes the physical power budget of a given radar system configuration, providing the received power as a function of assigned transmitter power and for a specific target at a given distance, after the attenuation of the transmitted signal due to propagation. This equation ties together the radar characteristics (e.g. transmitted power, antenna aperture), the target

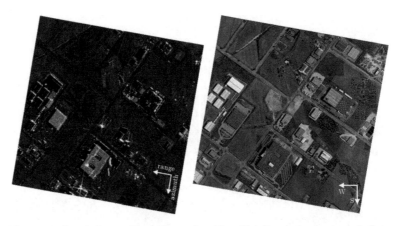

Figure 1.8. Resolutions achieved by modern Very High Resolution systems. Left: image with 1 m resolution captured by COSMO-SkyMed SAR over an industrial area close to the city of Matera in South Italy. Right image: The corresponding optical image from Google Earth.

properties (e.g. target radar cross section), the distance between target and radar (e.g. range) and the properties of the propagation medium (e.g. atmospheric attenuation). It represents a powerful tool for the radar designer, since it provides useful information about performance of the system without resorting to complex analysis and simulations. It is especially useful in early stages of the system design, when specific information about various components of the system (f.i., transmitted waveform, antenna design, signal processing algorithms, etc.) might not yet be available.

Let us consider the configuration depicted in Fig. 1.9 with a radar antenna which radiates the electromagnetic (e.m.) energy uniformly in all directions in the 3-Dimensional (3D) space, and a scattering object (target) placed at a (large) distance r from it.

Assuming now that P_T is the power transmitted by the radar antenna, the power density per surface unit incident on the target, see Fig. 1.9, is given by:

$$S_i\left(r\right) = \frac{P_T}{4\pi r^2} \qquad \left[\frac{\mathrm{W}}{\mathrm{m}^2}\right]. \qquad (1.9)$$

The power density (1.9) is constant in any direction, and depends only on distance r. In the real world, radar antennas are designed to concentrate the e.m. energy only in some portions of the 3D space. Such property of the radiating antenna to irradiate differently in different directions, is represented by the radar antenna gain $G_T\left(\theta, \phi\right)$, where θ and ϕ represent the polar angle (whose complementary angle is commonly referred to as elevation), and

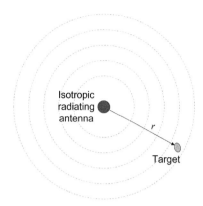

Figure 1.9. Isotropic radiating antenna.

the azimuth angle, respectively, in a spherical reference system centered on the radar antenna, as reported in Fig. 1.10.

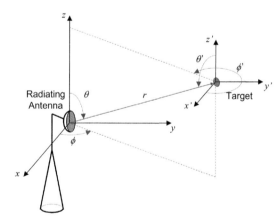

Figure 1.10. Radiating antenna in a spherical reference system. Monostatic configuration.

In directions where the gain is greater than 1, the antenna radiates a spatial power density greater than that which an isotropic radiator would radiate (considering the same P_T); vice versa happens when the gain is less than 1. In radar applications, the antenna will be designed so such that the gain in the directions (identified by the (θ, ϕ) pair) where one wants the antenna to radiate preferably, will be much greater than one; vice versa one will try to keep it much less than 1 in the other directions. Typical antenna gains in radar applications can be of the order of 30-40 [dB].

Consider now an anisotropic radar antenna whose gain in the direction (θ, ϕ) along the connecting line radar-target, is given by

$G_T(\theta, \phi)$. In this case the power density incident on the target is given by:

$$S_i(r, \theta, \phi) = \frac{P_T}{4\pi r^2} G_T(\theta, \phi) \qquad \left[\frac{W}{m^2}\right]. \qquad (1.10)$$

When the e.m. wave radiated by the radar antenna illuminates the target with the power density (1.10), the secondary currents induced in the target to satisfy the boundary conditions and the conditions of continuity of the e.m. incident field, in turn induce an e.m. field reflected (scattered) by the target.

The way in which the target scatters the secondary e.m. field depends on its orientation with respect to the direction of origin of the incident e.m. field. In particular, the target will scatter an e.m. field in a generic direction (θ', ϕ') that depends on the mutual radar-target geometry, on the size of the target, and on its physical constitution (mainly, on materials from which it is made, and its roughness). In the context of radar systems, to determine the scattered power in an arbitrary direction, a new parameter is introduced: the so-called radar cross-section (RCS) σ. It has the dimension of m^2, and it allows to write the power scattered by the target in the following way:

$$P_S(r, \theta, \phi, \theta', \phi') = \frac{P_T}{4\pi r^2} G_T(\theta, \phi) \sigma(\theta, \phi, \theta', \phi') \qquad [W]. \quad (1.11)$$

Hence, given a certain incident power density, the reflected power depends on the direction of arrival of the incident e.m. radiation θ and ϕ, and on scattering direction θ' and ϕ', which represent the polar angle, and the azimuth angle in a spherical reference system centered on the target (see Fig. 1.10). In case of a monostatic radar systems, the impinging and the backscattering directions both identify the direction joining radar and target, and the dependence on angles θ, ϕ, θ', and ϕ' and on distance r is usually omitted:

$$P_{BS} = \frac{P_T G_T \sigma}{4\pi r^2} \qquad [W]. \qquad (1.12)$$

The e.m. wave backscattered by the target reaches the radar antenna with a power density given by:

$$\frac{P_{BS}}{4\pi r^2} = \frac{P_T G_T \sigma}{\left(4\pi r^2\right)^2} \qquad \left[\frac{W}{m^2}\right]. \qquad (1.13)$$

The power received by the radar antenna can be expressed in the following way:

$$P_R = \frac{P_T G_T \sigma}{\left(4\pi r^2\right)^2} A_R \qquad [\text{W}], \qquad (1.14)$$

where A_R is the antenna Effective Area (when it works as a receiving antenna). The effective area of the antenna can be expressed as function of the gain of the receiving radar antenna G_R and of the working wavelength:

$$A_R = \frac{\lambda^2}{(4\pi)} G_R \qquad \left[\text{m}^2\right]. \qquad (1.15)$$

Then, substituting (1.15) into (1.14) we get:

$$P_R = \frac{P_T G_T G_R \lambda^2 \sigma}{(4\pi)^3 r^4} \qquad [\text{W}]. \qquad (1.16)$$

Note that in case of a bi-static radar the term r^4 must be substituted with $r_1^2 r_2^2$, where r_1 and r_2 represent, respectively, the distance between the transmitting antenna and the target, and the distance between the target and the receiving antenna.

In case of a monostatic radar, where the same antenna works as the transmitting and receiving antenna, the gains involved in (1.16) are equal ($G = G_T = G_R$), so that we get:

$$P_R = \frac{P_T G^2 \lambda^2 \sigma}{(4\pi)^3 r^4} \qquad [W], \qquad (1.17)$$

which represents the so-called Radar Equation. It represents the power associated with the received signal backscattered by a target having a RCS σ, at a distance r from the radar, when a power P_T is transmitted, using a working carrier at frequency $f = c/\lambda$ (c is the speed of light), and when the radar antenna gain in the direction of the target is G.

The radar equation (1.17) is very useful from a practical point of view. A first important characteristic of (1.17) is that the power of the received signal drops off very quickly with range, as $1/r^4$. This circumstance means that the signals backscattered by a target can be detected only if they have an appreciable power when they are in presence of the receiver noise and interfering signals. In fact, as the radar works in the real World, other signals are received or are present at the receiver, in addition to the signal that was transmitted by the radar itself. Neglecting for the sake of simplicity the presence of interfering/disturbing signals (f.i., other echoes, other electromagnetic signals in the same frequency bandwidth), it has to be considered the noise at the receiver, usually referred as thermal noise, which is usually modelled as an Additive White Gaussian Noise (AWGN). According to this hypothesis, the noise power at the output of a receiver is given by:

$$P_N = kT_e \mathcal{G} \mathcal{B}_N \qquad [W], \qquad (1.18)$$

where k is the Boltzmann constant, T_e and $\mathcal{G}$ are the noise equivalent temperature and the gain of the receiver; $\mathcal{B}_N$ is the receiver Equivalent Noise BandWidth (ENBW), which represents the (monolateral) bandwidth of an ideal lowpass filter, with the same gain of the actual receiver, producing the same noise power as that of the actual receiver.

We can compute now the Signal to Noise Ratio (SNR) at the receiver:

$$\text{SNR} = \frac{P_R}{P_N}, \qquad (1.19)$$

which is a fundamental parameter to evaluate the performance of a radar system.

In some cases, it is possible to increase the SNR. In addition to increasing the transmitted power and the gain of the antenna, when this is possible, one could act by "summing" coherently or incoherenty the received signals. These kinds of summing, called "integration" in the case of classical radar signal [1], and "image compression" (or image formation) in the case of radar imaging (e.g. SAR), allow to work reliably with radar systems, also when the radar antenna transmits tens or hundreds of Km apart from the target, and the received signal is very weak.

In the following two Subsections 1.4.1 and 1.4.2, we will introduce the radar equation in the two different cases of concentrated and of distributed targets, respectively.

1.4.1 Concentrated target

Consider the geometry reported in Fig. 1.11, where a radar system in presence of a concentrated target is presented. In such case, the transmitted power, concentrated substantially in the main lobe of the radiating radar antenna, is only partially intercepted by the target through its RCS. It is quite intuitive to understand that the larger the physical dimension of the target, the larger is the portion of the transmitted power that is intercepted by the target, and backscattered by it.

The target physical dimension is not the only element which contribute to the value of the RCS σ. Other elements contributing

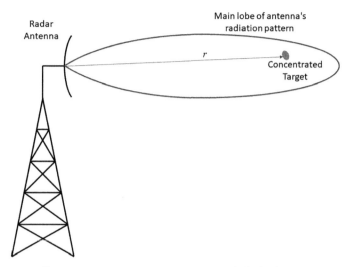

Figure 1.11. Two dimensional geometry of a monostatic Radar in presence of a concentrated Target.

to, are the size of the target relative to the working wavelength, the signal mutual orientation with respect to the incident wave direction (aspect angle), the target material and roughness, the working frequency, and the polarization.

In some cases, for particular shapes of the target, it is possible to evaluate analytically the RCS. This is important in itself, of course, and because it allows, once the radar signals have been received, to calibrate their values in the presence of known and reproducible targets response.

Among the targets for which it is possible to evaluate the RCS, there are spherical targets, planar targets, and the so-called corner reflectors. We will consider below in this section the case of monostatic radar systems, where the same antenna works both as transmitter and receiver, and the considered signals travel on the blue (dark gray in print version) line of Fig. 1.11. The analysis can be easily extended also to the case of bi-static radar systems.

Start with a spherical target of radius a. Its RCS in case of a sphere larger than at least two times the working wavelength λ, so we are in the so-called Optical region, is independent on λ, and is given by [1]:

$$\sigma \cong \pi a^2, \qquad a > 2\lambda \qquad (1.20)$$

where the equality is verified asymptotically (with respect to a) for $a \gg \lambda$.

In case of a small sphere in terms of the wavelength, in particular for $a < 0.1\lambda$, so we are in the so-called Raileigh region, the RCS is given by:

$$\sigma \cong \pi a^2 9 (ka)^4, \qquad a < 0.1\lambda \qquad (1.21)$$

where $k = 2\pi/\lambda$.

For spheres whose radius a satisfies the condition $0.1\lambda < a < 2\lambda$, so in the so-called Mie region [2], the RCS oscillates in the interval $\left(0.3\pi a^2, 3.6\pi a^2\right)$. It means that the same sphere, whose physical transversal dimension is always πa^2 can appear as 3.3 times smaller or, 3.6 times larger, only using different wavelengths in the interval $0.1\lambda < a < 2\lambda$. This can be explained considering resonant propagation modes, constructively or destructively interfering with each other, that can be induced during the scattering. It is important to note that the RCS of the sphere is not dependent on the direction of arrival of the incident wave.

Consider now a planar target whose area is A. It is easy to imagine that its RCS will strongly depend on its orientation with respect to the direction of the incident wave. In particular, if the normal of the target surface is aligned with the incident direction of the wave (normal incidence), the target will intercept the maximum possible portion of the incident power density and, consequently, the RCS of the target will assume its maximum value (with the same physical surface). Opposite reasoning if the normal of the target surface should be orthogonal to the incident direction of the wave. In this latter case, the planar surface would appear as its edge, and would intercept a minimal portion of the incident power density.

The RCS of planar (and smooth) target whose area is A, large in terms of the working wavelength, in case of normal incidence, is given by:

$$\sigma \cong \frac{4\pi A^2}{\lambda^2}, \qquad 2\pi A^{1/2} \gg \lambda. \qquad (1.22)$$

Such RCS is $4\pi A/\lambda^2$ times the actual area A. For planar surface whose area is approximately larger than $(0.3\lambda)^2$, the RCS of the planar target is always larger than the physical actual area. For instance, for a planar surface whose area is equal to $(10\lambda)^2$ the RCS (remember, a sort of equivalent area) is about 31 dB larger than the actual area (more than 1000 times larger!). In other words, the planar target for normal incidence, acts as an electromagnetic mirror. This effect can be explained, also in this case as in the case of spherical targets, with interference of different backscattering contributions.

Now, if a target is thus "visible" in certain directions, it will be less so in others. To better understand this statement, consider the particular case of a square planar target, whose side is a. Its RCS, for a monostatic radar, for an arbitrary incidence angle θ, is given by:

$$\sigma \cong \frac{4\pi a^4}{\lambda^2} \left[\frac{\sin(ka\sin\theta)}{ka\sin\theta} \right]^2 = \frac{4\pi a^4}{\lambda^2} \mathrm{sinc} \left(\frac{2a\sin\theta}{\lambda} \right)^2, \qquad 2\pi a \gg \lambda,$$
(1.23)

where $\mathrm{sinc}(x) = \sin(\pi x)/\pi x$. For normal incidence ($\theta = 0$), (1.23) reduces to (1.22), as $a^2 = A$. It is easy to verify that RCS (1.23) decays quickly with θ, so that unless the target is perfectly aligned for normal incidence, its RCS is usually very small. An expression similar to (1.23) can be obtained for the radar cross section can be obtained of planar circular targets.

As mentioned previously, having targets whose RCS expression is a-priori know can be very useful to calibrate the received signals and their following processed signals. In the field of imaging radars, especially in the case of airborne/spaceborne applications, the distance radar-target can be of tens/hundreds of Km. For such kind of applications, it would be useful to have calibration targets whose RCS does not significantly depend on the incidence angle (at least in a range of values of it), and which is sufficiently high to guarantee a return signal of sufficient power. Spherical targets satisfy the first requirement, but not the second, and planar targets satisfy the second, but not the first. A class of target which put the two requirements together, are the so-called corner reflectors.

The simplest case of corner reflector is the dihedral one, constituted by two planar surfaces forming an orthogonal angle to each other, shown in Fig. 1.12.

The peculiarity of such kind of targets is that an incident wave coming from an arbitrary direction belonging to a plane orthogonal to the two planar surface constituting the corner reflector, as long as it belongs to a range of values, is reflected back in the same direction of the incident wave. In the bottom part of Fig. 1.12 the cut of the dihedral corner reflector on a plane orthogonal to both planar surfaces is reported, and two incident directions $\alpha_1 = \pi/8$ and $\alpha_2 = \pi/4$ are considered. In both cases, as it can be easily inferred, the wave is reflected in the same (opposite) direction of the incident wave; this continues to apply as long as the angle α is within the range $(0, \pi/2)$. The portion of the incident wave that is intercepted and therefore reflected by the reflector depends on the length of the segments (A_1, B_1) and (A_2, B_2), which changes with α. In other words, the corner reflector appears of different dimension (an effective area) according to different values of α, and its maximum value is obtained for $\alpha = \pi/4$.

The limit of a dihedral corner reflector, at least in the case or airborne/spaceborne applications, is that in order to have a specular reflection, the incident direction must belong to the plane

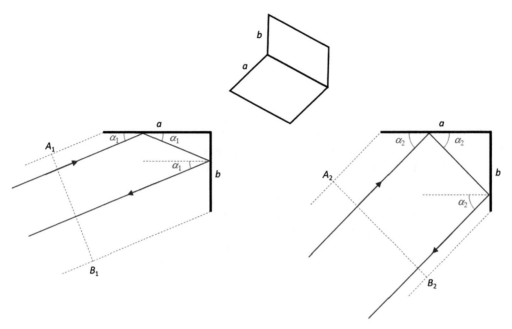

Figure 1.12. 3D geometry of a dihedral corner reflector for two different incidence angle (top). 2-D geometry, showing incident and reflected waves, for two different incidence directions, characterized by $\alpha_1 = \pi/8$ and $\alpha_2 = \pi/4$ (bottom left and right).

orthogonal to the two planar surfaces constituting the corner reflector, which is impractical. The trihedral corner reflectors, constituted by three planar surfaces forming an orthogonal angle to each other, overcome this limitation. As long as the direction of incidence belongs to a cone in three-dimensional space, the incident wave is reflected specularly. The trihedral corner reflector most used in radar imaging applications is the triangular trihedral one, shown in Fig. 1.13.

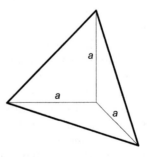

Figure 1.13. Triangular trihedral corner reflector.

The maximum value of the RCS of the trihedral corner reflector shown in Fig. 1.13, whose side is a is:

$$\sigma \cong \frac{4\pi a^4}{3\lambda^2}, \qquad \sqrt{3}\pi a \gg \lambda. \qquad (1.24)$$

Such value remains approximately constant (and in any case within 3 dB of reduction) as long as the direction of the incident wave belongs to a cone of aperture of approximately 20°, whose axis of symmetry is the bisector of the trihedral.[1]

Also for the triangular trihedral corner reflectors, similarly to the planar targets, the RCS appear much larger of actual projected area. The advantage with respect to planar targets is that the RCS value, as discussed above, is stable for about 40° of different incidence conditions. This allows to use the triangular trihedral corner reflectors very often for calibration purposes in airborne/spaceborne radar imaging missions.

1.4.2 Distributed target

Consider the geometry reported in Fig. 1.14, where a radar system in presence of a distributed target is shown. In such case, the transmitted power, concentrated substantially in the main lobe of the radiating radar antenna, is almost totally intercepted by the extended target.

In this case the target is usually an extended surface corresponding to the portion illuminated by the main lobe of the antenna. A case of interest for the purposes of this book is the case of a radar antenna, mounted on an aircraft or on a spacecraft, looking to the side with respect to the flight track, as discussed in Section 1.3 and shown in Fig. 1.15. This kind of radar system is called a Real Aperture Radar (RAR) or, a Side Looking Aperture Radar (SLAR). The footprint can be approximated by a rectangular with sides Y and X (indeed, its shape is more similar to an ellipse), whose dimension grows with the distance. The larger the distance of the surface from the radar, the larger Y and X are.

In this context it is convenient to express the RCS of the distributed target in the following way:

$$\sigma = A_d \sigma_0, \qquad (1.25)$$

where A_d is the area inside the antenna footprint of size XY, which is simultaneously illuminated by the transmitted pulse, and σ_0 is

---

[1]The bisector of the trihedral is a straight line such that its projections on the three planes of the trihedral are in turn bisectors of the right-angles identified on the trihedral faces.

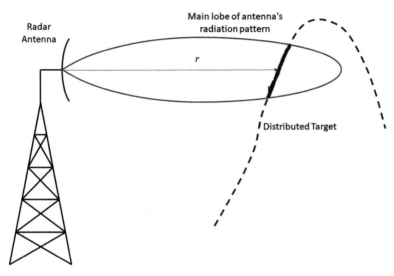

Figure 1.14. Two dimensional geometry of a radar in presence of a distributed Target (for instance, a relief or a mountain).

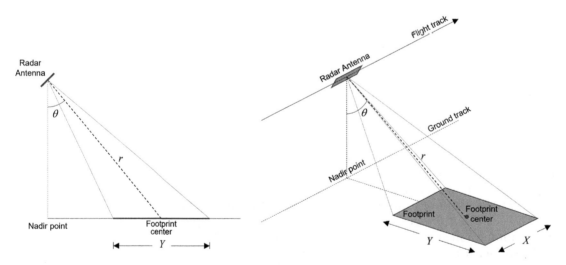

Figure 1.15. A Side Looking Aperture Radar and its footprint on the ground. View from behind (left); Lateral view (right).

the RCS (NRCS), where the normalization is carried out with respect to the surface extent. The term σ_0 embeds information also about the orientation of the illuminated surface with respect the direction of the incident wave. In our analysis, it is supposed that the surface is constituted by a homogeneous material. Other normalization options are discussed in Chapter 3, more specifically Section 3.2.

According to (1.25), the radar equation (1.17) becomes:

$$P_R = \frac{P_T G^2 \lambda^2 A_d \sigma_0}{(4\pi)^3 r^4} \qquad [W].\qquad (1.26)$$

It has to be considered that the area A_d grows with r, because of the growth with r of X. This means that the real dependence of the received power on r passes from $1/r^4$ relevant to concentrated targets to $1/r^3$ relevant to distributed ones.

For the noise power the relation (1.18) continues to be valid, so that the SNR becomes:

$$\text{SNR}_{\text{Dist}} = \frac{P_T G^2 \lambda^2 A_d \sigma_0}{(4\pi)^3 r^4 K T_e B}.\qquad (1.27)$$

1.5 Interferometry

In a SAR image, ground scatterers are discriminated in a 2D domain: azimuth (x) and range (r). Sampling of these variables on a discrete grid (x', r') is carried out by guaranteeing that the pixel size (distance between two samples) is lower than the resolution. As shown in the book, this is equivalent to stating that both the azimuth and range sampling frequencies has to fulfill the Nyquist criterion. In particular, the azimuth and range sampling frequencies must be higher than the Doppler and transmitted bandwidths, respectively.

A given pixel, say (x', r') provides a contribution which is integrated along the azimuth and range resolution cell. Along the vertical direction the radar integrates the contributions of scatterers with azimuth and range belonging to $(x' - \rho_a/2, x' + \rho_a/2)$ and $(r' - \rho_r/2, r' + \rho_r/2)$. Supposing that the azimuth is an ideal straight line, if a cylindrical reference system (x, r, θ) with θ being the angle between the nadir and the scatter line of sight (that is the range direction) it is clear that with a single SAR image there is no possibility discriminating θ.

The situation is depicted in Fig. 1.16: Here it is shown the geometry of a SAR in the plane orthogonal to the flight trajectory, that is to the azimuth.

In this plane the SAR system is able to discriminate the target range; however, the range alone is not sufficient to discriminate θ and consequently the vertical localization of the target. All points located in the circular crown determined by the range and range resolution would provide the same range measurement associated

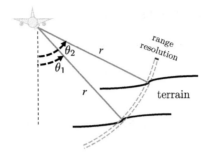

Figure 1.16. SAR geometry in the plane orthogonal to the flight track (azimuth): indeterminacy of the look angle and vertical location of a scatterer with a single antenna.

with the selected range pixel. The higher (numerically lower) the range resolution the narrower the crown. Stated differently, the natural 3D reference system of the radar is a cylindrical one with the axis corresponding to the flight trajectory where a point can be localized by specifying the projection along the axis (azimuth), the distance from the axis (the range) and the angle with respect to a fixed direction orthogonal to the azimuth, generally the angle with respect to the nadir referred to as look angle. With a single SAR antenna we can determine the range and the azimuth, with rather accurate localization, but the look angle remains unspecified within the illuminated beam.

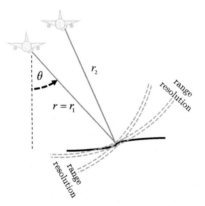

Figure 1.17. SAR geometry in the plane orthogonal to the flight track (azimuth): vertical localization with a two antenna (interferometric) system.

In order to achieve 3D sensitivity, that is to discriminate also the look angle to the target, SAR Interferometry exploits a second antenna that looks at the scene from a different view angle, see Fig. 1.17. This angular diversity in the viewing angle to the target,

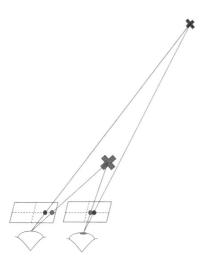

Figure 1.18. Binocular vision exploited for the depth sensitivity: an object (target) is judged less or more distant depending on the variation of its position in the image planes corresponding to the two eyes.

allows measuring two ranges to the target: The target localization is carried out by intersecting the two circular crowns corresponding to the measurement of the two ranges corresponding to the selected pixel (with the given accuracy/resolution). Said differently, if there are more vertical locations with the same range r_1, only one of this corresponds to the second range measurement r_2. Actually, what matters for the vertical localization is the range difference $r_2 - r_1$, which is hereafter referred to as δr. This concept, explained in more details in Chapter 6, is here simply explained by exploiting the analogy with the visual system which reconstructs the depth (i.e. achieves 3D localization) comparing the images of an object achieved from "slightly different" angles, that is by using a binocular vision as shown in Fig. 1.18. Eyes are capable to discriminate the angular localization of the objects: with a single eye it is more difficult, at limit impossible without any memory cognition of the context in which objects are located, to evaluate the depth of the scene, i.e. the distance of the targets from the eye. To overcome this limitation the binocular vision exploits two eyes: a target is less or more distant from the two-eyes system depending on the variation of its position in the image planes corresponding to the two eyes. In Fig. 1.18 the red (mid gray in print version) target provides projections in the image planes which are located at very different positions: one close to the right border for the image plane corresponding to the left eye, and one close to the center for the image plane corresponding to the right eye. Differently, the green (gray in print version) target, which is located more far than the red (mid gray in print version) target from the two-eyes system provides projections almost located at the midway between the center and right border in both the image planes. Accordingly, in the visual system the depth is sensed by comparing the positions of the same target in the two image planes and by evaluating, in our brain, how much the projection of the target moves from one image plane to the other. In the same way InSAR systems determine the vertical location of the targets by evaluating the variation of the range, referred to also as range or path difference, to the target.

Chapter 6 provides all the analysis and mathematical formulation to explain the importance of how, and to which accuracy SAR Interferometry (InSAR) allows the determination of the scene DEM. Here we want rather to point out that the height sensitivity of the system, that is the ratio between the path difference variation and the variation of the height, is related to the ratio between the component of the vector connecting the two SAR antennas, referred to as (spatial) baseline, orthogonal to the target los and the range of the target. Typically this ratio is low, on the order of $10^{-3} - 10^{-4}$ for satellite systems. This fact prevents the use of a radargrammetric system, i.e. a system that measure the range variation based on the amplitude or intensity (square amplitude) shift among the different acquisitions whose accuracy could be a fraction, between 1/10 and 1/100, of the range resolution.

InSAR systems exploit specifically the phase variation related to the oscillations of the transmitted radiation. Differently from the sunlight, which is characterized by large frequencies but random variations on the μm wavelength scale, the transmitted pulse for a radar is coherent: this means that oscillations are well controlled. The coherence concept is mathematically described in more details in the next chapters. However, to summarize, the phase control allows measuring path differences, which are fractions of the distance between two oscillation maxima (or minima), that is the wavelength (cm at microwaves). In practice InSAR exploits the phase interference between the signal at the two antennas to accurately measure the path difference. The capability of the interference to measure the height can be explained also by referring to the Young experiment, reported in Fig. 1.19, and used to explain the wave nature of the light. If an incoherent light is passed through a thin hole, the effect is to select a narrow frequency band, thus selecting an almost coherent spectral component of the light. Such component illuminates two vertically shifted holes which radiate in turn two coherent radiations. These radiations interfere each other in a constructive, destructive, or intermediate way. In the figure it is show that constructive interference allows identifying angular directions, represented as red (mid gray in print

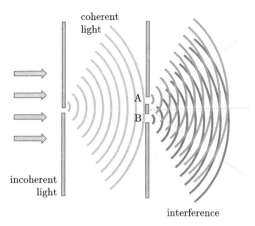

Figure 1.19. Young experiment showing the interference pattern between the coherent radiation (light) emitted by two vertically shifted sources.

version) dotted lines, which provide the height sensitivity. Constructive, destructive and in between interference are used in the same way in SAR Interferometry to estimate the ground topography. An example is provided in Fig. 1.20 where it is shown the overlap between a SAR image and the interferogram (phase difference wrapped in the $[-\pi, \pi]$ interval) collected by the F-SAR DLR airborne system. The latter is an image where sharp color variations correspond to a destructive interference, i.e. a phase difference which in modulus is π. This image shows that the interferometric lines corresponding to fringes provide a qualitative indication of the topography. The reader can refer to Chapter 6 to pass from qualitative analysis addressed so far, to a more quantitative description, achieving knowledge of the steps necessary to generate an interferogram and to convert it in a DEM.

1.6 Differential SAR interferometry

In the previous section it has been discussed the capability of an interferometric system to achieve the measurement of the range difference to an accuracy of the order of a fraction of the wavelength. In that case a range difference is induced by the angular imaging diversity, which is necessary to achieve the sensitivity to the height. One can think that the same capability can be exploited to measure the range difference induced by other physical mechanisms. In Differential SAR Interferometry (DInSAR), the range difference is induced by the movement of a target between two successive passes of the sensor. The situation is depicted in Fig. 1.21.

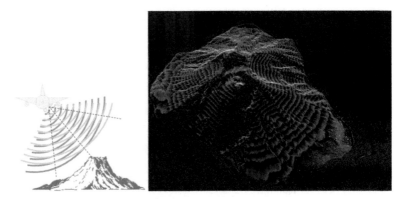

Figure 1.20. Left image: schematic representation of the wavefronts interference from two antennas, here mounted for example under the fuselage on an airborne. Right image: F-SAR DLR airborne interferogram over the Stromboli Island in South Italy.

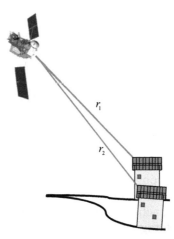

Figure 1.21. The Differential SAR Interferometry (DInSAR) principle. A SAR system is supposed to observe the scene in two subsequent passes and the target is supposed to be affected by deformations: a subsidence in the figure. If the pass is repeated on the same orbit, the range difference shows sensitivity to the deformation.

The methodology has been developed to a level that has boosted the investments on a global scale, especially for satellite systems whose multitemporal features and imaging regularity are intrinsic to the orbiting mechanisms, SAR technology is able to pierce clouds, with intrinsic and selective orbital revisiting, to provide regular and systematic observations. The DInSAR principle is rather simple: if a sensor passes over the same scene exactly repeating the same orbit, that is without inducing an imaging angular diversity (zero-baseline), the measured range differences are related either to the movement of the target or to a delay induced by the propagation in the atmosphere of the radiation from the target to the scene. The latter contribution is referred to as Atmospheric Phase Delay (APD) or Atmospheric Phase Screen (APS). What matters here is obviously only component of the APS variable over the scene, because a constant range difference (i.e. a constant phase difference for an interferometric system) is irrelevant. A rather spatially variable APS contribution is generally induced by the low atmosphere (troposphere), and this contribution is generally limited to the order of a few oscillations, that is a low multiple number of the wavelength. If the physical deformation phenomenon on the ground induces deformation which are several times larger than the wavelengths, f.i. large magnitude earthquakes, it can be directly measured by the DInSAR system. Similarly to the InSAR system, the DInSAR exploits the phase information, that is the high frequency (GHz) oscillations, that is centimeter wavelength in space, to accurately measure the range variation. If the passes are not repeated on the same orbits, the angular diversity will provide a range variation (i.e. phase) component, induced by the topography, additional to the one associated with the deformation. The additional component can be subtracted (hence the differential adjective) starting from an external DEM. More precisely the DEM is exploited together with the orbital information to evaluate the topographic range variation induced by the nonzero baseline. This operation is also referred to as zero baseline steering. In Fig. 1.22 it is shown the overlay of the amplitude image and the differential interferogram achieved by DInSAR analysis of two COSMO-SkyMed images taken before (April 4th 2009) and after (April 12, 2009) the earthquake that hit the city of L'Aquila on April 6th 2009. The sensor wavelength is approximately 3.11 cm and each fringe corresponds to half the wavelength, that is 1.55 cm approximately.

DInSAR technology was conceived and initially development at NASA-JPL at the end of eighties [3]. The clear demonstration of the potentialities of this powerful technology were however clearly demonstrated for the first time for the Landers earthquake [4] opening a new era for SAR. DInSAR concepts are described in more details in Chapter 6. Many applications involving its use for single or, generally, a few pairs of images relevant to the same area drew the lines for the development of the modern multitemporal DInSAR techniques.

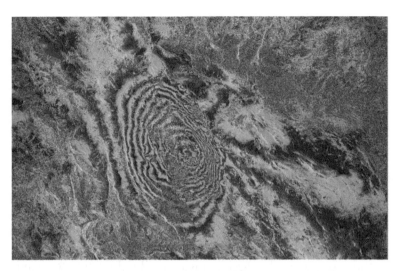

Figure 1.22. Overlay of the amplitude image with the DInSAR phase interference pattern for the L'Aquila earthquake in 2009 in center Italy.

1.7 Advanced differential SAR interferometry

As already stated SAR unique feature in Earth Observation is the systematic acquisition capability. This feature has led to the possibility to acquire stack of interferometric images over tens to hundreds of passes. Advanced Differential Interferometry (A-DInSAR), also referred to as MultiTemporal DInSAR (MT-DInSAR), represents a class of algorithms aimed at the coherent data analysis of such stacks to improve the performance of classical single pair (2-pass) DInSAR.

As explained in the previous section, DInSAR is able to capture large deformation signal associated with earthquakes but its applicability to weaker signals, as those occurring in part on volcanoes, but also to subsidence or displacements of slow moving landslides are impaired by the presence of atmosphere. A-DInSAR techniques exploit multiple interferometric coupling over a given period of observation as shown schematically in Fig. 1.23, together with an appropriate signal model, to estimate and compensate the signal associated with the atmospheric disturbance (APS) from the useful deformation signal. In doing this it allows also to pass from the measurement of single deformation variations to deformation time series. The possibility to model also the baseline diversity, spatial and temporal, allows at the same time to operate coherent analysis also at the highest possible resolution, to achieve accurate localization of the scatterers, even on scale of buildings, and measure at the same time its possible deformation signals. The method

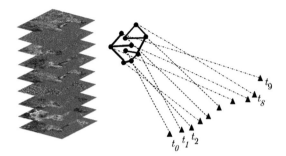

Figure 1.23. Schematic representation of the Advanced Differential Interferometric SAR (A-DInSAR) technique combining multiple acquisition to monitor ground deformations.

that for the first time showed the possibility to achieve such a fine interferometric analysis able to monitor scatterers able to retain over the time scattering stability was the Persistent Scatterers (PS) technique. With this approach it was demonstrated the possibility to model the phase response of ground scatterers capable to maintain the back-scattering properties over year even if subject to deformations. The implications of such data processing method for application to risk monitoring have been dramatic stimulating many Companies toward the offer of monitoring services. On the wave of the development of such a technology, similar techniques have been developed for the analysis of interferometric stacks at different degrees of spatial resolution. In addition to the Persistent Scatterers Interferometry (PSI) methods working at full resolution, methods based on the Stacking of Coherent Interferograms (SCI), such as the Small Baseline Subset (SBAS) approach, implementing spatial averaging (interferometric multilook) has been derived over the years. Typically a two scale processing strategy is adopted: data are first analyzed at low resolution to estimate and compensate the atmosphere and the small scale deformation, then a full resolution analysis is carried out to achieve accurate scatterer localization and high resolution monitoring. Fig. 1.24 provides an example showing the difference between the full resolution, i.e. large or fine scale analysis, and reduced resolution, i.e. small or rough scale analysis. The data correspond to a DInSAR processing of 28 COSMO-SkyMed images, with 3 m spatial resolution, acquired from January 2017 to September 2019. In the low resolution analysis the resolution was degraded to 30 m.

Figure 1.24. Example of measured deformation by a sequence of low and full resolution analysis. Left image: measured deformation mean velocity points at low resolution (small scale DInSAR analysis) over the whole image frame. Lower right image: a zoom of the small scale DInSAR analysis in the Campi Flegrei area defined by the white box in the left image. Upper right image: measured deformation mean velocity points at full resolution (large scale DInSAR analysis). The color scale is shown in the rightmost part of the image. All measured points have been superimposed to the Google map of the area.

1.8 Tomographic SAR for 3-dimensional and multi-dimensional analysis

SAR is a side looking system that is able to discriminate range distances. Such an imaging mechanism induces heavy geometric distortions in the presence of topographic variations.

Fig. 1.25 shows on the left an image of the financial area of Pudong in Shanghai (China) characterized by the presence of several skyscrapers, including the World Financial Center: the second tallest building (492 m) in the area with a singular corkscrew shape characteristic (right building in the right image). On the left it is shown the SAR image achieved with 3 m ground range resolution. The World Financial Center with its 492 m height should occupy more than 400 pixels, that is the 80% of the range extension of the image. Despite its size and the clear visibility of the adjacent skyscraper indicated by the green (gray in print version) circular

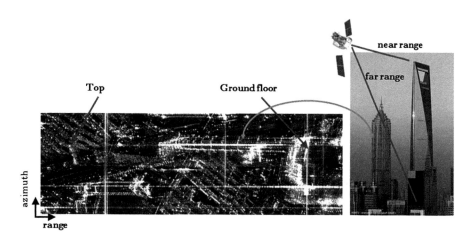

Figure 1.25. Graphical explanation of the distortion effect induced by the scene topography.

arrow, the World Financial Center is hardly visible. In the image with red (mid gray in print version) arrows have been depicted the top and the ground floor to show the typical distortion phenomenon in which the top of the building is sensed before, i.e. in near range, with respect to the ground floor which is located in far range. The effect is an equivalent folding of the vertical structures (as skyscrapers) toward the radar in the range dimension. This folding generates in urban areas the frequent presence of multiple contributions in a given image pixel: the phenomenon is known as "layover" and is the main reason of failure of interferometric approaches to localize (and monitor in the case of multitemporal analysis) all the scattering contributions. The failure is due to the assumption of interferometric methods to have one scatterer in each pixel. The situation is even worse when imaging vertical dense scattering layers as forest areas with lower frequency SAR systems which are able to penetrate the surface. To overcome the problem of inability of interferometric approaches to separate multiple vertical scatterers the tomographic SAR (TomoSAR) technique can be used. The principle is shown in Fig. 1.26 and found on the idea to exploit multiple antennas along the vertical direction to synthesize an array that is able to sharpen the beam of a single antenna. The principle is similar to the azimuth direction and the multiple antennas can be also synthesized with multiple passes, provided that the backscattering properties of the scene remain unchanged and that proper compensation of atmospheric disturbances is carried out. A schematic representation of the TomoSAR, which is able to profile the backscattering along the vertical direction is shown in Fig. 1.26.

Figure 1.26. Schematic representation of Tomographic SAR (TomoSAR): a vertical antenna is synthesized to sharpen the beam of a single antenna (red (mid gray in print version)), thus allowing to profile the vertical backscattering through a steering of the resulting fine beam (blue (dark gray in print version)).

TomoSAR can be combined with Differential Interferometry to allow separation of scatterers interfering in the same pixels, as well as monitoring of possible deformations (3D+time monitoring): the methodology is known as Differential TomoSAR (D-TomoSAR). TomoSAR and D-TomoSAR are the subject of Chapter 8.

2

Radar principles: ranging and Doppler

This chapter recalls the key principles of classical surveillance radars. The content does not provide a comprehensive treatment of the many developments of the recent years in surveillance radars but, rather, revisits the basic arguments and notions that are useful for the comprehension of the imaging radar concepts described in the subsequent chapters. Readers can refer to [5], [6], for a more complete address to classical radars.

Classical radar can exploit coherent and incoherent radiation. Because of the relevance in terms of capabilities for surveillance radars but, more important, with regard to the need to follow the path for understanding the mechanisms for high resolution capabilities of SAR, only coherent radars are of interest in this chapter.

Coherent radars can exploit continuous or interrupted transmission: the former are referred to as Continuous Wave (CW) radars, the latter as pulsed radars. Pulsed radars have typically a single antenna operating both in transmitting (TX) mode during the pulse transmission and in the receiving (RX) mode in the echo listening interval. Conversely, CW radars need an antenna dedicated to the transmission and another one exploited for the echo reception.

Finally, radars can operate in a monostatic or bistatic configuration: in the first case the TX and RX antenna (if different from the TX one) are generally almost colocated; in the second case they are typically placed far away to achieve, depending on the applications, slight or significant angular diversity in the target line of sight with respect to the monostatic case.

If not explicitly stated, pulsed monostatic radars are considered because of their importance in satellite high-resolution radar imaging.

Multi-Dimensional Imaging with Synthetic Aperture Radar
https://doi.org/10.1016/B978-0-12-821655-2.00006-1

2.1 Target ranging with a rectangular pulse

The key working principle of radars for detecting the presence of objects and for measuring their distance (range) from the illuminating antenna phase center is the transmission of a proper probing signal and the reception of the echoes backscattered by the target. This operation is technological implemented, as explained in the previous chapter, by exploiting an antenna radiating an electromagnetic field, see Fig. 2.1, which impinges on the target. The interaction with the target generates a scattered field, whose backscattered quota is received by the antenna in the echo listening interval and recorded by the receiver for subsequent processing.

Figure 2.1. Radar basic scheme.

With reference to a pulsed radar system let

$$p_{BP}(t) = a(t) \cos{(2\pi f_0 t + \varphi(t))} \tag{2.1}$$

be the real valued Radio-Frequency (RF) transmitted signal at carrier frequency f_0 (i.e., wavelength $\lambda = c/f_0$ with c being the speed of the light), with envelope $a(t)$, generally characterized by a short support, and phase $\varphi(t)$: the pedix "BP" highlight the band-pass nature of the signal. Let also $\mathcal{E}_T = \mathcal{E}_p/2$, be the transmitted energy, with $\mathcal{E}_p$ being the energy of the envelope $a(t)$.

As described in Appendix A.2, it is more convenient to consider the complex valued low pass equivalent (LPE) representation, say $p(t)$, of the band pass signal $p_{BP}(t)$ given by:

$$p(t) = a(t)e^{j\varphi(t)} \tag{2.2}$$

Such signal is related to the band pass signal $p_{BP}(t)$ by:

$$p_{BP}(t) = \Re\left\{p(t)e^{j2\pi f_0 t}\right\} \tag{2.3}$$

with $\Re$ being the operator extracting the real part of the argument.

Unless otherwise stated, in the following we refer to the LPE. Moreover, note that $p(t)$ and $a(t)$ share the same energy $\mathcal{E}_p = 2\mathcal{E}_T$. In addition to this we also assume $p_u(t)$ to be the normalized, i.e. unitary energy, version of $p(t)$, so that:

$$p(t) = \sqrt{\mathcal{E}_p}\, p_u(t) = \sqrt{2\mathcal{E}_T}\, p_u(t) \tag{2.4}$$

Additionally, we denote with T_κ the "κ dB pulse duration", i.e., the pulse time extension corresponding to the generic κ decibel peak power attenuation ($\kappa \geqslant 0$), i.e.:

$$T_\kappa : |t| \geqslant \frac{T_\kappa}{2} \implies 20 \log_{10}\left[\frac{|p(t)|}{|p(0)|}\right] \leqslant -\kappa \tag{2.5}$$

where, without loss of generality, we have assumed the amplitude of $p(t)$ to be an even function attaining its maximum value in $t = 0$. The (largest) pulse duration, denoted by T, is the null-to-null pulse extension (support) and, thus, represents its "∞dB duration", i.e., $T_\infty = T$.

The simplest form of the transmitted pulse is

$$p_u(t) = \frac{1}{\sqrt{T}} \Pi\left(\frac{t}{T}\right) \tag{2.6}$$

where $\Pi(x)$ is the rectangular pulse, defined as

$$\Pi(x) = \begin{cases} 1 & \text{for } |x| \leqslant 1/2 \\ 0 & \text{elsewhere} \end{cases} \tag{2.7}$$

It is worth noting that, being the signal in (2.6) a "flat" pulse over all the own support, we have $T_\kappa = T \; \forall \kappa \geqslant 0$.

The received signal $s(t)$, backscattered by a stationary (i.e., without relative motion with respect to the radar) target, which can be assimilated to a point scatterer (point like target) located at range r in the volume irradiated by the radar, is an attenuated and delayed version of the transmitted signal, i.e.:

$$s(t) = q(\sigma, r)\, p(t - \tau) \exp\left(-j2\pi f_0 \tau\right) \tag{2.8}$$

where the delay corresponds to the round trip time:

$$\tau = 2r/c \tag{2.9}$$

and the scaling coefficient $q(\sigma, r)$ accounts for the attenuation induced by the range propagation (through the variable r) and for the amplification/attenuation (with respect to an isotropic scatterer) of the power returning to the antenna, related to the scattering characteristic of the target. This characteristic is "quantified" by the target radar cross section (RCS) σ and depends on the geometric and electromagnetic properties of the target and on its orientation with respect to the impinging radiation, see Chapter 1. Otherwise stated, numerically we have: $|q|^2 \propto \sigma$ and $|q|^2 \propto r^{-4}$, see Chapter 1. It is worth underlining that the exponential term at the right hand side is the phase change associated with the effect of the delay τ on the carrier. Note that the carrier is suppressed in the LPE representation: the phase change must be, therefore, explicitly introduced in the expression of the received echo when this is derived from the LPE signal representation, see Appendix A.2. Addressed in several sections of this book, this phase term plays a fundamental role in many important radar applications, especially in radar interferometric configurations.

Eq. (2.8) can also be expressed in terms of the spatial variables by using (2.9) and by the substitution $t = 2r'/c$:

$$s(r') = q(\sigma, r) p(r' - r) \exp\left(-j\frac{4\pi}{\lambda}r\right) \qquad (2.10)$$

where

$$p(r') = A\Pi\left(\frac{r'}{\rho_R}\right) \qquad (2.11)$$

with:

$$A = \sqrt{\frac{\mathcal{E}_p}{T}} \qquad (2.12)$$

and

$$\rho_R = \frac{cT}{2} \qquad (2.13)$$

Note that the functions in (2.10) and (2.11) should be indicated with symbols different from (2.8) and (2.4), respectively. To avoid weighting down the notation we let the difference being discriminated by the dependence on the time or range variables.

Eq. (2.8), with (2.6) and (2.4) (or, equivalently, (2.10) and (2.11)) shows that T (ρ_R) is the temporal (spatial) extension of the response of a point like target: it measures the spreading of the return from a point scatterer induced by the spreading of the pulse transmitted by the radar, i.e. by the band-limited nature of the radar. To better highlight this point, we rewrite (2.10) as:

$$s(r') = p(r') \otimes q(\sigma, r)\delta(r' - r) \exp\left(-j\frac{4\pi}{\lambda}r\right) \qquad (2.14)$$

where $\delta(r')$ is the Dirac delta function and $\otimes$ denotes the convolution operator. Eq. (2.14) highlights that a point like target located at a distance r from the radar, mathematically described by the Dirac delta function, produces a response (target echo) spread by the system, which acts as a filter with impulse response function $p(r')$ for the impulsive input, on a region (support) with extension ρ_R, centered in $r' = r$. As better explained in the following sections, the spreading of the target echo relates to the capability of the system in resolving (separating) the responses of multiple targets. The quantity ρ_R measures such a capability, more specifically, it provides the range resolution, i.e., the minimum separation between two point like targets easily (say "visually") distinguishable in the received signal. Lower resolutions, i.e. higher values of ρ_R, indicate larger spreading and hence a reduced capability of the system in discriminating the response by multiple targets.

An ideal system would map the response of a point like target with a Dirac delta function, i.e., should have a resolution $\rho_R \rightarrow 0$. According to (2.13), this means that the transmitted pulse should have $T \rightarrow 0$ or, equivalently, an infinite bandwidth. In fact, for the basic, i.e., rectangular pulse in (2.6), the resolution ρ_R can be rewritten as

$$\rho_R = \frac{c}{2B_t} \tag{2.15}$$

with $B_t = 1/T$ being the 3 dB pulse bandwidth.

In practice, however, T is lower bounded by both the system requirement on the transmitted energy $\mathcal{E}_p$ and the system limitation on the transmitted peak power $\mathcal{P}_p$, being such three parameters related as

$$T = \frac{\mathcal{E}_p}{\mathcal{P}_p} \tag{2.16}$$

In fact, on one hand, $\mathcal{E}_p$ is lower bounded by the minimum value of Signal-to-Noise ratio (SNR) required to avoid the overwhelming of the received echo by the unavoidable noise, which have been so far neglected. More specifically, referring to the terminology of radar applications, the transmitted energy cannot be reduced below certain bound without prejudicing the target detection performances. On the other hand, $\mathcal{P}_p$ is upper bounded by the limitations of the transmitter circuitry, particularly the radio frequency apparatus. As a consequence, according to (2.16), for a given required value of $\mathcal{E}_p$, T cannot be arbitrarily reduced by operating an increase of $\mathcal{P}_p$.

To emphasize the target echo with respect to the noise contribution, or better, for an effective implementation of the detection stage typically presents in surveillance radars, the received signal undergoes a processing after reception. Similarly to the basic processing typically implemented also in digital communication systems, the radar processing involves in fact a matched filtering operation [7]. More precisely, by matching the filter to the normalized (unitary energy) pulse, the output signal $y(t)$ results to be the mutual correlation between $p_u(t)$ and the shifted version of $s(t)$

$$y(t) = s(t) \otimes p_u^*(-t) \tag{2.17}$$

To highlight the filtering effect on the response of the target, the expression (2.8) of $s(t)$, which neglects the noise contribution, can be substituted in (2.17). This leads to

$$y(t) = q(\sigma, r)\sqrt{\mathcal{E}_p}\, r_u(t - \tau) \exp\left(-j2\pi f_0 \tau\right) \tag{2.18}$$

where

$$r_u(t) = p_u(\xi) \otimes p_u^*(-t) \tag{2.19}$$

is the autocorrelation function (ACF) of $p_u(t)$, i.e., of the transmitted (normalized) pulse. The case of rectangular pulse in (2.6) provides:

$$r_u(t) = \Lambda\left(\frac{t}{T}\right) \tag{2.20}$$

where $\Lambda(x)$ is the triangular pulse defined as

$$\Lambda(x) = \begin{cases} 1 - |x| & \text{for } |x| \leqslant 1 \\ 0 & \text{elsewhere} \end{cases} \tag{2.21}$$

Some remarks about the processing in (2.18) are now in order.

First of all, it can be shown [6,7] that the matched filter carrying out the above processing, which is sketched in Fig. 2.2, allows maximizing the Signal to Noise Ratio on the output sample at $t = \tau$, i.e.

$$\text{SNR} = \frac{|y(\tau)|^2}{\mathcal{N}_0 \mathcal{G} \mathcal{B}_N} \tag{2.22}$$

where $\mathcal{G}$ and $\mathcal{B}_N$ are the filter gain and Equivalent Noise Bandwidth (ENBW), respectively, whereas $\mathcal{N}_0$ is Power Spectral Density (PSD) of the (LPE) thermal noise contribution added in the receiver.

Secondly, it is worth underlining that the role played by the pulse $p_u(t)$ in the unprocessed (i.e., original) received echo $s(t)$ is played by the ACF $r_u(t)$ in the processed signal $y(t)$. Being $r_u(0) = 1$, it follows that $|y(\tau)|^2$ in (2.22) (i.e. the intensity of the signal in

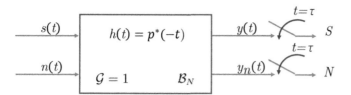

Figure 2.2. Basic processing on the receive line.

(2.18)) measures the energy of the received echo which is proportional to the transmitted Energy and the target RCS. As better explained in the next section, the matched filter operation allows improving the resolution capability of the system, especially when using pulses with proper "shapes".

2.2 Echoes from multiple targets

The radar concepts introduced in the previous sections for the case of a single target can be straightforwardly extended to the case of multiple targets. Under the linearity assumption corresponding to the vast majority of situations, the signal backscattered by multiple point like target with RCSs σ_k and located at range distances r_k from the radar, $k = 1, \ldots, K$, can be expressed as:

$$s(t) = \sum_{k=1}^{K} q_k \sqrt{\mathcal{E}_p}\, p_u(t - \tau_k) \exp\left(-j2\pi f_0 \tau_k\right) \qquad (2.23)$$

where $q_k = q(\sigma_k, r_k)$ and $\tau_k = 2r_k/c$ are the coefficient and the time delay corresponding to the generic kth target, respectively.

Accordingly, the output signal of the matched filter is given by:

$$y(t) = \sum_{k=1}^{K} q_k \sqrt{\mathcal{E}_p}\, r_u(t - \tau_k) \exp\left(-j2\pi f_0 \tau_k\right) \qquad (2.24)$$

where $r_u(t)$ is the ACF in (2.19).

As already discussed, the capability of the radar in separating the echos scattered by different (point like) targets is quantified by the resolution parameter which is strictly related to the spread of the target echo induced by the band-limited characteristics of the system. This concept can be now better understood by looking at the expressions of the unprocessed and processed signal in (2.23) and (2.24), respectively. Both are the superposition of K pulses, each one centered in the round trip time associated with the corresponding target position. To resolve the targets, their echoes should be "somehow" distinguishable: this is certainly possible when they are not overlapped, i.e., if the distance between each target pair is at least equal to temporal spread of the echoes. However, differently from the pulses $p_u(t)$ in the unprocessed signal $s(t)$, the pulses $r_u(t)$ appearing in the processing output $y(t)$ in (2.24) are characterized by a shape which is peaked at the pulse center: this characteristic is a direct consequence of the properties of the ACF. By inspecting the processed signal, the discrimination of the targets and identification of the peaks is thus possible even if the echos are partially overlapped, provided that the peaks are still distinguishable. In other words, while the resolution attainable from the unprocessed signal is

$$\rho_R = \frac{cT}{2} \tag{2.25}$$

where T is the width (i.e., the null-to-null extension) of $p_u(t)$, the resolution achievable from the processed signal is

$$\rho_r = \frac{cT_\kappa}{2} \tag{2.26}$$

where T_κ is the κ dB duration of $r_u(t)$ and thus lower than its null-to-null extension. The specific T_κ duration can be properly set according to the shape of the processed pulse. Hereafter, we refer to ρ_R as the *raw resolution* and ρ_r as the *system resolution*, to better emphasize the fact that they are attained, respectively, *before* and *after* the system processing. Summarizing, differently from the raw resolution, which is always associated with the (maximum) extension of the transmitted pulse, the system resolution depends on the considered κ dB extension of its ACF. For this reason the system resolution in (2.26) is also referred to as κ dB system resolution or simply κ dB resolution.

To better explain this concept, an example on simulated data is provided in the following. In Fig. 2.3 it is shown the processed echo of a point like target located at a range of 3 Km illuminated by a rectangular pulse whose temporal width corresponds to a raw resolution $\rho_R = 3$ m. It is worth noting, also according to the expressions (2.6) and (2.20) of the rectangular pulse and its autocorrelation function, respectively, that the pulse width of the processed echo is doubled, thus leading to a null-to-null spatial extension equal to $2\rho_R$. In Fig. 2.4 it is shown the plot corresponding to Fig. 2.3 after conversion in dB: by limiting the horizontal axis to the interval of extension ρ_R and centered in the target location, it is highlighted that the 6 dB system resolution ρ_r equals the raw resolution ρ_R. It can be shown that, by transmitting the rectangular pulse in (2.6), the 6 dB system resolution $\rho_r = \rho_R$ is actually the minimum separation allowing the resolution of two point like targets. To this aim, let us consider the simple scenario composed by three point like targets, whose starting relative separation is ρ_R and is then increased of the 10%. The processed signal, displayed in Fig. 2.5, shows that a separation exactly equal to ρ_R does not generate peaks in the output thus impairing the easy localization of the three targets. It is sufficient a small (10%) increase of the distance with respect to ρ_R for generating three distinct peaks in the output of the correlator.

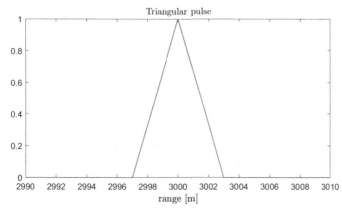

Figure 2.3. Response of a target to an ideal rectangular pulse after the processing.

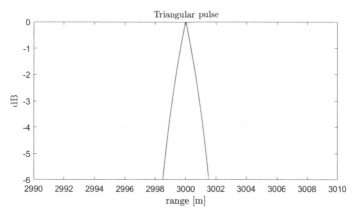

Figure 2.4. Response, in dB, of a target to an ideal rectangular pulse after the processing.

The above example has shown how the matched filter aids in "visually" separating multiple targets. Nevertheless, it has also shown that the use of the simple rectangular pulse in (2.6) does not

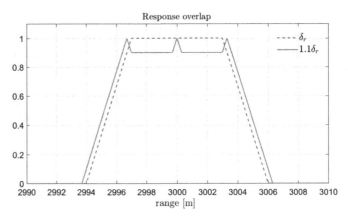

Figure 2.5. Example of processed received data for three targets separated by a distance corresponding to the resolution (dashed blue (dark gray in print version) line) and to the resolution increased by 10% (red (mid gray in print version) line).

allow improving the system resolution with respect to the raw one attainable before the processing of the received signal. To achieve an improved system resolution, while limiting or even reducing the transmitted peak power, the transmitted pulse should belong to the class of the so-called "sophisticated signal".

2.3 Chirp compression

We saw that the raw resolution ρ_R is the range mapping of the pulse time duration T, which is lower bounded by the maximum transmittable peak power. On the other hand, the system resolution ρ_r is attainable after the matched filter and can be defined as the range mapping of the κ dB duration T_κ of the ACF of the transmitted pulse. Such a duration represents the effective time extension of the ACF and usually is set as a very high percentage of the reciprocal of the transmitted frequency bandwidth B_t. In other words, while the raw resolution is expressed as in (2.25), i.e.:

$$\rho_R = \frac{cT}{2} \tag{2.27}$$

the system resolution is given by:

$$\rho_r \approx \frac{c}{2B_t} \tag{2.28}$$

Accordingly, the ratio between the two resolutions:

$$\frac{\rho_R}{\rho_r} \approx B_t T \tag{2.29}$$

almost equals the time-bandwidth product of the transmitted pulse. Therefore, a pulse shaped such as to provide $B_t T \gg 1$ leads to an improvement of the final system resolution: the matched filter allows "to compress" the low (i.e., numerically large) raw resolution ρ_R in (2.27) in the high (i.e., numerically small) system resolution ρ_r in (2.28). In this case, the processing implemented by the matched filter is referred to as pulse compression, and the ratio (2.29) between the raw and system resolutions is the compression ratio: the higher the compression ratio, the higher the gain of range resolution after the pulse compression.

Simple pulses, as the rectangular one previously exploited, have a time-bandwidth product $B_t T \approx 1$ and, thus, are not suitable for achieving high system resolutions when large values of T are required. Pulses belonging to the class of the so-called sophisticated signal, characterized by large time-bandwidth product ($B_t T \gg 1$), can be instead profitably exploited. Among them, the referred to as "chirp" is a frequency modulated pulse whose instantaneous frequency increases (up-chirp) or decreases (down-chirp) linearly with the time. The (real) band-pass and (complex) LPE representations of a chirp signal are given by

$$p_{BP}(t) = \sqrt{\frac{\mathcal{E}_p}{T}} \cos\left(2\pi f_0 t \pm \pi \alpha t^2\right) \Pi\left(\frac{t}{T}\right) \qquad (2.30)$$

and

$$p(t) = \sqrt{\frac{\mathcal{E}_p}{T}} \exp\left(\pm j\pi \alpha t^2\right) \Pi\left(\frac{t}{T}\right) \qquad (2.31)$$

respectively. The sign of the quadratic phase term accounts for the frequency variation ($+$ for the up-chirp and $-$ for the down-chirp). The factor α is the chirp-rate, a positive real number determining the degree of modulation and, thus, determining the transmitted bandwidth. The latter, according to the Carson's rule [7], can be written as:

$$B_t \approx f_{\max} - f_{\min} = \alpha T \qquad (2.32)$$

where $f_{\min}$ and $f_{\max}$ are the minimum and the maximum values of the instantaneous frequency, respectively (see Appendix A.2). Fig. 2.6 shows an example with a transmitted up-chirp.

Eq. (2.32) shows that, for a given value of T, α allows tuning B_t so to achieve the required system resolution in (2.28).

The matched filter operation carried out in the receiver, actually implements the so-called chirp compression step. To show this we develop the expression of the ACF $r_p(t)$ of the pulse in (2.31) by applying the factorization property in Appendix A.3. This

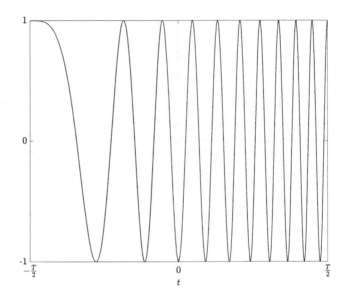

Figure 2.6. Example of transmitted chirp pulse.

leads to:

$$r_p(t) = \frac{\mathcal{E}_p}{T} \Pi\left(\frac{t}{2T}\right) \exp(\mp\pi\alpha t^2) \mathcal{F}_\xi \left\{ \Pi\left(\frac{\xi - t/2}{T - |t|}\right) \right\} \Bigg|_{f=\mp\alpha t} \qquad (2.33)$$

where $\mathcal{F}_\xi$ is the Fourier transform operator A.1 from the ξ to the f domains.

Further developing (2.33) leads to:

$$r_p(t) = \mathcal{E}_p \Lambda\left(\frac{t}{T}\right) \operatorname{sinc}\left[B_t\left(1 - \frac{|t|}{T}\right)t\right] \qquad (2.34)$$

where

$$\operatorname{sinc}(x) = \frac{\sin(\pi x)}{\pi x} \qquad (2.35)$$

The function in (2.34) is, thus, sinc-like shaped within the time interval $[-T, T]$: its nulls are achieved at

$$t_k = \pm\frac{T}{2}\left(1 \pm \sqrt{1 - \frac{4k}{B_t T}}\right) \qquad (2.36)$$

for $k \in \{0, \ldots, \lfloor\frac{B_t T}{4}\rfloor\}$, where $\lfloor x \rfloor$ is the greatest integer lower than x. Being the transmitted chirp characterized by a high time-bandwidth product ($B_t T \gg 1$), the null positions in (2.36) in the interval $(-T/2, T/2)$ corresponding to the values of k as small as to make reliable the approximation

$$\sqrt{1 - \frac{4k}{B_t T}} \approx 1 - \frac{2k}{B_t T} \qquad (2.37)$$

can be in turn approximated as

$$t_k \approx \pm \frac{k}{B_t} \qquad (2.38)$$

The time extension of the main lobe is thus given by

$$\Delta t \approx \frac{2}{B_t} \qquad (2.39)$$

and, at least within the interval where the approximation in (2.38) is reliable, the compressed chirp is well approximated by the (more simple) expression

$$r_p(t) \approx \mathcal{E}_p \text{sinc}\,(B_t t) \qquad (2.40)$$

Eqs. (2.39) and (2.40) confirm that the system resolution improves by increasing the transmitted bandwidth.

It can be shown that the value of system resolution in (2.28) corresponds more precisely to the 4 dB extension of the main lobe of the compressed chirp.

Most common is, however, the 3 dB system resolution, which results to be

$$\rho_r = 0.89 \frac{c}{2B_t} \qquad (2.41)$$

The chirp compression of the received signal $s(t)$ can be carried out in the time domain

$$y(t) = \int s(\xi) p^*(\xi - t) d\xi \qquad (2.42)$$

or more efficiently in the frequency domain

$$Y(f) = S(f) P^*(f) \qquad (2.43)$$

being $P(f)$, $S(f)$, and $Y(f)$ the Fourier transforms of $p(t)$, $s(t)$, and $y(t)$, respectively. The evaluation of $P(f)$ is thus in order. By referring without loss of generality to the up-chirp in (2.31), we have:

$$P(f) = \int_{-\infty}^{\infty} p(t) e^{-j2\pi ft} = \int_{-\infty}^{\infty} \Pi\left(\frac{t}{T}\right) e^{j\left(\pi\alpha t^2 - 2\pi ft\right)} dt$$

$$(2.44)$$

The integral in (2.44) cannot be evaluated in closed form. However, a simple and generally good approximated expression can be found by resorting to the Stationary Phase Approximation (SPA), see Appendix A.4, which is largely used to determine the Fourier transform of a frequency modulated signal with an arbitrary phase.

Application of the SPA to the integral in (2.44) leads to the following approximation for the spectrum of the transmitted signal:

$$P(f) \simeq \sqrt{\frac{j}{\alpha}} \exp\left(-j\frac{\pi}{\alpha}f^2\right) \Pi\left(\frac{f}{B_t}\right) \tag{2.45}$$

for sufficiently large α (i.e., for an high compression ratio) and for values of the frequency far from $f = \pm B_t/2$, corresponding to the values of stationary point where the envelope of the chirp is discontinuous. The amplitude term in (2.45) is often neglected. The pair $p(t) \longleftrightarrow P(f)$ can be hence rewritten as:

$$p(t) = e^{+j\frac{B_t}{T}\pi t^2} \Pi\left(\frac{t}{T}\right) \longleftrightarrow P(f) \simeq e^{-j\frac{T}{B_t}\pi f^2} \Pi\left(\frac{f}{B_t}\right) \tag{2.46}$$

where the time-frequency symmetry is better highlighted: an up-chirp in the time domain corresponds to a down-chirp in the frequency domain with reciprocal chirp-rate and vice-versa. According to (2.43) and (2.45), the chirp compression in the frequency domain is, thus, the compensation of the quadratic phase term. It is performed by the matched filter $P^*(f)$ that, having $P(f)$ a constant amplitude, is equivalent to the inverse filter $P^{-1}(f)$. In other words the chirp compression can be interpreted as the operation of recovering the "spectrum of the back-scattering" within the "system bandwidth" B_t. As a last comment, we remark that (2.45) is an approximation of the chirp spectrum. Actually, depending on the product $B_t T$, the spectrum is characterized by the presence of amplitude oscillations, which are emphasized at the edges of the bandwidth. The phenomenon, known as Gibbs oscillation, is widely treated in the existing literature [8] and outside of the scope of this book.

2.4 Multiple pulses

Actually, a pulsed radar transmits a signal constituted by a series of N uniformly separated pulses, usually referred to as *pulse train*, so that the scene is scanned regularly over the time. The time separation T_I between subsequent pulses is referred to as Inter-Pulse Period (IPP) or Pulse Repetition Interval (PRI) [5], whereas the inverse $1/T_I$ is the Pulse Repetition Frequency (PRF).

The (real) band pass transmitted signal is thus:

$$\check{p}_{BP}(t) = \sum_{n=-N/2}^{N/2-1} p_{BP}(t-t_n) \tag{2.47}$$

where $p_{BP}(t)$ is the basic pulse, $t_n = nT_I$ is the time instant corresponding to the center of the nth pulse and, without loss of generality, an even number of N pulses has been considered. Moreover, we refer to a *coherent pulse train*, which is characterized by pulses with the same carrier frequency and known initial phase: the total time span NT_I of the signal in (2.47) is thus referred to as Coherent Processing Interval (CPI).

Although the transmitted train is usually constituted by chirp pulses, for sake of simplicity and without loss of generality we prefer carrying out the subsequent discussion by considering again rectangular pulses. Accordingly, the signal in (2.47) specializes as:

$$\check{p}_{BP}(t) = \sqrt{\frac{\mathcal{E}_p}{T}} \sum_{n=-N/2}^{N/2-1} \Pi\left(\frac{t-t_n}{T}\right) \cos[2\pi f_0(t-t_n)] \tag{2.48}$$

where all pulses are assumed to be starting with the same phase value: this is obtained by shifting, for each of them, the carrier with the known phase offsets $\varphi_n = 2\pi f_0 t_n$.

The corresponding LPE is given by:

$$\check{p}(t) = \sqrt{\frac{\mathcal{E}_p}{T}} \sum_{n=-N/2}^{N/2-1} \Pi\left(\frac{t-t_n}{T}\right) \tag{2.49}$$

where the exponential terms associated with the phase offsets φ_n have been neglected for sake of simplicity. It is worth noting that this fact does not affect the generality of the discussion, since known phase offsets can be always compensated at receiver stage.

The PRI should be set with the constraint of avoiding the overlap of the returns from the scene over different illuminating intervals. Letting $\tau_{max} = 2r_{max}/c$ to be the maximum delay, i.e., corresponding to the farthest scatterer, one should choose $T_I > \tau_{max}$ so that the echo from the farthest range r_{max} should be received before the transmission of the next pulse. This requirement may in some cases be too stringent, especially when the scene is far away from the transmitter as in the case of satellite imaging radars debated in the following chapters. A less stringent requirement is achieved by referring to the spanned range interval. Letting $\tau_{min} = 2r_{min}/c$ be the minimum delay corresponding to the nearest scatterer, the following lower limit on the PRI must be satisfied:

$$T_I \geqslant \tau_{max} - \tau_{min} = 2\frac{r_{max} - r_{min}}{c} \qquad (2.50)$$

Condition in (2.50) is tantamount to guarantee that the returns from the scene can be contained within a time interval between two subsequent pulses.

A remark is now in order. Let us consider the presence of a relative motion between the radar and the target: without loss of generality, we can assume that it is due to a moving target in the field of view of a stationary radar. In the case of a single pulse system, neglecting this movement means neglecting the target displacement occurring during the round trip time of the transmitted pulse. However, being the target speed typically by far much lower than the pulse speed, as first approximation the stationary target assumption can be considered to some extent reliable. In the case of a pulse train, instead, the target displacement which accumulate during the CPI cannot be anymore neglected: the stationary target assumption is in this case generally unacceptable. A better approximation, in this case, is that of considering different target positions at the time instants corresponding to the pulses transmission, keeping the stationary assumption just for the round trip time of each pulse: this condition will be relaxed in the next section.

Accordingly, let us denote with $R(t_n)$ the distance of the target at time t_n relevant to the nth transmitted pulse and with $\tau(t_n) = 2R(t_n)/c$ the corresponding delay. From now on the use of a capital letter for the distance emphasizes the dependence of the range over the time. The received signal can be therefore expressed as:

$$s(t) = q\sqrt{\frac{\mathcal{E}_p}{T}}\sum_{n=-N/2}^{N/2-1}\Pi\left(\frac{t - t_n - \tau(t_n)}{T}\right)\exp\left(-j2\pi f_0\tau(t_n)\right) \qquad (2.51)$$

It is worth noting that the signal in (2.51) should be considered as a one-dimensional function with respect to the variable t spanning the whole echo time interval, which is given by:

$$t \in \left[-\frac{NT_I}{2} + t_{d_{min}}, \frac{(N-2)T_I}{2} + t_{d_{max}}\right]$$

However, since the echos corresponding to consecutive pulses are not overlapped, an equivalent two-dimensional representation of the received signal could be exploited by collecting the N echos as so many functions of the time, provided that the position $t - t_n \rightarrow t$

is done. Accordingly, the equivalent two-dimensional representation of the signal in (2.51) results to be:

$$s(t, t_n) = q \sqrt{\frac{\mathcal{E}_p}{T}} \, \Pi\left(\frac{t - \tau(t_n)}{T}\right) e^{-j2\pi f_0 \tau(t_n)} \qquad n = -N/2, \, .., \, N/2 - 1$$

(2.52)

where the axes t and t_n correspond to two different temporal scales: the former, t, is referred to as "fast time", since it is relevant to the pulse propagation; the latter, t_n, instead, is referred to as "slow time", since it is associated with the pulse repetition. The two-dimensional representation in (2.52), typically adopted for the imaging radars, would have a graphical correspondence where all the transmitted pulses in (2.49) would be aligned along the slow time axis (see Fig. 2.7).

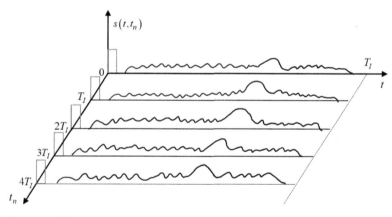

Figure 2.7. 2D arrangement of the received signal as a function of the slow time t_n and of the fast time t.

The received signal is processed with the filter matched to the basic pulse of the transmitted signal. The one-dimensional and two-dimensional representation of the output are thus:

$$y(t) = q \sum_{n=-N/2}^{N/2-1} r_p(t - t_n - \tau(t_n)) \exp(-j2\pi f_0 \tau(t_n)) \qquad (2.53)$$

and

$$y(t, t_n) = q r_p(t - \tau(t_n)) \exp(-j2\pi f_0 \tau(t_n)) \qquad (2.54)$$

respectively, where $r_p(t)$ is the ACF of the basic pulse.

Similarly to the case of the transmission of a single pulse, the received signal can be referred to the spatial coordinates. The two-dimensional representation in (2.54) can thus be rewritten as

$$y(r', t_n) = q r_p \left(r' - R(t_n) \right) \exp \left(-j \frac{4\pi}{\lambda} R(t_n) \right) \qquad (2.55)$$

where r' is the variable associated with the fast time, i.e., $t - t_n \rightarrow t = 2r'/c$, and, with a slight abuse of notation, we have posed $r_p(2r'/c) \rightarrow r_p(r')$.

Eq. (2.55) shows that, in general, the signal received from a generic scatterer migrates in range with a slow-time dependent distance law $R(t_n)$, which impacts the position of the return, i.e. the argument of the ACF, as well as the phase factor in the carrier. The latter will be relevant for the frequency analysis along the slow time (Doppler).

2.5 Doppler effect

Beside the possibility to detect the target and determine its position, coherent radar systems allow also measuring instantaneously the radial component of the velocity of a moving target. This goal can be achieved by exploiting the so-called Doppler effect, which consists in a frequency shift between the transmitted and the received radiation induced by the motion of the target with respect to the radar. As shown in the following, the variation of the distance of the scatterer induces not only a variation of pulse centering and phase shift in the received echo, but also a more complex modification of the signal.

We begin by considering a stationary coherent pulsed radar, transmitting the train of N pulses:

$$\check{p}(t) = \sum_{n=-N/2}^{N/2-1} p(t - t_n) \qquad (2.56)$$

where $p(t)$ is the basic pulse.

2.5.1 Effect of the movement of a target on a single echo

In the previous section, we accounted for the moving target by assuming that it stops during the time interval between the transmission and the reception of each pulse (Start-Stop approximation). A deeper analysis of the Doppler effect requires, however, relaxing this assumption. We start from the description of the motion of the target. For sake of simplicity, let us assume a rectilinear and uniform motion of the target described by the velocity vector $\boldsymbol{v}$, see Fig. 2.8. In the more general case of a nonuniform motion, this can be in most of the cases an assumption valid in the temporal interval associated with the pulse transmission and reception. The target equation of motion along the range can be expressed as:

$$R(t) = r - v_r t \qquad (2.57)$$

where r is the range position of the target at the reference time $t = 0$ and v_r is the radial component of the target velocity vector $\boldsymbol{v}$. The latter, sometimes also referred to as range rate, is:

$$v_r = \boldsymbol{v} \cdot (-\hat{\boldsymbol{r}})$$

where, by referring again to Fig. 2.8, $\hat{\boldsymbol{r}}$ is the unit vector of radar line of sight (LOS), which is the direction connecting the TX/RX antenna and the target, oriented toward the target. The target motion is represented by the blue (dark gray in print version) line in Fig. 2.9 for $v_r > 0$.

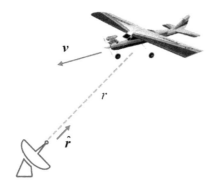

Figure 2.8. Doppler effect basic reference scheme.

We now aim to account for the motion of the target during the transmission of the pulse, as well as during the travel of the pulse from the transmitter to the target. We therefore rewrite the n-th pulse in (2.56) as a superposition of contributions which are distributed in the $[-T/2, T/2]$ interval around t_n, that is:

$$p(t - t_n) = p(t) \otimes \delta(t - t_n) = \int_{-T/2}^{T/2} p(\xi)\delta[t - (t_n + \xi)]d\xi \qquad (2.58)$$

Eq. (2.58) allows interpreting the transmitted radiation as the superposition of elementary contributions $p(\xi)\delta[t - (t_n + \xi)]$ transmitted at times $t_n + \xi$, with $\xi \in [-T/2, T/2]$ and $n = -N/2, \ldots,$

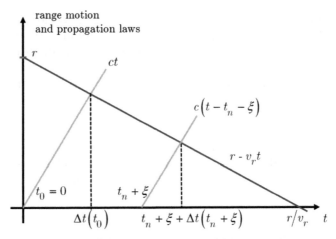

Figure 2.9. Representation of the one way forward motion (along the line of sight) of the pulse (red (mid gray in print version) line) and of the target (blue (dark gray in print version) line) for a positive v_r.

$N/2 - 1$. Neglecting amplitude factors inessential for this analysis, thanks to the linearity we can express the received echo $s_n(t)$ corresponding to the n-th pulse backscattered by a target as the superposition of the response to the generic Dirac pulse $\delta[t - (t_n + \xi)]$ in (2.58). The latter will be delayed by a quantity, say $\tau(t_n + \xi)$, that depends on the transmitted time $t_n + \xi$ due to the target motion. We have therefore:

$$s_n(t) = \int_{-T/2}^{T/2} p(\xi)\delta[t - (t_n + \xi) - \tau(t_n + \xi)]d\xi \qquad (2.59)$$

To evaluate the time delay $\tau(t_n + \xi)$ we express it as twice the time $\Delta t(t_n + \xi)$ required by the elementary contribution $\delta[t - (t_n + \xi)]$ to reach the moving target. Fig. 2.9 shows the motion of the target along the range described by (2.57) and the range travelled by pulses sent at $t_0 = 0$ (reference time) and at a generic t_n. This figure shows that Δt can be achieved by considering the intersection between the motion of the pulse and the motion of the target. From Fig. 2.9, considering the Dirac pulse at $t_n + \xi$, we therefore have:

$$c\Delta t = r - v_r(t_n + \xi + \Delta t) \qquad (2.60)$$

which provides:

$$\Delta t = \Delta t(t_n + \xi) = \frac{[r - v_r(t_n + \xi)]}{c + v_r} \qquad (2.61)$$

Finally, considering the round trip delay we have:

$$\tau(t_n + \xi) = \frac{2[r - v_r(t_n + \xi)]}{c + v_r} \tag{2.62}$$

In (2.62), the presence of the target velocity at the denominator is a consequence of the movement of the target during the time needed to each elementary contribution $\delta[t - (t_n + \xi)]$ to reach the target. As a matter of fact, for a target proceeding toward the radar, the velocity increases with respect to the light speed thus reducing the delay; the opposite, i.e. a velocity reduction is pertinent to a target receding from the radar.

Eq. (2.62) can be recast in the more convenient form:

$$\tau(t_n + \xi) = \frac{1}{1 + \beta} [\tau_{SS}(t_n) - 2\beta\xi] \tag{2.63}$$

where $\beta = v_r/c$ and

$$\tau_{SS}(t_n) = \frac{2}{c} R(t_n) = \frac{2}{c}(r - v_r t_n) \tag{2.64}$$

The time delay $\tau_{SS}(t_n)$ in (2.64), which differently from $\tau(t_n + \xi)$ does not depend on the time ξ running on the pulse duration interval, is the delay achieved by assuming, as in the previous section, that the target is not moving in the time interval between the transmission and reception of the pulse. The latter assumption is referred to as Start-Stop approximation.

The use of the time delay expression in (2.63) allows rewriting the delayed elementary contribution transmitted at $t_n + \xi$ as:

$$\delta[t - (t_n + \xi) - \tau(t_n + \xi)] = \frac{1 + \beta}{1 - \beta} \delta\left[\xi - \left(\frac{1 + \beta}{1 - \beta}\right)\left(t - t_n - \frac{\tau_{SS}(t_n)}{1 + \beta}\right)\right] \tag{2.65}$$

where the time-scaling property of δ, $\delta(at) = \delta(t)/|a|$, has been used. Substituting (2.65) in (2.59), we finally have:

$$s_n(t) = p\left[\left(\frac{1 + \beta}{1 - \beta}\right)\left(t - t_n - \frac{\tau_{SS}(t_n)}{1 + \beta}\right)\right] \tag{2.66}$$

where an inessential amplitude factor (which for small β is close to 1) has been neglected.

Let us rewrite (2.66) as:

$$s_n(t) = p\left[a\left(t - t_n - \tau_n\right)\right] \tag{2.67}$$

where

$$a = \frac{1 + \beta}{1 - \beta} \tag{2.68}$$

and

$$\tau_n = \frac{\tau_{SS}(t_n)}{1+\beta} = \frac{1}{1+\beta}\left[\frac{2}{c}(r-v_r t_n)\right] \tag{2.69}$$

Eq. (2.67) shows that the fast time of the echo backscattered by the moving target is scaled by a and delayed by τ_n. According to (2.68), $a > 1$ for $v_r > 0$ and $a < 1$ for $v_r < 0$: the fast time is, thus, compressed or expanded by the target motion if the target is approaching to or departing from the radar, respectively. Finally, (2.69) shows that the time delay τ_n is a scaled version, by $(1+\beta)^{-1}$, of the delay corresponding to $\tau_{SS}(t_n) = 2(r - v_r t_n)/c$. As for the delay scaling, it is worth noting that the delay decreases for a target proceeding toward the sensor while it increase for a target receding from the radar.

2.5.2 Analysis on the band pass signal

To further analyze the effect of the target movement on both the envelope and the carrier of the received signal, let us specialize the results of the previous subsection for the (real) BP representation of the involved signals.

Following the notation in (2.3), the relationship between the elementary BP pulse and its LPE is:

$$p_{BP}(t) = \Re\{p(t)e^{j2\pi f_0 t}\} \tag{2.70}$$

independent of the adopted waveform. A train of N pulses of length T is supposed to be transmitted. According to (2.67), the nth received echo is:

$$s_{BP}(t,n) = \Re\{p[a(t-t_n-\tau_n)]e^{j2\pi f_0 a(t-t_n-\tau_n)}\} \tag{2.71}$$

where an inessential amplitude factor has been omitted.

Eq. (2.71) shows that the envelope of the received echo is delayed by τ_n and scaled in time by a, thus achieving a scaled length T/a.

As for the carrier, it results:

$$e^{j2\pi f_0 a(t-t_n-\tau_n)} = e^{j2\pi f_r(t-t_n)}e^{-j\varphi_r} \tag{2.72}$$

where

$$f_r = af_0 = f_0 + \frac{2\beta}{1-\beta}f_0 \tag{2.73}$$

is the received carrier frequency and

$$\varphi_r = 2\pi f_0 a\tau_n = \frac{1}{1-\beta}\frac{4\pi}{\lambda}(r-v_r t_n) \tag{2.74}$$

is the received phase shift associated to the time delay τ_n.

The difference $f_d = f_r - f_0$ between the received and the transmitted carrier frequencies represents the so-called Doppler shift. According to Eq. (2.73), it is given by:

$$f_d = \frac{2\beta}{1-\beta} f_0 = \frac{1}{1-\beta} \frac{2v_r}{\lambda} \qquad (2.75)$$

which represents a positive or negative Doppler shift according to the sign of the v_r. More precisely, by the definition of v_r it follows that targets approaching the radar have a positive Doppler shift, whereas targets receding from the radar show a negative Doppler shift. Moreover, the factor $(1-\beta)^{-1}$ amplifies positive Doppler shift, while a reduction is achieved for the opposite case.

The expression of signal received over multiple pulse is therefore given by the superposition of the echos in (2.67):

$$s_{BP}(t) = \Re \left\{ \sum_{n=-N/2}^{N/2-1} p\left[a(t - t_n - \tau_{SS}(t_n)/(1+\beta))\right] \right.$$
$$\left. \exp\left[j2\pi (f_0 + f_d)(t - t_n) - \frac{4\pi}{\lambda} \frac{(r - v_r t_n)}{1-\beta} \right] \right\} \qquad (2.76)$$

Eq. (2.76) is the most general expression for the received signal in the presence of a moving target derived according to the classical mechanics. Frequently the above expression is approximated by assuming $\beta \ll 1$ as follows:

$$s_{BP}(t) = \Re \left\{ \sum_{n=-N/2}^{N/2-1} p\left[t - t_n - \tau_{SS}(t_n)\right] e^{j2\pi(f_0 + f_d)} \right.$$
$$\left. e^{-j\frac{4\pi}{\lambda}(r - v_r t_n)} \right\} \qquad (2.77)$$

As shown in the subsequent chapters, the Doppler shift plays an important role in radars, especially in radar imaging. The presence of the Doppler shift has a twofold effect. On one side its estimation allows measuring the target velocity (at least the radial component). The use of a pair of radars allows even determining two velocity components: specifically the component of the three dimensional velocity vector in the plane determined by the line of sight unit vectors associated with the two radars. On the other hand the Doppler shift induces a reduction of the signal power level at the output of the matched filter. Investigation of the latter aspect is instructive and therefore carried out below.

2.5.3 Analysis on the matched filter output

As described in Section 2.1, the basic radar processing consists of filtering the received signal with the filter matched to the transmitted pulse and then sampling where the output amplitude is maximized. The presence of the target is usually declared if this sample overcomes a properly set threshold: in this case the target position is provided by the range corresponding to the time delay corresponding to the sampling time, see Fig. 2.10.

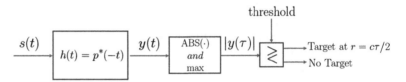

Figure 2.10. Block scheme of basic radar processing.

A Doppler shift generated by the relative movement between target and sensor generates a mismatch between the target echo and the matching pulse, thus reducing the level of the output sample. This condition may impair even the detection of the target when, as happens in real cases, the thermal noise is superimposed to the signal at receiver stage. To better understand this concept, let us go back to the LPE representation of the signals and, for sake of simplicity, let us consider the echo corresponding to $n = 0$ only in (2.77), that is:

$$s_0(t) = p\,(t - \tau_{SS})\exp[j2\pi(f_d t - f_0\tau_{SS})] \tag{2.78}$$

where $\tau_{SS} = \tau_{SS}(0) = 2r/c$.

The amplitude of the output of the matched filter is given by:

$$
\begin{aligned}
|y_0(t)| &= \left|s_0(t) \otimes p^*(-t)\right| \\
&= \left|\int p(\xi)p^*(\xi - (t - \tau_{SS}))\exp(j2\pi f_d\xi)\,d\xi\right| \tag{2.79} \\
&= \chi_p(t - \tau_{SS}, f_d)
\end{aligned}
$$

where

$$\chi_p(t, f) = \left|\int p(\xi)p^*(\xi - t)\exp(j2\pi f\xi)\,d\xi\right| \tag{2.80}$$

is the ambiguity function of the pulse $p(t)$, defined as the amplitude of the time-frequency autocorrelation function [5]. It is a two-dimensional (time-frequency) function such that:

$$\chi_p(t, f) \leqslant \chi_p(0, 0) \tag{2.81}$$

i.e., with an absolute maximum at $t = 0$ and $f = 0$, where its main lobe is centered. Accordingly, Eq. (2.79) highlights that the mismatch between the received echo and the transmitted pulse introduced by the Doppler shift causes a reduction in the output amplitude peak, which in general could even not be reached at $t = \tau_{SS}$. In the most realistic case, which accounts also for the presence of the additive thermal noise, the mismatch can impair not only the target localization (selection of the maximum) but also the target detection (overcoming of the threshold).

This drawback can be circumvented by tailoring the matched filter to the Doppler shift, i.e. by means of the Doppler-dependent filter:

$$h_D(t, f_D) = p^*(-t) \exp(-j2\pi f_D) \tag{2.82}$$

whose Doppler-dependent output has amplitude:

$$|y_0(t, f_D)| = \chi_p(t - \tau_{SS}, f_D - f_d) \tag{2.83}$$

Eq. (2.83) shows that the ambiguity function can be seen as the postprocessing response of the target to the transmitted pulse: its main lobe is centered in τ_{SS} and f_d so that, by jointly maximizing the amplitude of the output over the time-frequency domain, it is possible to avoid the attenuation of the peak and in the meantime to estimate both the range position and the radial velocity of the target, related to τ_{SS} and f_d, respectively.

A further consideration on the ambiguity function is now in order. The cuts of (2.80) for $f = 0$ and $t = 0$ are

$$\chi_p(t, 0) = \left| \int p(\xi) p^*(\xi - t) \xi \right| = \left| \int |P(f)|^2 \exp(-j2\pi ft) df \right| \tag{2.84}$$

and

$$\chi_p(0, f) = \left| \int p(\xi) p^*(\xi) \exp(j2\pi f\xi) d\xi \right| = \left| \int |p(t)|^2 \exp(j2\pi ft) dt \right| \tag{2.85}$$

respectively, where the right hand side of (2.84) is the (amplitude of the) ACF of the pulse $p(t)$ expressed as inverse FT of its energy spectral density. According to the duality property of the FT, Eqs. (2.84) and (2.85) show that, independent from the shape of the transmitted pulse, the extension of the main lobe of the ambiguity function along the time and the frequency direction is given by the reciprocal of the transmitted bandwidth B_t and duration T, respectively: such extensions provide a measure of the time (range) and Doppler (radial velocity) system resolutions achievable with the transmitted pulse. The former, as defined in Section 2.2, is the minimum temporal separation between two point like targets "visually" distinguishable and is driven by transmitted bandwidth: it improves by increasing B_t; the latter is the minimum measurable difference between the Doppler shifts of two point like targets and is driven by the transmitted duration: it improves by increasing T. This seems to suggest that sophisticated pulses, which can be transmitted with very high values of both T and B_t, allow achieving very high level of resolution both in time and Doppler domains. However, such a result is only partially true, since it must be considered in the light of the uncertainty principle, which expresses the following property of the ambiguity function:

$$\iint \chi_p(t, f) dt df = \mathcal{E}_p \tag{2.86}$$

where $\mathcal{E}_p$ is the energy of the transmitted pulse. This property constrains the narrowing of the main lobe in one direction to be counterbalanced by i) its widening in the other direction and/or ii) an increase of the contribution coming from the side lobes. The first situation is almost intuitive and characteristic of the basic pulses, as the rectangularly shaped, where the increase of the time length produces a reduction of the transmitted bandwidth. The last situation is typical of pulses with a high $B_t T$ product, as the chirped pulse, where very high values of T (and therefore of B_t) can even make the value of the local maxima associated with the side lobes indistinguishable from the absolute maximum where the main lobe is centered.

At this point, it is instructive to analyze the ambiguity function of the two pulses mentioned in the above discussion and exploited in the previous sections. As for the rectangular pulse, the use of the factorization property in Appendix A.3 leads to:

$$
\begin{aligned}
\chi_p(t, f) &= \left| \Pi\left(\frac{t}{2T}\right) \mathcal{F}_\tau\left\{ \Pi\left(\frac{\tau - t/2}{T - |t|}\right) \right\} \right| \\
&= \left| \Lambda\left(\frac{t}{T}\right) \operatorname{sinc}\left[fT\Lambda\left(\frac{t}{T}\right) \right] \right|
\end{aligned}
\tag{2.87}
$$

where an irrelevant amplitude term has been omitted. The amplitude (2.82) of the output of the filter matched with respect both the time and Doppler is thus:

$$|y_0(t, f_D)| = \left| \Lambda\left(\frac{t - \tau_{SS}}{T}\right) \operatorname{sinc}\left[(f_D - f_d)T\Lambda\left(\frac{t - \tau_{SS}}{T}\right) \right] \right| \tag{2.88}$$

Eq. (2.88) reduces to the modulus of (2.18) for $f_D = f_d$ and shows that a Doppler mismatch of f_d, as in (2.79), causes a reduction of the peak (achieved at $t = \tau_{SS}$ independently from f_d) by the factor $\text{sinc}(f_d T)$: a mismatched Doppler frequency shift higher than $1/T$, which in this case is also the bandwidth B_t of the transmitted pulse, leads therefore to a significant attenuation of the response of the target. Finally, as anticipated, time, and Doppler resolutions cannot be improved both: increasing one of them decreases the other one, being $B_t = 1/T$ for the rectangular pulse.

It is well known that typical lengths of the transmitted rectangular shaped pulse allow achieving poor values of the time (range) resolution and that better results are provided by chirped pulses. In this case, with reference to an up-chirped pulse with chirp-rate α, the factorization property in Appendix A.3 leads to:

$$
\begin{aligned}
\chi_p(t, f) &= \left| \Pi\left(\frac{t}{2T}\right) \mathcal{F}_\tau \left\{ \Pi\left(\frac{\tau - t/2}{T - |t|}\right) \right\}_{|f \leftarrow f + \alpha t} \right| \\
&= \left| \Lambda\left(\frac{t}{T}\right) \text{sinc}\left[(fT + tB_t)\Lambda\left(\frac{t}{T}\right) \right] \right|
\end{aligned}
\tag{2.89}
$$

where $B_t = \alpha T$ and, once again, an irrelevant amplitude term has been omitted. The amplitude of the output of the matched filter when the received echo is associated with a target characterized by a time delay τ_{SS} and a shift Doppler f_d is thus:

$$
\begin{aligned}
|y_0(t, f_d)| = &\left| \Lambda\left(\frac{t - \tau_{SS}}{T}\right) \right. \\
&\left. \text{sinc}\left[((f_D - f_d)T + (t - \tau_{SS})B_t)\Lambda\left(\frac{t - \tau_{SS}}{T}\right) \right] \right|
\end{aligned}
\tag{2.90}
$$

Eq. (2.89) shows a coupling of Doppler and time given by the term $fT + tB_t$: beside the main lobe centered at $t = 0$ and $f = 0$, this coupling leads to the presence of side lobes centered at $t = -f/\alpha$ for each value of f, whose peaks and extensions are modulated by the triangular function of the time. Consequently, although both resolutions can in principle be improved by transmitting a very long chirped pulse, very high values of T produce side lobes whose peaks are comparable to that of the main lobe (in accordance to the uncertainty principle). This aspect is much more appreciated assuming that the length of the transmitted pulse is so high that $|t - \tau_{SS}| \ll T$, thus simplifying (2.90) into:

$$
|y_0(t, f_D)| \simeq \left| \text{sinc}\left[(f_D - f_d)T + (t - \tau_{SS})B_t \right] \right|
\tag{2.91}
$$

Eq. (2.91) clearly shows that a Doppler mismatch, which can significantly reduce the amplitude of the output signal when a short rectangular pulse is transmitted, see (2.88), impairs even the detection of the target, in case of a long chirped pulse produces a shift of the time corresponding to the maximum with respect to the time corresponding to the range target position. This leads to a wrong localization of the target, although its detection turns out to be preserved.

The last part of this section is dedicated to the problem of measuring the Doppler shift. As already pointed out, the Doppler shift is typically much lower than the bandwidth of the transmitted signal. For instance, referring to the rectangular pulse, for a target showing a radial velocity of 100 m/s, at X-Band (i.e., $\lambda = 1.55$ cm), the Doppler shift is about 13 KHz. For a radar achieving range resolutions of the order of 1 m, i.e. a transmitted pulse bandwidth B_t of 150 MHz, the ratio f_d/B_t is $10^{-4} \div 10^{-5}$. A more viable approach to the measurement of the Doppler shift can be carried out by exploiting the variation of the phase of the signal scattered back by the target over multiple subsequent pulses, i.e. the interpulse phase variation. We refer to the two-dimensional LPE representation of the signal in (2.77):

$$s(t, t_n) = Aq \, \Pi \left(\frac{t - \tau_{SS}(t_n)}{T} \right) e^{j2\pi f_d t - j\frac{4\pi}{\lambda}r + j2\pi\frac{2v}{\lambda}t_n} \qquad (2.92)$$

Eq. (2.92) shows that the target velocity is coupled also to the slow time t_n in the phase factor. Accordingly, provided that the delay in the amplitude is compensated (Range Migration Compensation (RCM)), a Fourier transform along t_n would allow detecting a peak located at frequency $f_d = 2v/\lambda$. It is clear that the resolution of the measurement of the Doppler is provided by the inverse of the slow time interval where the Fourier transform has been carried out, i.e., along the CPI $T_a = NT_I$, generally much larger than the transmitted pulse duration T. In conclusion, the slow time analysis allowed by the transmission of the pulse train provides a Doppler resolution equal to $1/T_a$.

2.6 Radar waveforms

Radar systems can use different transmitted waveforms. The choice of a particular waveform is related to the specific application field. The cost and complexity associated with the hardware and software implementation of the system is one of the main factors to be taken into account for the decision process.

Radar waveforms can be Continuous Waveforms (CW) or pulsed waveforms, modulated or unmodulated.

Pulsed radars transmit short high-power bursts and register returned echoes. CW radars emit a continuous waveform and thus avoid high-power bursts and the related constraints. CW radars use two antennas, one to transmit and the other to receive. As explained in the following this characteristic allows avoiding blind ranges, in which the echo receive time window has to be interrupted for the transmission of the pulses. On the other hand, coupling between the transmitting and receive antennas is an issue for CW radar. The simplest type of CW radars is the one that detects moving targets.

2.6.1 Continuous wave

Pulsed radar divides the PRI interval in two portions, one exploited for the signal transmission and one for the echo reception. As seen in the previous sections, the exploitation of sophisticated signals allows increasing the transmitted pulse duration T in (2.6) without jeopardizing the range resolution. To reduce the transmitter peak power, and hence the transmitter complexity, it could be convenient to increase as much as possible the pulse duration. However, upper limitations on T are present due to the influence on the range coverage: for a single antenna TX/RX radar, an increase of T inevitably impacts the minimum range defined in (2.50), that is the region, referred to as "blind range", where it is impossible to acquire data due to the use of the channel for the transmitting operation. Moreover, to account the channel occupation for the transmission, (2.50) should be replaced by

$$T_I \geqslant T + \tau_{max} - \tau_{min} = T + 2\frac{r_{max} - r_{min}}{c} \qquad (2.93)$$

where T_I is the PRI, τ_{max} and τ_{min} are the minimum and maximum echo delay corresponding to the near and far range, respectively. Eq. (2.93) clearly shows that, for a fixed PRI (i.e. a fixed T_I), an increase of T inevitably impacts the range coverage. Above contraindications on the transmitted pulse widening of pulsed radars inevitably pose a lower bound on the transmitted peak power, thus generally limiting the possibility to adopt simple technological solutions for the transmitter. Continuous Wave (CW) Radar allows overcoming such limitations, generally at the expenses of a degradation of the quality of the signal. In fact, CW radar transmits a sinusoidal signal at a fixed frequency f_0, of the form:

$$p_{BP}(t) = A \cos(2\pi f_0 t) \qquad (2.94)$$

where A^2 is the peak power.

With respect to (2.30), the absence of the amplitude (envelope) modulation entails the use of the whole time for transmitting, thus improving the ratio between the average and the peak transmitted power with respect to a pulsed radar where this ratio is lowered by the factor T/T_I. Obviously, an additional antenna must be exploited for the echo reception thus raising issues of isolation to prevent the coupling between the TX and RX antennas. The residual coupling generally leads to a degradation of performances of the radar in terms of the lowest detectable RCS level. In return, simple solid state technology can be used to simplify the apparatus and achieve also cost gains.

If a target is present at range r in the volume irradiated by the radar antenna, the TX CW signal will be backscattered according to its radar section σ:

$$s(t) = Aq \cos\left(2\pi f_0(t - \tau) + \varphi_R\right) \tag{2.95}$$

where q is the previously explained amplitude attenuation factor (depending, among others, on σ), φ_R is the phase term introduced by the target e.m. backscattering and τ is the time delay associated with r as in (2.9). In the case of a bistatic radar system, r should be in principle replaced by $(r_1 + r_2)/2$, where r_1 and r_2 are the ranges to the TX and RX antennas. For CW radar generally $r_1 \simeq r_2 = r$.

The received signal can be therefore rewritten as:

$$s(t) = Aq \cos\left(2\pi f_0 t - \frac{4\pi}{\lambda}r + \varphi_R\right) \tag{2.96}$$

will differ from the transmitted signal because it will have a lower amplitude and a phase shift with respect to the transmitted signal related to the frequency and distance traveled.

In the case of antenna and target whose distance r does not change over time, the phase shift term will be constant over time, and the received signal will always be a sinusoidal signal at the frequency f_0. In case this distance changes over time, the phase shift term will vary over time according to the Doppler effect described above in Section 2.5. If we consider the particular case of whose mutual radar-target distance varies linearly over time with a rate $v_r = dr(t)/dt \ll c$, the received signal will be characterized by a constant frequency, greater or lower than the frequency of the signal transmitted, depending on whether this distance decreases or increases over time, respectively. We therefore have:

$$s(t) = Aq \cos\left(2\pi(f_0 + f_d)t + \varphi_R\right) \tag{2.97}$$

The transmitted and received signals exhibit therefore different frequencies. In case the mutual distance $r(t)$ decreases, $f_d = 2v_r/\lambda$ is positive; the received signal has a frequency larger than the transmitted one, while in case the mutual distance increases with time, f_d is negative, and the received signal has a frequency smaller than the transmitted one. Consequently, targets that move with any velocity perpendicularly the direction joining radar antenna and target are characterized by $f_d = 0$ (case of circular motion of the target around the radar antenna).

The Doppler frequency can be detected and measured according to a circuit constituted by a mixer-heterodyne and an amplifier. In particular, a reduced power replica of the transmitting signal (spilled by the transmitting signal) and the receiving signal, at working frequencies f_0 and $f_0 + f_d$, respectively, are used as inputs of the mixer, where high frequencies and low frequency signal components are produced by multiplying the two signals. The low frequency part of the signal contains both a constant signal (the stationary part related to the possible presence of stationary targets), and the signal at the Doppler frequency. The amplifier eliminates high frequency components and the constant signal, while it amplifies the Doppler signal. The cutoff frequencies of the amplifier state the minimum value of Doppler frequency and the maximum value of Doppler frequency that can be detected. A block scheme of the CW transmitter/receiver is reported in Fig. 2.11.

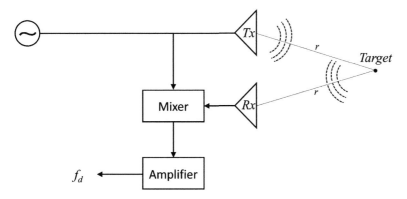

Figure 2.11. Block scheme of the CW transmitter/receiver.

CW radars are used in several civil and military application fields where movement detection and velocity estimation is required.

CW radars in the form previously presented (transmitted signal at constant frequency) cannot be used for target range estimation. To extend the CW radars capability to this purpose, a frequency modulation must be introduced. A radar system whose transmitted signal is frequency modulated as in:

$$s(t) = A \cos \left(2\pi f_0 t + 2\pi \Delta f \int_0^t m(u)du + \varphi_T \right) \qquad (2.98)$$

is called a FM-CW radar. In (2.98) $m(t)$ is the modulating signal such that max $|m(t)| \leqslant 1$, and Δf is the maximum frequency deviation of the FM signal. The instantaneous frequency of signal (2.98) is given by:

$$f_T(t) = f_0 + m(t)\,\Delta f \qquad (2.99)$$

Different modulating signals $m(t)$ can be used. We consider the case of a linearly frequency modulated signals according to a train of triangular wafeforms of period T. The instantaneous frequency (2.98) generated by such signal $m(t)$ of period T that linearly grows from -1 to 1, is shown with the blue (dark gray in print version) line in Fig. 2.12. The signal (2.98) has an approximate bandwidth $B_t = 2\Delta f$ which can also be graphically inferred by the total variation of the instantaneous frequency of Fig. 2.12.

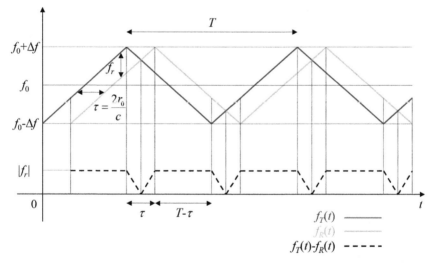

Figure 2.12. Instantaneous frequency of transmitted signal (blue (dark gray in print version) line), and of received signal (red (mid gray in print version) line); frequency of mixing signal (black dash line). Case of a stationary point target.

If the radar signal is backscattered by a stationary point target at distance r_0, the received signal will be a replica of the transmitted signal delayed of $\tau = 2r/c$, and it will have the instantaneous frequency shown with the red (mid gray in print version) line in Fig. 2.12, given by:

$$f_R(t) = f_T(t - \tau) = f_0 + \Delta f\, m(t - \tau) = f_0 + \Delta f\, m\left(t - 2\frac{r}{c}\right) \quad (2.100)$$

The information about the target distance is related to the difference between the transmitted and received instantaneous frequencies, reported with the black line in Fig. 2.12. Such difference, when $\tau \ll T$, will be almost always equal to:

$$f_r = \frac{4\Delta f}{T} \frac{2r}{c} \quad \Longrightarrow \quad r = \frac{cTf_r}{8\Delta f} = \frac{cTf_r}{4B_t} \tag{2.101}$$

The accuracy of the range measurement is directly related to the accuracy of the measurement of the frequency measure:

$$\rho_r = \frac{cT}{4B_t}\delta f_r \tag{2.102}$$

Consequently, as the frequency accuracy is inversely proportional to the frequency sweep, i.e., approximately $2/T$, the range resolution is given by:

$$\rho_r = \frac{cT}{4B_t} \frac{2}{T} = \frac{c}{2B_t} \tag{2.103}$$

the same value obtained in Eq. (2.15) for the range resolution in the case of a pulse radar.

The derivation of (2.100)-(2.103) has been done under the hypothesis of a single stationary point target.

The extension to the case of more targets is straightforward. For instance, in the case of two targets at two different distances r_{0_1} and r_{0_2}, two different frequencies f_{r_1} and f_{r_2} will be generating by mixing transmitted and received signals. If these two frequencies are different from each other more than δf_r, the two targets will be detected as different ones.

In the case of nonstationary targets the presence of the Doppler frequency term f_d must be considered. In particular, the instantaneous frequency of the transmitted signal (2.98) and of the received one, given by:

$$f_R(t) = f_T(t - \tau) = f_0 + f_d + \Delta f\, m(t - \tau) = f_0 + f_d + \Delta f\, m\left(t - 2\frac{r}{c}\right) \tag{2.104}$$

are represented in Fig. 2.13.

The mixing of the two waveforms provides two beat frequency values f_{b_1} and f_{b_2}:

$$\begin{aligned} f_{b_1} &= f_r - f_d \\ f_{b_2} &= f_r + f_d \end{aligned} \tag{2.105}$$

The frequency value f_r depending on target distance r_0 and the frequency values f_d depending on target radial velocity v_r, can be

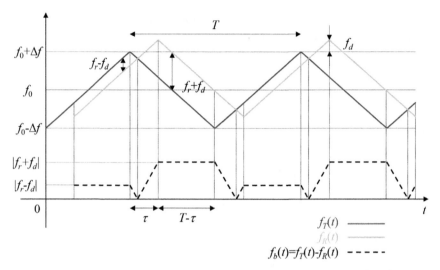

Figure 2.13. Instantaneous frequency of transmitted signal (blue (dark gray in print version) line), and of received signal (red (mid gray in print version) line); frequency of mixing signal (black dash line). Case of a moving (nonstationary) point target.

easily obtained by adding or subtracting the two beating frequencies f_{b_1} and f_{b_2}. Consequently, values of r_0 and v_r are:

$$r = \frac{cT}{4B_t}\left(\frac{f_{b_1} + f_{b_2}}{2}\right)$$

$$v_r = \frac{\lambda}{2}\left(\frac{f_{b_1} - f_{b_2}}{2}\right)$$

(2.106)

CW radars can be adopted also in remote sensing, especially for short range imaging applications due to the advantage to eliminate blind ranges. With respect to pulsed radar imaging systems, the main difference is related to the range compression stage that, as discussed in this section, requires a different processing strategy. An in-depth study of this issue can be found in [9].

2.6.2 Stepped frequency

The range resolution of a radar system is determined by the bandwidth of the transmitted signal. Thus, ultra-high range resolution radars require the generation of ultra-wideband signals, which is strongly limited by the requirement of a very high sampling rate and by the increasing hardware demands. To solve this problem, a stepped frequency approach can be used.

Stepped Frequency radars allow the reduction of the instantaneous bandwidth and then of the sampling rate requirements, while maintaining high range resolution. They use a large bandwidth waveform (Stepped Frequency Waveform or SFW) synthesized by generating samples of its Fourier transform. The governing principle is to transmit multiple subbandwidth signals and synthesize a large-bandwidth echo through signal processing. A high resolution estimate of the target range profile can be recovered by means of Inverse Discrete Fourier Transformation (IDFT) [10]. The drawback is that the acquisition is slowed down, thus requiring a distribution of the transmission of the total bandwidth over multiple PRI.

The SFW transmitted by the radar is composed of a series of N narrow-band and hence long duration pulses, with central frequencies f_n with $n = 0, \ldots, N - 1$. The frequency f_n from pulse to pulse is stepped by a fixed frequency step Δf. Each group of pulses is referred to as a burst. Then the transmitted burst signal can be written as:

$$p_{BP}(t) = \sum_{n=0}^{N-1} A \cos\left(2\pi f_n t + \varphi_n\right) \Pi_{(0,T)}\left(t - nT_b/N\right) \qquad (2.107)$$

where A is the constant amplitudes of the pulses,

$$f_n = f_0 + n\Delta f \qquad n = 0, \ldots, N - 1, \qquad (2.108)$$

T_b/N is the PRI (i.e., $T_b = NT_I$ is the total time of the burst), $T \leqslant T_b/N$ is the duration of each narrow band pulse and $\Pi_{(0,T)}(t)$ is the (monolateral) rectangular pulse in the $(0, T)$ interval, defined as

$$\Pi_{(0,T)}(t) = \Pi\left(\frac{t - T/2}{T}\right) = \begin{cases} 1 & \text{for } 0 \leqslant t \leqslant T \\ 0 & \text{elsewhere} \end{cases} \qquad (2.109)$$

Consequently, the transmitted signal achieves a large global bandwidth equal to $(N - 1)\Delta f$. The time-frequency representation of the transmitted burst is reported in Fig. 2.14 for the case $T = T_b/N$ and in Fig. 2.15(a) for the case $T < T_b/N$.

We note that the use of a pulse duration time $T < T_b/N$ enables to use the system in an interrupted continuous wave (ICW) modality, by turning the transmitter/receiver off for a period, in such a way to obtain an increase in the SNR and a reduction of the transmitting/receiving antenna coupling respect to the CW modality. This can be achieved by switching off the receiver for a short blanking time in such a way to avoid the signals from very close, unwanted targets (e.g.disturbing reflections from walls in closed propagation environments and antenna leakage signals).

In practice, an ICW signal can be generated by switching on and off the transmitter and the receiver of a CW system in such a way

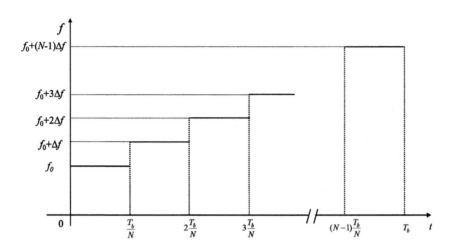

Figure 2.14. Time-frequency representation of the stepped frequency transmitted burst in the CV modality.

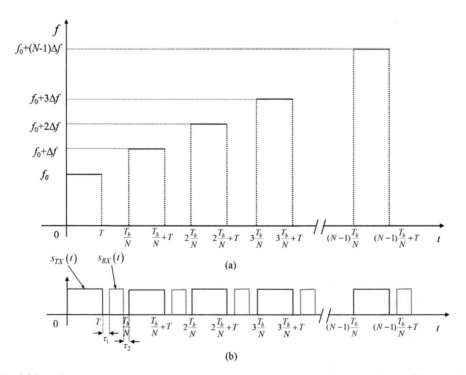

Figure 2.15. (a) Time-frequency representation of the stepped frequency transmitted burst in the ICW modality; (b) Gating functions for the transmitter (blue (dark gray in print version)) and receiver (red (mid gray in print version)) switching in the ICW modality.

that they are not simultaneously active. These time gating can be performed by multiplying the transmitted and received signals by the binary functions $s_{TX}(t)$ and $s_{RX}(t)$ shown in Fig. 2.15(b). The transmitter or receiver is off when the value of the binary functions $s_{TX}(t)$ or $s_{RX}(t)$ is 0 and on when the value is 1. Although we talk of "transmitted pulses" and "received pulses", the pulse length does not influence the radar range resolution. The setting of the system parameters, T_I (the PRI), N, T, and Δf, depends on the specific application (short or long range imaging), particularly on the extension of the scene, its distance from the radar and the required range resolution.

In the gating sequences shown in Fig. 2.15(b) two additional delays τ_1 and τ_2 (with $\tau_1, \tau_2 \geqslant 0$) are used to change the duty cycle of the square wave applied at the receiver. In particular, a proper setting of τ_1 and τ_2 allows to increase the extension of the blind ranges in the near and far zones.

Assume now that, for the sake of simplicity, a single stationary target is present in the region of interest at a distance r from the radar antenna, with $r_{min} < r < r_{max}$. The received signal corresponding to the n-th pulse of the burst (2.107) can be represented as:

$$s_{nBP}(t) = q A \cos\left(2\pi f_n \left(t - 2\frac{r}{c}\right) + \varphi_n\right) \Pi_{(0,T)} (t - 2r/c - nT_b/N)$$

(2.110)

where q is the usual attenuation factor. In order to have the receiver active when the n-th pulse echo reception starts, it must be $T + \tau_1 < 2r/c < T_b/N - \tau_2$.

The received signal is base-band converted by mixing each pulse $s_{nBP}(t)$ with the signals $\cos(2\pi f_n t + \varphi_n)$, so that the low-pass equivalent representation of the received signal $s_{BP}(t)$ is:

$$s(t) = \sum_{n=0}^{N-1} q A \exp\left(-j4\pi f_n \frac{r}{c}\right) \Pi_{(0,T)} (t - 2r/c - nT_b/N) \quad (2.111)$$

The baseband signal $s(t)$ is sampled at low frequency (audio) rates at the times $t_k = (k+1)T_b/N - \tau_2$, with $k = 0, ..., N - 1$, so that we obtain:

$$\begin{aligned}
s(t_k) &= \sum_{n=0}^{N-1} A q \exp\left(-j4\pi f_n \frac{r}{c}\right) \\
&\quad \Pi_{(0,T)} ((k-n)T_b/N + (T_b/N - \tau_2 - 2r/c)) = \\
&= A q \exp\left(-j4\pi f_k \frac{r}{c}\right) \qquad k = 0, ..., N - 1
\end{aligned}$$

(2.112)

as the rectangular window in the summation in (2.112) is equal to 1 only for $k = n$, and is always equal to 0 for $k \neq n$, since in order to sample the signal before the pulse echo is ended, it has to be $2r/c + T > T_b/N - \tau_2$, so that $T_b/N - 2r/c - T/2 - \tau_2 < T/2$.

By substituting expression (2.108) in (2.112), the samples of the received signal can be written as:

$$
\begin{aligned}
s(t_k) &= Aq \exp\left(-j4\pi f_0 \frac{r}{c}\right) \exp\left(-j2\pi \frac{\Delta f}{T_b/N} \frac{2r}{c} k \frac{T_b}{N}\right) = \\
&= a(r) \exp\left(-j2\pi \frac{\Delta f}{T_b/N} \frac{2r}{c} k \frac{T_b}{N}\right), \qquad k = 0, ..., N-1
\end{aligned}
\tag{2.113}
$$

The first term $a(r)$ in (2.113) is the product of amplitude terms and a phase shift depending on r and constant with k. The second term is a phasor with a frequency given by the multiplication of the rate of change of SFW frequency with the round-trip delay time and sampled in the times kT_b/N. This term represents a shift in frequency during the round-trip time. Thus, the scatterer range is converted into a frequency shift.

From (2.113) it can be noted that the time delay $2r/c$, and then the scatterer range r, can be estimated from the difference of the phases of $s(t_{k-1})$ and $s(t_k)$:

$$
\angle s(t_{k-1}) s^*(t_k) = 2\pi \Delta f \frac{2r}{c}.
\tag{2.114}
$$

For obtaining the target range profile, we observe that the signal samples $s(t_k)$ in (2.113) represent the samples of the Fourier transform of the scatterer reflectivity obtained by a single burst. Therefore, an N-IDFT can be applied to recover the scatterer range profile:

$$
\begin{aligned}
\frac{1}{N} \sum_{k=0}^{N-1} s(t_k) \exp\left(j\frac{2\pi ki}{N}\right) &= \\
&= \frac{1}{N} \sum_{k=0}^{N-1} a(r) \exp\left(-j\frac{4\pi k\Delta f r}{c}\right) \exp\left(j\frac{2\pi ki}{N}\right) = \\
&= \frac{a(r)}{N} \frac{1 - \exp\left(-j2\pi \left(\frac{2N\Delta f r}{c} - i\right)\right)}{1 - \exp\left(-j2\pi \left(\frac{2\Delta f r}{c} - \frac{i}{N}\right)\right)} = \\
&= \frac{a(r)}{N} \exp\left(-j\pi (N-1) \left(\frac{2\Delta f r}{c} + \frac{i}{N}\right)\right) \frac{\sin\left(\pi \left(\frac{2N\Delta f r}{c} - i\right)\right)}{\sin\left(\pi \left(\frac{2\Delta f r}{c} - \frac{i}{N}\right)\right)}.
\end{aligned}
\tag{2.115}
$$

Then, by setting $r_i = ic/(2N\Delta f)$, the amplitude of the signal (2.115) obtained after the N-IDFT can be written as:

$$|y(r_i)| = \frac{|a(r)|}{N} \left| \frac{\sin\left(\pi \frac{2N\Delta f}{c} (r_i - r)\right)}{\sin\left(\pi \frac{2\Delta f}{c} (r_i - r)\right)} \right|. \qquad (2.116)$$

We can observe that, according to (2.116), $|y(r_i)|$ is proportional to the amplitude of a Dirichlet function (often denoted as periodic sinc function) centered around the scatterer range value r, periodic with period $c/(2\Delta f)$, and with a first null in $r_i - r = c/(2N\Delta f)$. The 3 dB range resolution of the reconstructed range profile is then equal to:

$$\rho_r = \frac{c}{2N\Delta f} \qquad (2.117)$$

and is inversely proportional to the overall bandwidth of the SFW.

Due to the periodicity of the reconstructed range profile (2.116), an ambiguity in the determination of the range of the scatterer is present. This ambiguity is related to the fact that the frequency spectrum (2.115) is a discrete spectrum. The maximum range interval Δr_{unamb} that can be unambiguously reconstructed is related to the period of the function in (2.116) and is expressed by the following equation:

$$\Delta r_{unamb} = \frac{c}{2\Delta f} \qquad (2.118)$$

This result can be also derived starting from (2.114).

2.7 Stretch processing and sampling

Range compression has been described in Sections 2.1 and 2.3 for pulsed radars as a cross correlation operation, that is a convolution performed between the received echo and the transmitted signal, which can be easily implemented in the FT domain. FMCW radars perform, however, the range measurement by exploiting a beating between the received echo and the transmitted signal. The product of the beating is then analyzed in the spectral domain in order to achieve the measurement of the target range. As shown, the final time (delay) resolution is always the inverse of the transmitted bandwidth.

The procedure described for FMCW radar is termed as Spectral Analysis (SPECAN) or Stretch Processing. SPECAN can be employed also in pulsed radar to efficiently implement, when the transmitted pulse is a chirp signal, the matched filter operation.

The efficiency of the SPECAN approach is particularly evident in the cases where a high range resolution is required: We here address this topic.

The stretch processing presumes the transmission of a chirp signal, therefore we refer to (2.31) for the transmitted signal and for sake of simplicity to the up-chirp: the particularization for the down-chirp case is straightforward.

We assume a target located at a range corresponding to a delay τ; the received signal can be therefore expressed as:

$$s(t) = Aqe^{j\pi\alpha(t-\tau)^2} \Pi\left(\frac{t-\tau}{T}\right) \tag{2.119}$$

SPECAN proceeds by mixing the received signal in (2.119) with the conjugate version of the transmitted chirp phase component centered on a reference time, typically corresponding to the time delay τ_c associated with the scene center (mid-range). This operation can be easily analyzed by shifting back the received echo by τ_c before the mixing. We therefore have:

$$h_{\mathrm{mix}}(t) = s(t+\tau_c)e^{-j\pi\alpha t^2} \tag{2.120}$$

and hence, after simple manipulation:

$$h_{\mathrm{mix}}(t) = Aqe^{j\pi\alpha(\tau-\tau_c)^2}e^{-j2\pi\alpha t(\tau-\tau_c)}\Pi\left[\frac{t-(\tau-\tau_c)}{T}\right] \tag{2.121}$$

The signal generated by the beating in (2.121) highlights the fundamental concept of the SPECAN approach, i.e. that after the signal mixing the position of the target (referred to τ_c), i.e., $\tau - \tau_c$, and the fast time t are conjugate variables of a Fourier transformation. The coupling of delays and frequencies is a direct consequence of the peculiar shape of the ambiguity function described in Subsection 2.5.3.

The result of the mixing operation between the received echo and the reference chirp is shown in Fig. 2.16 with reference to a near range target, i.e. $\tau = \tau_{min}$ (green (gray in print version) line) and far range target, i.e. $\tau = \tau_{max}$ (red (mid gray in print version) line). The mixing operation with the conjugate chirp (blue (dark gray in print version) line) generates two constant frequencies f_{b1} and f_{b2}. Due to the peculiarity to set to zero the slopes of the echos instantaneous frequency response, the mixing is usually referred to as deramping. Accordingly, it is convenient interpreting the signal in (2.121) as a spectrum by letting $f = \alpha t$. Operating the substitution $h_{\mathrm{mix}}(t = f/\alpha) \leftarrow S(f)$ leads to

$$S(f) = Aqe^{j\pi\alpha(\tau-\tau_c)^2}e^{-j2\pi f(\tau-\tau_c)}\Pi\left[\frac{f-\alpha(\tau-\tau_c)}{B_t}\right] \tag{2.122}$$

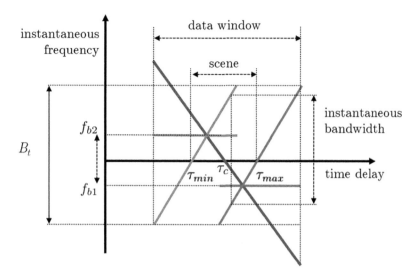

Figure 2.16. Time frequency diagram.

where at the denominator of the rectangular pulse in (2.122) it has been considered the relation between bandwidth, duration, and chirp rate of the transmitted signal.

But for the frequency offset $\alpha\tau$, which shifts for $\tau \neq \tau_c$ the spectral support from the reference zero frequency (i.e. $\alpha\tau_c$), the signal in (2.122) describes the spectrum of the response of a point target after the chirp compression: the spectral support is B_t and the linear phase contribution account for the location τ of the target with respect to the reference point τ_c; no distortion factors are present because the first phase factor is constant for a given point target. A simple inverse Fourier transformation can be therefore applied to achieve the focused data, i.e.:

$$s(t) = Aq \exp\left[j\pi\alpha(\tau - \tau_c)^2\right] \int_{-\infty}^{\infty} df \exp\left[-j2\pi f(t - (\tau - \tau_c))\right]$$
$$\Pi\left(\frac{f - \alpha(\tau - \tau_c)}{B_t}\right)$$

$$(2.123)$$

that is

$$s(t) = Aq\,B_t \exp\left[j\pi\alpha(\tau - \tau_c)^2 - j2\pi\alpha(\tau - \tau_c)(t - (\tau - \tau_c))\right]$$
$$\text{sinc}\left[B_t(t - (\tau - \tau_c))\right]$$

$$(2.124)$$

By simply rearranging the phase factors we have:

$$s(t) = Aq B_t \exp\left[-j\pi\alpha(\tau - \tau_c)^2 + j2\pi\alpha t(\tau - \tau_c)\right]$$
$$\mathrm{sinc}\left[B_t(t - (\tau - \tau_c))\right] \tag{2.125}$$

It is worth noting that, in writing (2.123), (2.124), and (2.125) a slight abuse of notation has been done, being $s(t)$ usually exploited for denoting the received echo in (2.119).

The last operation of the SPECAN processing is a multiplication again by the quadratic phase factor exploited in (2.120) i.e.:

$$y(t) = s(t) \exp\left[-j\pi\alpha t^2\right] = Aq B_t \exp\left[-j\pi\alpha(t - (\tau - \tau_c))^2\right]$$
$$\mathrm{sinc}\left[B_t(t - (\tau - \tau_c))\right] \tag{2.126}$$

The phase multiplication in (2.126) implements the correction of the linear (in t) phase factor in (2.125) thus avoiding the shift of the support of the spectrum in (2.124) and is called Residual Video Phase (RVP). Such a multiplication also zeroes the phase of the peak of the mainlobe of the target echo. Comparing the result achieved by SPECAN with that of the standard correlation processing for the chirp in (2.34) it is possible to notice the disappearance of the triangular function, which was further neglected in Section 2.3, but the appearance in (2.126) of a quadratic phase factor. Such a phase factor can be neglected as well within the mainlobe of the focused response, i.e. $|t - (\tau - \tau_c)| \leqslant 1/B_t$ when $T B_t \gg 1$.

SPECAN, or stretch processing, is an approach applicable uniquely to linear frequency modulated systems, pulsed or continuous wave, allows achieving pulse compression, and connected advantages, on the same basis of the classical chirp correlator processor. Anyway, stretch processing has specific feature which allows taking advantage on the signal sampling aspect when the pulse duration and scene extension are one large and the other small and viceversa. We start by referring to Fig. 2.16 in which B_t represent the transmitted bandwidth, i.e. the frequency span covered by the transmitted signal, and hence by the received echos. The Figure quite well illustrates the concept that, depending on the scene extension (i.e., $T_s = \tau_{max} - \tau_{min}$), a bandwidth reduction can be achieved after the deramping operation. To achieve quantitative evaluation of the sampling gain we start by observing that the Nyquist condition for sampling of the transmitted signal would require:

$$F_s \geqslant B_t \tag{2.127}$$

With reference to the scheme reported in Fig. 2.17, the signal after the deramping can be interpreted as the spectrum of the scene, see (2.122). The relation for converting time to frequency is:

$$f = \alpha t \qquad (2.128)$$

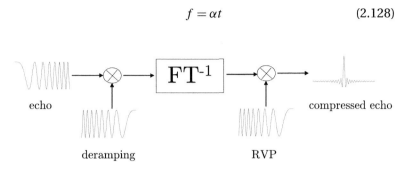

Figure 2.17. Stretch processing.

Therefore sampling the input at a rate F_s^{in}, i.e. $t_k = k/F_s^{in}$ is tantamount to sampling the spectrum at a rate F_s^{in}/α, i.e. $f_k = k\alpha/F_s^{in}$. To avoid time domain aliasing it is required that the inverse of the sampling rate is larger than the time support, i.e. of the scene size (in time):

$$F_s^{in}/\alpha \geqslant (\tau_{max} - \tau_{min}) = T_s \qquad (2.129)$$

The minimum input sampling frequency to avoid aliasing is therefore:

$$F_{s_{min}}^{in} = \alpha T_s = \frac{T_s}{T} B_t \qquad (2.130)$$

According to the above analysis, if $T \gg T_s$, i.e. the chirp duration is much larger than the scene duration, the received signal can be undersampled by a factor corresponding to the ratio between the scene duration and the chirp duration. This is the reason why stretch processing is exploited in very high resolution, i.e. very large bandwidth systems in which the chirp duration is large and the illuminated scene is typically small. Interesting to note is that, whatever the fixed time, αT_s is the "instantaneous bandwidth" of the signal that need to be sampled, whereas the total bandwidth for whatever echo is distributed over a larger time T. So the "trick" exploited is to correctly sample the instantaneous bandwidth. Such a subsampling is not possible with the standard Frequency domain filtering because the signal subsampling would generate a frequency aliasing. Finally it is natural to ask which is the output data sampling. To this end we again refer to the signal stage in Fig. 2.17 after deramping, i.e. prior to the FT operation, which represent the stage in which the signal represents the spectrum of the scene. The spanned frequency interval is α times the signal support (i.e. the data window in Fig. 2.16, $T + T_s$). Accordingly the output sampling frequency F_s is given by:

$$F_s = \alpha(T + T_s) = B_t \frac{(T + T_s)}{T} \tag{2.131}$$

thus guaranteeing the correct total bandwidth sampling as $T_s/T \geqslant 1$.

A last remark is dedicated to the terminology adopted in the radar context. Radars exploiting the stretch processing generally by implementing the deramping operation directly at the receive stage, i.e. onboard for imaging radars which are described in the rest of the book. In such a case it is frequent to term the radar as a "deramp-on receive radar".

2.8 MATLAB® examples

2.8.1 Doppler analysis

In this example, the frequency shift induced by a relative motion among the target and the radar will be analyzed.

The code starts with the definition of system parameters characteristics of the transmitted carrier and of the modulated pulse:

```
%% Physical Constants
c = physconst('LightSpeed');    %Speed of light

%% Carrier Parameters
f0 = 9.6e+09;                   %Central frequency [Hz]
lambda = c/f0;                  %Wavelenght [m]

%% Pulse Parameters
B = 72.5e6;                     %Transmitted bandwidth [Hz]
T = 40e-6;                      %Pulse duration [s]
alpha = (B/T)*(B*T > 1);        %Chirp rate (0 for ...
   nonmodulated pulse)

%% Sampling parameter
fs = 1.5*B;                     %Sampling frequency [Hz]
```

Based on these settings, the range resolution `dr` and the range sampling step `rs` result in:

```
dr = c/2/B;                     %Range resolution
rs = c/2/fs;                    %Range step
```

The receiver range window within the transmission of a single pulse is designed by imposing an opening of the receiver window at a time corresponding to the distance of the closest target sensed, i.e. located at the *near range*, which, for this example is set at `rnear=1e3 [m]`. The duration of the receiver window, corresponding to a distance equal to the *range swath* has been designed to be equal to twice the duration of the transmitted pulse:

```
%% Scene Extent
rnear = 1e3;                    %Near range [m]
Rswath = 2*c*T/2;              %Range swath [m]
rfar = rnear + Rswath;         %Far range [m]
```

The observed target has been simulated to be located at the center of the observed range, moving with a radial velocity of 100 m/s (see Fig. 2.18):

```
%% Target Parameters
rt = rnear + Rswath/2;          %Range position of the ...
   target at t=0
vt = 100;                       %Radial component of ...
   target velocity [m/s]
```

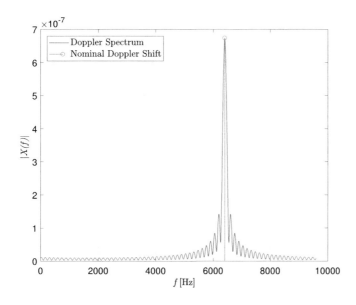

Figure 2.18. Doppler Spectrum of a simulated moving target.

The observation time `Tobs` of the radar is also set to address the sensing of the moving target in the *slow time*: in this example, it is specifically designed to limit the effect of the migration of the ACF in the range direction induced by the distance variation in the *slow time*. In particular, it is set as the time needed to for a target moving with velocity `vt` to travel a distance corresponding to half of a range step. The repetition of the pulses is provided by the PRF which is set accordingly to the expected nominal Doppler shift fd:

```
%% Observation Parameters
Tobs = (rs/2)/vt;              %Observation time [s]
fd = 2*vt/lambda;             %Nominal Doppler shift [Hz]
PRF   = 1.5*fd;                %Pulse Repetition ...
     Frequency [Hz]
```

The target echo is generated by means of the echo_gen.m function available in the book repository. The function is fed by the target and pulse parameters(rt, lambda, T, B, rout), as well as by the further parameters for the Doppler analysis (vt, Tobs, ... PRF). The output of the echo simulation is a matrix in which the columns correspond to the samples of the received echo in the *fast time* domain, the rows to the subsequent echoes associated to the transmission of multiple pulses along the *slow time* domain

```
%% Generation of Target Echo
rout = (rnear:c/2/fs:rfar);       %Range support of the ...
     received echo
[x] = echo_gen(rt, lambda, T, B, 'rout', rout, 'vt', vt, ...
     'Tobs', Tobs, 'PRF', PRF);
```

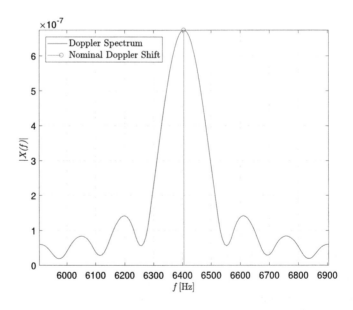

Figure 2.19. Zoom of the Doppler Spectrum on the nominal Doppler Shift.

The analysis of the Doppler shift involves the computation of the echo Fourier Transform. A zero-padding operation in the time domain is considered to achieve finer sampling in the frequency domain. The estimated spectrum, as provided in Fig. 2.19 is well centered around the nominal Doppler shift, which, for the parameters of this example is about $f_d = 6404$ [Hz].

```
%% Doppler Spectrum Analysis
%zero padding
nc = length(rout);              %Number of columns
nr = length(x(:, 1));           %Number of rows
ovs = 20;                       %Oversampling factor in ...
    Doppler domain
nro = nr*ovs;                   %Number of rows after ...
    zero-padding
xz = zeros(nro, nc);            %Zero-padding initialization
xz(1:nr, :) = x;                %zero-padding of the ...
    simulated echo
xf = fft(xz, [], 1);            %FFT of the zero-padded ...
    simulated echo
xfm = sum(abs(xf)/nc, 2);       %Average along the range ...
    direction, (columns)
af = (0:1:nro-1)/(nro)*PRF;     %Output frequency axis

%visualization of the Doppler Spectrum
figure('Name', 'Doppler Spectrum', 'NumberTitle', 'off');
plot(af, xfm), grid on
xlabel('{\it f} [Hz]','Interpreter','latex'), ...
    ylabel('{\it $\mid$X(f)$\mid$ }','Interpreter','latex')
hold on; stem(fd, max(xfm), 'r')
legend({'Doppler Spectrum', 'Nominal Doppler ...
    Shift'},'Interpreter','latex', 'Location', 'northwest')
```

3

Scene characterization

Imaging radars transmit an electromagnetic wave in the radio frequency band, and register the echo signals backscattered toward the radar by the observed scene. Backscattered signals contain information on the geometrical and electromagnetic properties of the scene, that can be retrieved by means of proper signal processing. The understanding of the scattering mechanism between the electromagnetic wave and the scene is a fundamental step for extracting information from the received signals.

There are many models describing the interaction between the incident electromagnetic wave and a scene on the Earth surface, which are depending on the operating frequency and on the geometrical and electromagnetic properties of the scene. In this chapter, we discuss some quantities useful to representing an electromagnetic wave and few simple scattering models, whose range of validity with respect to frequency, view angles, polarization, and surface characteristic, is matching to the typical imaging radars parameters and applications.

3.1 Electromagnetic wave polarization

A radar system transmits a vector electric field $\mathbf{E}$ and measures the vector electric field backscattered from the target, according to the electromagnetic wave equation [11].

For imaging radars, the target is represented by the illuminated ground scene characterized by permittivity $\varepsilon(x, y, z)$ and conductivity $\sigma_c(x, y, z)$, and by a random height profile $z(x, y)$, as shown in Fig. 3.1. When the transmitted wave impinges on the interface surface, it is in part backscattered toward the upper medium and in part it can penetrate into the lower medium till to a penetration depth d_p depending on the operating wavelength. For unvegetated scenes and for the typical frequencies used by imaging radars, d_p assumes usually small values, so that the electromagnetic parameter maps of the scene can be considered two-dimensional (2D), i.e. $\varepsilon(x, y)$ and $\sigma_c(x, y)$ [11].

Multi-Dimensional Imaging with Synthetic Aperture Radar
https://doi.org/10.1016/B978-0-12-821655-2.00007-3

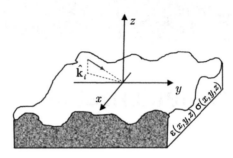

Figure 3.1. Observed ground target.

Assume that the radar system transmits a completely polarized monocromatic plane wave propagating along the direction $\hat{k}_i = \hat{r}$ usually referred to as wave vector (see Appendix A.6). The polarization of the wave is defined as the time–space variation of the electric field vector in the plane perpendicular to its propagation direction [12].

Let us define a coordinate system $(\hat{h}, \hat{v}, \hat{k}_i)$, where $\hat{h}$ and $\hat{v}$ denote the orthogonal directions of vertical and horizontal linear polarizations, respectively. The reference coordinate system used for the wave representation is entirely arbitrary, although the plane in which polarization is described is always perpendicular to the direction of propagation. This reference plane is fixed in space, so that the polarization properties are described by considering the apparent motion of the electric field vector as the wave moves through the plane.

In remote sensing, when the viewing direction is off-nadir, the horizontal axis is defined to be parallel to the Earth's surface (i.e. perpendicular to the plane of incidence), while the vertical axis is defined as being perpendicular to the horizontal one and to the propagation direction (i.e. it is in the plane of incidence and it needs not be "vertical" with respect to the Earth's surface). When viewing towards the nadir, the choice is completely arbitrary and there is no distinction between horizontal and vertical at the ground surface.

If we assume that the transmitted wave is a completely polarized monocromatic wave, the field incident on the ground surface can be represented as:

$$\mathbf{E}^i = E_h^i e^{-jkr} \hat{h} + E_v^i e^{-jkr} \hat{v} \tag{3.1}$$

where E_h^i and E_v^i are the LPE (see Appendix A.2) horizontal and vertical (complex valued) components of the time varying (see Appendix A.6) incident electric field:

$$e^i(t,r) = |E_h^i| \cos(\omega t - kr + \angle E_h)\hat{\boldsymbol{h}} + |E_v^i| \cos(\omega t - kr + \angle E_v)\hat{\boldsymbol{v}},$$

(3.2)

and k is the propagation constant $k = 2\pi/\lambda$.

We can write Eq. (3.1) in the form of a complex vector:

$$\mathbf{E}_{hv}^i = \begin{bmatrix} E_h^i \\ E_v^i \end{bmatrix},$$

(3.3)

which is often referred to as the Jones vector representation.

Horizontal polarization is usually defined as the case in which the electric vector is perpendicular to the plane of incidence, i.e. the plane containing the direction of propagation of the wave $\hat{\boldsymbol{k}}_i$ and the unit vector normal to the surface, while vertical polarization component is orthogonal to both horizontal polarization and direction of propagation, and corresponds to the case in which the electric field vector is in the plane of incidence.

The polarization of a plane wave describes the shape traced by the tip of its electric field (3.2) in a plane orthogonal to the direction of propagation, as the wave passes through that plane. In the general case the wave (3.1) is elliptically polarized. In the special case when E_h^i and E_v^i have a phase difference of $n\pi$, with n any integer, the wave is said to be linearly polarized. When the two amplitudes $|E_h^i|$ and $|E_v^i|$ are the same and the relative phase difference is either $-\pi/2$ or $\pi/2$ the wave is circularly polarized. A representation of an elliptically polarized wave is given in Fig. 3.2.

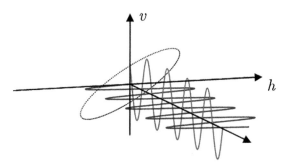

Figure 3.2. Horizontally and vertically polarized components of an elliptically polarized wave.

If instead of using the basis $(\hat{\boldsymbol{h}}, \hat{\boldsymbol{v}})$, we can introduce an alternative representation as a linear combination of two arbitrary orthonormal polarization states $\mathbf{E}_m^i$ and $\mathbf{E}_n^i$ for which:

$$\mathbf{E}^i = E_m^i \hat{\boldsymbol{u}}_m + E_n^i \hat{\boldsymbol{u}}_n$$

(3.4)

Table 3.1 Polarization states in terms of normalized Jones vector E.

Polarization	Linear horizontal	Linear vertical	Linear 45°	Circular left-handed	Circular right-handed
Jones vector $(\hat{h}, \hat{v})$ basis	$\begin{bmatrix} 1 \\ 0 \end{bmatrix}$	$\begin{bmatrix} 0 \\ 1 \end{bmatrix}$	$\frac{1}{\sqrt{2}}\begin{bmatrix} 1 \\ 1 \end{bmatrix}$	$\frac{1}{\sqrt{2}}\begin{bmatrix} 1 \\ j \end{bmatrix}$	$\frac{1}{\sqrt{2}}\begin{bmatrix} 1 \\ -j \end{bmatrix}$
Jones vector $(\hat{l}, \hat{r})$ basis	$\frac{1}{\sqrt{2}}\begin{bmatrix} 1 \\ 1 \end{bmatrix}$	$\frac{1}{\sqrt{2}}\begin{bmatrix} -j \\ j \end{bmatrix}$	$\frac{1}{2}\begin{bmatrix} 1 & -j \\ 1 & j \end{bmatrix}$	$\begin{bmatrix} 1 \\ 0 \end{bmatrix}$	$\begin{bmatrix} 0 \\ 1 \end{bmatrix}$

with the standard basis vectors $\hat{u}_m$ and $\hat{u}_n$ in general orthonormal, i.e. $\hat{u}_m^H \hat{u}_n = 0$ and $\hat{u}_m^H \hat{u}_m = 1$, $\hat{u}_n^H \hat{u}_n = 1$, with ^{H} denoting the Hermitian operator, and both orthogonal to the propagation vector.

The Jones vector $\mathbf{E}_{mn}^i$ in the basis $(\hat{u}_m, \hat{u}_n)$ may be defined as:

$$\mathbf{E}_{mn}^i = \begin{bmatrix} E_m^i \\ E_n^i \end{bmatrix}. \tag{3.5}$$

A circular polarization basis $(\hat{l}, \hat{r})$, where $\hat{l}$ and $\hat{r}$ are two left-hand and right-hand circularly polarized unit vectors, can be also considered, since they are an orthonormal set under an hermitian inner product.

The Jones vectors of different polarizations states are reported in Table 3.1 for different polarization basis [13].

The state of polarization of a wave is determined by the electric field components, f.i. E_h^i and E_v^i. For completely polarized waves these components are constant over the time. For partially polarized narrow-band waves, the base-band signals $E_h^i(t)$ and $E_v^i(t)$ become slowly varying functions of time, and are usually assumed to be ergodic random processes. The polarization characteristics of a partially polarized wave are described by the wave covariance matrix $\mathbf{C}_p$, defined as:

$$\mathbf{C}_p = \begin{bmatrix} \langle E_h(t)E_h^*(t)\rangle & \langle E_h(t)E_v^*(t)\rangle \\ \langle E_v(t)E_h^*(t)\rangle & \langle E_v(t)E_v^*(t)\rangle \end{bmatrix} \tag{3.6}$$

where * denotes conjugate and $\langle \cdot \rangle$ denotes the time averaging operation which, under the ergodic assumption, is equivalent to the statistical mean operation. It is interesting to note that when the wave is completely polarized, we have that the (LPE representations) components of the vector field do not depend on the time.

Therefore we have:

$$\det(\mathbf{C}_p) = \langle E_h(t)E_h^*(t)\rangle\langle E_v(t)E_v^*(t)\rangle - \langle E_h(t)E_v^*(t)\rangle\langle E_v(t)E_h^*(t)\rangle$$
$$= E_h E_v^* E_v E_h^* - E_h E_v^* E_v E_h^* = 0$$

$$(3.7)$$

where $\det()$ indicates the determinant. The determinant of a square matrix is also equal to the product of the eigenvalues. Accordingly, the validity of (3.7) implies the zeroing of one of the two eigenvalues of the polarization coherency matrix. Letting λ_1 and λ_2 the two eigenvalues, it is possible to define the two quantities:

$$p_1 = \frac{\lambda_1}{\lambda_1 + \lambda_2} \qquad p_2 = \frac{\lambda_2}{\lambda_1 + \lambda_2}, \qquad (3.8)$$

so that we can associate an entropy to the polarization coherence matrix:

$$H_p = -p_1 \log p_1 - p_2 \log p_2 \qquad (3.9)$$

and hence the case of completely polarized wave corresponds to zero entropy.

Finally the following degree of polarization can be defined:

$$d_p = 1 - H_p \qquad (3.10)$$

which is one in the case of complete polarization. The case of absence of polarization evidently corresponds to $p_1 = p_2 = 1/2$, that is $\lambda_1 = \lambda_2$. This can be achieved when the matrix $\mathbf{C}_p \propto \mathbf{I}$ (with $\mathbf{I}$ being the identity matrix) thus leading to a condition of uncorrelation between the vertical and horizontal electric field components. Under this condition it is not possible to predict by a linear transformation with a limited mean square error one component of the field starting from the other component.

3.2 Scattering matrix, scattering coefficient, and radar cross section

Let us now consider a wave incident on a surface delimiting a given volume. The total electric field $\mathbf{E}$ in the upper space is the sum of the incident field $\mathbf{E}^i$ and the scattered field $\mathbf{E}^s$, which carries information about the target geometry and its electromagnetic properties.

The scattering geometry can be described in a wave system denoted as Forward Scattering Alignment (FSA) or in an antenna system denoted as Backward Scattering Alignment (BSA) (see

Fig. 3.3). In the FSA, the directions of the vertical and horizontal unit vectors $\hat{v}$ and $\hat{h}$ are always defined in such a way that the positive directions of the unit vectors $\hat{k}_i$ and $\hat{k}_s$ orthogonal to them are oriented in the same direction of wave propagation for both the incident and scattered signals. On the contrary, in the BSA, the positive direction of the unit vectors $\hat{k}_i$ and $\hat{k}_s$ points toward the target for both the incident and scattered signals. The conversion between FSA and BSA is straightforward. In the following we consider the FSA convention.

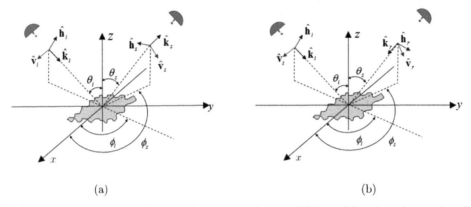

(a)   (b)

Figure 3.3. Coordinate systems for the (a) forward scattering alignment (FSA) and (b) backward scattering alignment (BSA).

A completely polarized wave incident upon a deterministic simple target produces a completely polarized scattered wave, and the scattering is told to be coherent.

The scattered field can be found by solving the wave equation under the far-field approximation, obtaining the horizontal and vertical components of the scattered field (in the FSA convention):

$$\begin{bmatrix} E_h^s(r) \\ E_v^s(r) \end{bmatrix} = \frac{e^{jkr}}{r} \mathbf{S} \cdot \mathbf{E}^i = \frac{e^{jkr}}{r} \begin{bmatrix} S_{hh} & S_{hv} \\ S_{vh} & S_{vv} \end{bmatrix} \cdot \begin{bmatrix} E_h^i \\ E_v^i \end{bmatrix} \qquad (3.11)$$

with $\mathbf{E}^s = E_h^s \hat{h} + E_v^s \hat{v}$, r the distance between the observation point and the scattering point, and $\mathbf{S}$ a 2×2 complex matrix, denoted as the scattering matrix, which in the FSA convention is also known as the Jones matrix, while in the BSA convention it is also recognized as the Sinclair matrix. It is well known that plane waves are characterized by a decay proportional to the distance; moreover the phase changes linearly with the distance through the wavenumber. This behavior in the scattered fields is accounted for by ratio in (3.11) before the scattering matrix, so that the latter becomes independent on the distance and rather represents a property of the target. In particular, each element of the scattering matrix S_{pq}, with $p, q \in h, v$, represents the complex scattering coefficient under the q-polarization incidence and p-polarization scattering. The four complex scattering amplitudes in the matrix **S** characterize the scattering behavior of the illuminated target for the four possible combinations of the h and v polarization orientations of the incident and scattered fields. Each of them is a function of not only the target shape, size, orientation, permittivity, and conductivity, but also of the illumination and scattering angles, (θ_i, ϕ_i) and (θ_s, ϕ_s), respectively (see Fig. 3.3), that is:

$$S_{pq} = S_{pq}(\hat{k}_i, \hat{k}_s) = \lim_{r \to \infty} r e^{-jkr} \frac{E_p^s}{E_q^i} \tag{3.12}$$

where the limit denotes that r is in the far-zone of the scattering target.

In general, S_{hv} and S_{vh} are not simply related, but in the backscatter direction $(\hat{k}_i = -\hat{k}_s = \hat{k}_r)$, the reciprocity theorem of electromagnetic scattering [11] states that:

$$S_{hv} = S_{vh}. \tag{3.13}$$

Radar systems can be fully polarimetric systems, when they can measure the entire scattering matrix **S**; they can be single polarization (single-pol) or dual-polarization (dual-pol) systems when they measure only one or two elements of the scattering matrix, respectively. Conventionally, co-pol refers to scattering with the same polarization for incidence and scattering, while cross-pol refers to different polarization for incidence and scattering.

In most conditions encountered in Earth observation the principle of reciprocity is satisfied, then only three complex measurements are given in a fully polarimetric system. It is therefore often convenient to define a target vector, instead of a matrix, which in a linear basis $(\hat{h}, \hat{v})$ is defined as:

$$\boldsymbol{k}^T = \begin{bmatrix} S_{hh} & S_{vv} & S_{hv} \end{bmatrix}^T. \tag{3.14}$$

From the target vector a number of other matrices can be defined. The most common is the covariance matrix $\mathbf{C}_s$, defined as:

$$\mathbf{C}_s = \left\langle \boldsymbol{k}^T \boldsymbol{k}^{T\,H} \right\rangle = \left\langle \begin{bmatrix} S_{hh} \\ S_{vv} \\ S_{hv} \end{bmatrix} \begin{bmatrix} S_{hh}^* & S_{vv}^* & S_{hv}^* \end{bmatrix} \right\rangle =$$
$$= \begin{bmatrix} \langle |S_{hh}|^2 \rangle & \langle S_{hh} S_{vv}^* \rangle & \langle S_{hh} S_{hv}^* \rangle \\ \langle S_{vv} S_{hh}^* \rangle & \langle |S_{vv}|^2 \rangle & \langle S_{vv} S_{hv}^* \rangle \\ \langle S_{hv} S_{hh}^* \rangle & \langle S_{hv} S_{vv}^* \rangle & \langle |S_{hv}|^2 \rangle \end{bmatrix} \tag{3.15}$$

where the brackets $\langle \cdot \rangle$ are denoting ensemble averaging, which is usually approximated by averaging over a set of measurements (usually pixels).

This definition is similar to the one of covariance matrix, introduced in Eq. (3.6), as they both measure the statistical interrelationship between the different channels. It is well known that covariance matrix allows, in the zero mean case, the complete (joint) statistical characterization of random variables. Moreover, covariance is a measure of degree of predictability between random variables. Accordingly the matrix in (3.15) allows characterizing the joint statistical behavior of the elements of the scattering matrix.

Another important definition is the polarimetric radar cross-section (RCS) σ_{pq}. It is defined as the ratio between the scattered power in a given direction, in the polarization p, as if the scattered field were uniform in all directions and the incident power density in the polarization q, that is:

$$\sigma_{pq} = 4\pi |S_{pq}|^2. \tag{3.16}$$

The RCS is more useful for single-polarization radar where phase information is not critical.

For a distributed target, like the ground surface, the RCS normalized to the area A illuminated by the radar antenna, denoted as Normalized Radar Cross Section (NRCS) or scattering coefficient, is of interest. It is a normalized dimensionless number, defined as:

$$\sigma_{pq}^0 = \frac{4\pi}{A} |S_{pq}|^2 \tag{3.17}$$

Moreover, being the distributed scattering of interest generated generally by a rough surface, the quantity in (3.17) is a random variable whose fluctuation is referred to as speckle. The following quantity is thus considered:

$$\sigma_{pq}^0 = \frac{4\pi}{A} E[|S_{pq}|^2] \tag{3.18}$$

in place of (3.17), with $E[]$ being the statistical averaging.

The NRCS σ_{pq}^0, related to the modulus of a quantity referred to as the complex reflectivity γ_{pq} by:

$$\sigma_{pq}^0 = 4\pi E[|\gamma_{pq}|^2] \tag{3.19}$$

The complex reflectivity γ_{pq} is usually expressed as a function of the coordinates (x, y) of the elementary scattering area on the observed surface, and is also referred as reflectivity function.[1] It is defined, except for a constant, as the ratio between the phasor of the electric field in the polarization p re-radiated from the point (x, y) towards the radar (E_p^s), and the phasor of the electric field in the polarization q incident at the same point (E_q^i):

$$\gamma_{pq}(x, y) = \frac{E_p^s(x, y)}{E_q^i(x, y)}. \tag{3.20}$$

The reflectivity function γ_{pq}, differently from the NRCS σ_{pq}^0, retains the phase information of the scattered signal.

The RCS is a quantity introduced in the context of radar surveillance. In remote sensing, the scattering can be concentrated for targets as metallic objects, dihedral elements of buildings, etc., or distributed as in the case of roads, forest, and natural surfaces. Accordingly, in this field it is usual to refer to the normalized radar cross section, i.e. a sort of "scattering density" with respect to the illuminated area, or to γ_{pq} which is related to σ_{pq}^0 via (3.19), and is commonly referred to as reflectivity pattern.

A last remark is related to the area of normalization in (3.17) for which different options have been proposed in the literature [14]. In radar remote sensing, typically the area refers to the illuminated area on the ground, that is the ground resolution area after the focusing operation. A different option is the NRCS per unit effective surface area, which is defined as the area perpendicular to the beam instead of the illuminated ground area: this quantity is still referred to as backscattering coefficient but the symbol typically used is γ^0. Finally, a less used option is the radar brightness, defined as the average RCS per unit area in the radar plane, i.e. slant-range and azimuth. This quantity, referred to as radar brightness, commonly indicated with β^0, is however rarely used because leads to measurements that are dependent on the sensor.

The computation of the scattering matrix **S**, and then of the RCS, involves integration of the total electric field over the target volume. Closed-form solutions exist for many canonical targets, f.i. a scatterer of perfectly spherical shape, for which the scattering coefficients expression can be found by using Mie theory. Explicit expressions for **S** also exist for symmetric geometries such as cylinders, spheroids, and so on.

For a large scattering body with complicated geometry, numerical methods have to be used (f.i. the method of moments (MoM),

---

[1] In the literature, following the terminology adopted in optics, the term "reflectivity" refers to the square modulus of γ_{pq}, i.e. to the dimensionless quantity representing the percentage of scattered power density with respect to the incidence one.

finite element method (FEM), and finite difference time difference), or high-frequency methods such as geometrical optics and physical optics.

3.3 Scattering matrix of canonical objects

Polarimetric radars measure the full scattering matrix to collect the information regarding the scatterer. The four elements of the scattering matrix can be measured by first transmitting a wave with one polarization and receiving echoes in two orthogonal polarizations simultaneously, and then transmitting a wave with a second polarization and, again, receiving echoes with both polarizations simultaneously. If the radar operates in the backscatter mode, reciprocity holds so that $S_{hv} = S_{vh}$ and only three independent elements of the scattering matrix have to be determined.

The polarization response of a target is strictly related to the target characteristics. Canonical targets can be used for polarimetric radar calibration. For instance, the polarimetric response in a linear polarimetric basis of a thriedal corner reflector must have the following characteristics (see Table 3.2):

- cross polarization components are not present ($S_{hv} = S_{vh} = 0$);
- horizontal and vertical backscattering cross sections are the same ($|S_{hh}| = |S_{vv}|$);
- horizontal and vertical copolarized components are in phase $\underline{/S_{hh} S_{vv}^*} = 2l\pi$, with l integer.

For a dihedral corner reflector, instead, the scattering matrix has the following structure (see Table 3.2):

- cross polarization components are not present ($S_{hv} = S_{vh} = 0$) when it is illuminated by a purely horizontally or vertically polarized wave;
- horizontal and vertical backscattering cross sections are the same ($|S_{hh}| = |S_{vv}|$);
- horizontal and vertical copolarized components are out of phase $\underline{/S_{hh} S_{vv}^*} = (2l + 1)\pi$, with l integer.

Table 3.2 gives few examples of the scattering matrix structure of some other canonical scattering targets.

To evaluate the scattering matrix of a canonical object rotated by an angle θ in the plane orthogonal to the propagation direction with respect to the position in which the scattering matrix is known, the following relation can be used:

$$\mathbf{S}(\theta) = \begin{bmatrix} \cos\theta & \sin\theta \\ -\sin\theta & \cos\theta \end{bmatrix} \begin{bmatrix} S_{hh} & S_{hv} \\ S_{vh} & S_{vv} \end{bmatrix} \begin{bmatrix} \cos\theta & -\sin\theta \\ \sin\theta & \cos\theta \end{bmatrix} \qquad (3.21)$$

Table 3.2 Structure of the scattering matrices of canonical targets.

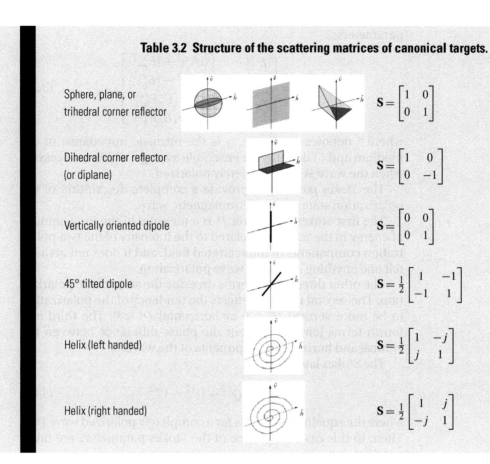

Sphere, plane, or trihedral corner reflector	$\mathbf{S} = \begin{bmatrix} 1 & 0 \\ 0 & 1 \end{bmatrix}$
Dihedral corner reflector (or diplane)	$\mathbf{S} = \begin{bmatrix} 1 & 0 \\ 0 & -1 \end{bmatrix}$
Vertically oriented dipole	$\mathbf{S} = \begin{bmatrix} 0 & 0 \\ 0 & 1 \end{bmatrix}$
45° tilted dipole	$\mathbf{S} = \frac{1}{2} \begin{bmatrix} 1 & -1 \\ -1 & 1 \end{bmatrix}$
Helix (left handed)	$\mathbf{S} = \frac{1}{2} \begin{bmatrix} 1 & -j \\ j & 1 \end{bmatrix}$
Helix (right handed)	$\mathbf{S} = \frac{1}{2} \begin{bmatrix} 1 & j \\ -j & 1 \end{bmatrix}$

3.4 Stokes parameters and Mueller matrix

Eq. (3.11) relates the h-polarized and v-polarized components of field $\mathbf{E}^s$ scattered from an object to the h-polarized and v-polarized components of the field $\mathbf{E}^i$ incident on it.

For the case of coherent scattering from a single deterministic target, the scattered wave is completely polarized, that means $|E_v^i|$, $|E_h^i|$ and $\measuredangle E_v^s E_h^{s*}$ are constants or at least slowly varying functions of time. As opposed to coherent scattering, incoherent scattering from random or distributed targets or from random surfaces produces partially polarized waves, that can be studied considering the second-order moment of the random field $\mathbf{E}^s$, which is in power units.

Then, for describing partially polarized waves by means of observable power terms, and not by amplitudes and phases, it is convenient to use the Stokes vector $\mathbf{I}^s$, consisting of the four Stokes parameters:

$$\mathbf{I}^s = \begin{bmatrix} I^s \\ Q^s \\ U^s \\ V^s \end{bmatrix} = \frac{1}{\eta} \begin{bmatrix} \langle |E_v^s|^2 + |E_h^s|^2 \rangle \\ \langle |E_v^s|^2 - |E_h^s|^2 \rangle \\ 2Re\langle E_v^s E_h^{s*} \rangle \\ 2Im\langle E_v^s E_h^{s*} \rangle \end{bmatrix} \tag{3.22}$$

where * denotes conjugate, η is the intrinsic impedance of the medium and $\langle \cdot \rangle$ denotes the ensemble average, which is necessary when the wave is not completely polarized.

The *Stokes parameters* provide a complete description of the polarization state of an electromagnetic wave.

The first Stokes parameter I^s is a measure of the total amount of energy in the wave. It is related to the intensity of the two polarization components of the scattered field, and it does not actually tell one anything about the wave polarization.

The other three Stokes terms describe the state of the polarization. The second term Q^s reflects the tendency of the polarization to be more vertical $Q^s > 0$ or horizontal $Q^s < 0$. The third and fourth terms jointly represent the phase difference between the vertical and horizontal components of the wave.

The Stokes law gives:

$$I^{s2} \geqslant Q^{s2} + U^{s2} + V^{s2} \tag{3.23}$$

where the equality only holds for a completely polarized wave [11]. Then, in this case, only three of the Stokes parameters are independent.

Another commonly used form of Stokes parameters is:

$$\breve{\mathbf{I}}^s = \begin{bmatrix} I_v^s \\ I_h^s \\ U^s \\ V^s \end{bmatrix} = \frac{1}{\eta} \begin{bmatrix} \langle |E_v^s|^2 \rangle \\ \langle |E_h^s|^2 \rangle \\ 2Re\langle E_v^s E_h^{s*} \rangle \\ 2Im\langle E_v^s E_h^{s*} \rangle \end{bmatrix} \tag{3.24}$$

Using Stokes vectors $\mathbf{I}^i$ and $\mathbf{I}^s$ to represent the incident and scattered wave, the incoherent scattering can be represented by:

$$\mathbf{I}^s = \bar{\mathbf{M}} \cdot \mathbf{I}^i \tag{3.25}$$

where $\bar{M}$ is the 4×4 real Mueller matrix defined as:

$$\bar{M} = \begin{bmatrix} \langle |S_{vv}|^2 \rangle & \langle |S_{vh}|^2 \rangle & Re\langle S_{vv} S_{vh}^* \rangle & -Im\langle S_{vv} S_{vh}^* \rangle \\ \langle |S_{hv}|^2 \rangle & \langle |S_{hh}|^2 \rangle & Re\langle S_{hv} S_{hh}^* \rangle & -Im\langle S_{hv} S_{hh}^* \rangle \\ 2Re\langle S_{vv} S_{hv}^* \rangle & 2Re\langle S_{vh} S_{hh}^* \rangle & Re\langle S_{vv} S_{hh}^* + S_{vh} S_{hv}^* \rangle & -Im\langle S_{vv} S_{hh}^* - S_{vh} S_{hv}^* \rangle \\ 2Im\langle S_{vv} S_{hv}^* \rangle & 2Im\langle S_{vh} S_{hh}^* \rangle & Im\langle S_{vv} S_{hh}^* + S_{vh} S_{hv}^* \rangle & Re\langle S_{vv} S_{hh}^* - S_{vh} S_{hv}^* \rangle \end{bmatrix} \tag{3.26}$$

Some useful considerations about the Mueller matrix can be done:

- For coherent scattering, only seven of the sixteen elements of the Mueller matrix are independent (degrees of freedom of Mueller matrix).
- For most natural targets, copolarization and cross-polarization components are mutually uncorrelated, so that the 2×2 upper-right block and the 2×2 lower-left block approach to zero.
- The upper-left block reflects the average power of the two polarizations and the lower-right block reflects the correlation between different polarizations.

Moreover, from Stokes parameters and Mueller matrix some target signatures can be inferred:

- copolarization signature:

$$\sigma_{co} = 4\pi \frac{1}{2} \mathbf{I}^{sT} \mathbf{I}^s \qquad (3.27)$$

- cross-polarization signature:

$$\sigma_x = 4\pi \frac{1}{2} \left(\mathbf{I}^{sT} \mathbf{I}^s - \mathbf{I}^{sT} \mathbf{I}^i \right) \qquad (3.28)$$

- polarization degree:

$$m_s = \frac{Q^{s2} + U^{s2} + V^{s2}}{I^{s2}} \qquad (3.29)$$

where $m_s \leqslant 1$, with $m_s = 1$ when the scattered wave is completely polarized;

- co-pol/cross-pol phase difference:

$$\phi_{vh} = \tan^{-1} \frac{M_{43} - M_{34}}{M_{33} + M_{44}} \qquad (3.30)$$

where M_{ij} denotes the ij element of the Mueller matrix $\bar{M}$.

3.5 Coherent polarimetric decomposition

The scattering matrix $\mathbf{S}$ defined in (3.11) can characterize the scattering target only when the incident and the scattered waves are both completely polarized waves. This requires the scattering object to be a coherent target. In this case target characterization can be performed by means of coherent decomposition of the scattering matrix.

The objective of coherent decomposition is to express the scattering matrix as the combination of the scattering matrices of simpler objects:

$$S = \sum_{i=1}^{K} c_i S_i \tag{3.31}$$

where the matrices S_i are scattering matrices of canonical objects and c_i are expansion coefficients representing the scattering power of the different objects.

In order to simplify the understanding of (3.31), it is desirable that the matrices S_i are linearly independent to avoid that a particular scattering behavior to be present in more than one matrix. Often, the less restrictive property of orthogonality of the matrices S_i is imposed:

$$\text{trace}\left(S_i S_j^H\right) = 0 \qquad \text{for } i \neq j \tag{3.32}$$

where trace is the sum of the diagonal element of a square matrix. The decomposition in (3.31) is not unique, in the sense that it is possible to find several sets $S_i, i = 1, ..., K$, in which the matrix S can be decomposed, but only few of them are convenient for the interpretation of the information contained in S.

A basis that is very convenient for representing polarimetric data is the "Pauli" basis, given by the four 2×2 matrices $\{S_a, S_b, S_c, S_d\}$ defined as follows:

$$S_a = \frac{1}{\sqrt{2}} \begin{bmatrix} 1 & 0 \\ 0 & 1 \end{bmatrix} \tag{3.33}$$

$$S_b = \frac{1}{\sqrt{2}} \begin{bmatrix} 1 & 0 \\ 0 & -1 \end{bmatrix} \tag{3.34}$$

$$S_c = \frac{1}{\sqrt{2}} \begin{bmatrix} 0 & 1 \\ 1 & 0 \end{bmatrix} \tag{3.35}$$

$$S_d = \frac{1}{\sqrt{2}} \begin{bmatrix} 0 & 1 \\ -1 & 0 \end{bmatrix}. \tag{3.36}$$

Since in monostatic configuration reciprocity holds and $S_{hv} = S_{vh}$, the Pauli basis can be reduced to a basis composed by the three matrices S_a, S_b, and S_c, so that the scattering matrix can be expressed by:

$$S = \begin{bmatrix} S_{vv} & S_{hv} \\ S_{hv} & S_{hh} \end{bmatrix} = \alpha S_a + \beta S_b + \gamma S_c \tag{3.37}$$

where

$$\alpha_p = \frac{S_{vv} + S_{hh}}{\sqrt{2}} \tag{3.38}$$

$$\beta_p = \frac{S_{vv} - S_{hh}}{\sqrt{2}} \tag{3.39}$$

$$\gamma_p = \sqrt{2}S_{hv} \tag{3.40}$$

Then, a target can be characterized by the target vector in the Pauli basis:

$$\boldsymbol{k}_p = \frac{1}{\sqrt{2}} \begin{bmatrix} S_{vv} + S_{hh} & S_{vv} - S_{hh} & 2S_{vh} \end{bmatrix}^T \tag{3.41}$$

For many applications this is a more useful representation since it helps to emphasize the phase difference between the co-pol terms.

The interpretation of the Pauli decomposition can be done on the base of the following considerations:

- The matrix $\mathbf{S}_a$ corresponds to the scattering matrix of a sphere, a plate or a trihedral, and refers to single- or odd-bounce scattering mechanism (see (3.33) and Table 3.2). Then, the intensity $|\alpha|^2$ represents the power of the wave returned by single- or odd-bounce scattering.
- The matrix $\mathbf{S}_b$ corresponds to the scattering matrix of a dihedral oriented at 0 degrees (see (3.34) and Table 3.2), and refers to double or even-bounce scattering mechanism. Then, the intensity $|\beta_p|^2$ represents the power of the wave returned by double- or even-bounce scattering.
- The matrix $\mathbf{S}_c$ corresponds to the scattering matrix of a dihedral oriented at 45 degrees (see (3.34) and Table 3.2), and refers to the scattering mechanism of the scatterers which are able to return the polarization orthogonal to the incident one. Then, one of the best examples is the volume scattering produced by the forest canopy. Then, the intensity $|\gamma_p|^2$ typically represents the power of the wave returned by volume scattering.

The information extracted using Pauli basis can be pictorially represented in the form of a color image, by associating a primary color of the RGB color system to the intensities of the coefficients α_p, β_p, and γ_p (f.i. $|\alpha_p| \rightarrow Blue$, $|\beta_p| \rightarrow Red$, $|\gamma_p| \rightarrow Green$).

Pauli image composition of fully polarimetric L-band acquisitions of the E-SAR airborne sensor of the German Aerospace Center (DLR) on the city of Drtesden (Germany) is shown in Fig. 3.4. Forested and vegetated areas can be easily recognized in the upper part of the image. Moreover, it can be noted that the individual buildings and building blocks exhibiting dihedrals oriented parallel to the azimuth (vertical) direction are characterized by high HH-VV (red) component, while buildings/blocks with an orientation angle with respect to azimuth direction are characterized by a higher cross-polar component (HV), and then appear colored in shades closer to green.

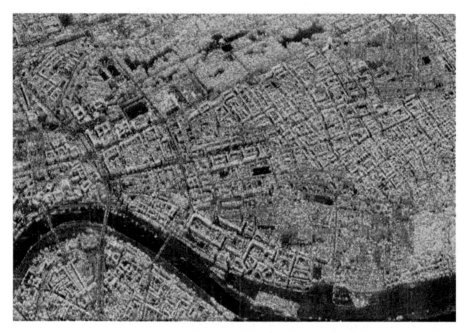

Figure 3.4. RGB coded Pauli representation of a fully polarimetric image acquired by the ESAR airborne sensor of DLR on the city of Dresden (Germany). Vertical direction is azimuth, horizontal direction is range.

Referring to the Pauli representation basis (3.41), the polarimetric coherency matrix, also called **T** matrix, can be defined as:

$$\mathbf{T} = \left\langle k_p k_p{}^H \right\rangle =$$

$$= \frac{1}{2} \left\langle \begin{bmatrix} S_{vv} + S_{hh} \\ S_{vv} - S_{hh} \\ 2S_{vh} \end{bmatrix} \begin{bmatrix} (S_{vv} + S_{hh})^* & (S_{vv} - S_{hh})^* & 2S_{hv}^* \end{bmatrix} \right\rangle = \quad (3.42)$$

$$= \frac{1}{2} \begin{bmatrix} \langle |S_{vv}+S_{hh}|^2 \rangle & \langle (S_{vv}+S_{hh})(S_{vv}-S_{hh})^* \rangle & \langle 2(S_{vv}+S_{hh})S_{vh}^* \rangle \\ \langle (S_{vv}-S_{hh})(S_{vv}+S_{hh})^* \rangle & \langle |(S_{vv}-S_{hh})|^2 \rangle & \langle 2(S_{vv}-S_{hh})S_{vh}^* \rangle \\ \langle 2S_{vh}(S_{vv}+S_{hh})^* \rangle & \langle 2S_{vh}(S_{vv}-S_{hh})^* \rangle & \langle 4|S_{hv}|^2 \rangle \end{bmatrix}$$

The **T** matrix (3.42) is by definition an Hermitian positive semidefinite matrix. Its diagonal elements are real valued and represent the backscattered power, while the offdiagonal elements are the complex cross correlations between the elements of k_p. Then, there exist nine independent real parameters in the polarimetric coherency matrix, which can provide a geometrical or physical interpretations of the radar scattering mechanisms [15] [16] [17]. This issue is addressed more in detail in Chapter 8.

3.6 Scattering models

Basically, there are two different interaction mechanisms that can take place when the transmitted electromagnetic wave impinges on the illuminated scene: surface scattering and volume scattering. Surface scattering is caused by dielectric discontinuities at all points on a surface boundary, while volume scattering is caused by dielectric discontinuities distributed throughout a volume. In the following sections, these scattering mechanisms are briefly described.

3.6.1 Surface scattering

Surface scattering is the most common scattering mechanism for natural scenes. The most basic model is given by a perfectly planar surface of infinite extension.

Let us consider a surface separating two homogeneous media with different electric or magnetic properties. When a plane electromagnetic wave propagating in the upper half-space is incident on it, a part of the energy will be radiated toward the upper half-space, and a part of the energy will be transmitted in the lower half-space. If the lower medium is homogeneous, or it can be considered as such, we have a surface scattering problem, because scattering takes place only on the separation surface.

In the case of a perfectly flat surface separating a vacuum upper half-space from a dielectric lower half-space with a complex relative dielectric constant $\epsilon = \epsilon_r + \sigma_c/(j\omega\epsilon_0)$, the radiated field is composed of a reflected plane wave propagating in the specular direction in the upper half-space with amplitude depending on the well-known Fresnel reflection coefficients

$$R_h = \frac{\cos\theta_i - \sqrt{\epsilon - \sin^2\theta_i}}{\cos\theta_i + \sqrt{\epsilon - \sin^2\theta_i}} \qquad R_v = \frac{\epsilon\cos\theta_i - \sqrt{\epsilon - \sin^2\theta_i}}{\epsilon\cos\theta_i + \sqrt{\epsilon - \sin^2\theta_i}} \qquad (3.43)$$

where θ_i is the incidence angle, and of a refracted or transmitted plane wave propagating in the lower half-space in the direction (see Fig. 3.5):

$$\theta' = \arcsin\left(\frac{\sin\theta_i}{\sqrt{\epsilon}}\right). \qquad (3.44)$$

From Fig. 3.5 it is clear that all the specular reflected waves have the same path lengths and then are satisfying the in-phase condition.

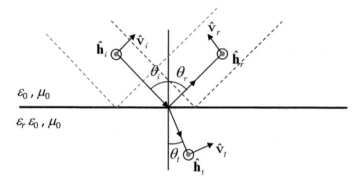

Figure 3.5. Reflection from a plane surface.

Then, when the surface is perfectly flat and has an infinite extent, the radiated field in the upper half-space is concentrated in the specular direction. If the surface is finite in extent, or if the infinite surface is illuminated by a finite-extent uniform plane wave, the maximum amount of reflected power still appears in the specular direction, but a lobe structure appears around this direction. The calculation of the radiated field can be done using physical optics. This component is often referred to as the coherent component of the scattered field. It will be very small in directions away from the specular one.

Assume now that the surface is rough, so that some deviations from the flat surface are present. It is clear from Fig. 3.6 that the two reflected waves have a path length difference equal to $2\Delta r = 2\Delta h \cos\theta_i$. This will give a phase difference of:

$$\Delta\varphi = \frac{4\pi}{\lambda}\Delta h \cos\theta_i. \tag{3.45}$$

If Δh is small respect to the wavelength, the phase difference is not significant and the reflected waves will sum up in-phase. Instead, if the phase difference is significant, the reflected waves will be not in-phase and the specular reflection will be reduced.

Another effect that occurs in the case of rough surfaces is that some of the energy is radiated in all directions different from the Fresnel reflection direction. The amount of this energy is depending on the magnitude of the surface roughness with respect to the wavelength of the incident field, and is associated to the scattered field (Fig. 3.7).

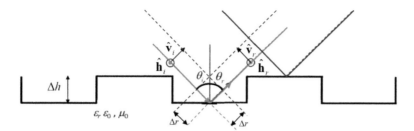

Figure 3.6. Reflection from a nonplanar surface.

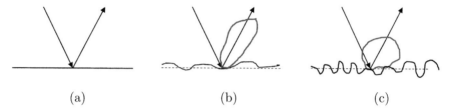

Figure 3.7. Scattering from a surface. (a) Plane interface: only specular scattering is present. (b) Moderately rough interface: specular and diffuse scattering are present. (c) Very rough interface: diffuse scattering is dominant.

A simple criterion that can be used to establish if a surface behaves as a rough surface is the Rayleigh condition, which regards a surface as rough if the phase difference (3.45) is greater than $\pi/2$, that means [18]:

$$\Delta h > \frac{\lambda}{8\cos\theta_i}. \tag{3.46}$$

When condition (3.46) is not verified, the surface is regarded as a smooth surface.

The component of the scattered power associated with the surface roughness is referred to as the incoherent component of the scattered field. It is predominant respect to the coherent component at angles significantly away from the specular direction.

A rough surface is described by a height function $z = f(x, y)$ which is not perfectly known, so that it is treated as a random process. The height profile of roughness is, then, characterized in terms of some statistical quantities, such as mean value, which is usually assumed to be zero, height standard deviation σ_z, height correlation length L, and height correlation function $C_z(x_1 - x_2, y_1 - y_2)$, which is equivalent to assign the roughness spectrum. The models applied for computing the scattered field from the rough surface are depending on these statistical parameters [19].

Then, the electromagnetic scattering from a natural surface consists of two components: a reflected or coherent component, resulting from the reflection of a smooth surface and concentrated around the specular direction, and a scattered or incoherent component, resulting from the diffuse scattering due to surface roughness. As roughness becomes larger, the coherent component becomes negligible and diffuse scattering is dominant (see Fig. 3.7).

Electromagnetic scattering from rough surfaces has been the subject of intensive studies. Many approaches have been developed and tested using empirical measurements. Anyway, the approaches that are of practical interest for natural surfaces rely on approximate models, and hence exhibit more or less restricted applicability constraints.

The most common approaches used to model radar scattering from rough surfaces are physical optics and small-perturbation model. Physical optics is mainly applied for small angles of incidence (less than 20°-30°), where reflection from regular facets dominates. Small-perturbation model, instead, is generally applied for larger angles of incidence, where scattering from the small-scale roughness dominates, and often is referred to as the Bragg model. This model treats the surface roughness as a small perturbation from a flat surface.

Physical optics

Physical Optics can be applied for modeling electromagnetic scattering for smooth surfaces, for which the radius of curvature for each point on the surface has to be large compared to the wavelength [20] [21]. It is based on the Kirchhoff approach, which exploits the tangent plane approximation, i.e. at any surface point electromagnetic fields are approximated by the fields that would be present on the tangent plane.

The surface smoothness assumption can be considered verified in a large band of operating frequencies if we consider the surface profile on a macroscopic scale, i.e. the profile that describes the morphology of the ground surface on a spatial scale larger than resolution.

The morphology variation of the surface on a small spatial scale, which describes the surface roughness, can be taken into account including a stochastic roughness model associated with a stochastic diffuse scattering component.

A very simple, general and flexible model based on Physical Optics is the so-called "facet model". It exploits a two scale model for describing the observed surface profile: the macroscopic, large scale, surface profile is approximated by means of a series of small plane facets, each one tangent to the effective surface, while the microscopic, small scale, profile is described by means of a stochastic model [22]. Coherent scattering components from each small plane facet can be computed exploiting the Fresnel reflection coefficient (3.43) and taking into account the radiation diagram of the facet. In this way, it is very easy to adopt the model to every smooth surface and to account for polarization, through the use of the appropriate expression of the Fresnel coefficients. The incoherent component, associated with the surface roughness, instead, can be included by introducing an appropriate phase stochastic factor and a weighting by a facet radiation diagram that are depending on the roughness degree of the facet.

The small perturbation model

The Small Perturbation Model (SPM) [20] can be applied when the surface roughness is small. The surface roughness can be considered "small" when the roughness parameters, such as the height standard deviation σ_z and the roughness correlation length L are small with respect to the radar wavelength (f.i. $4\pi\sigma_z/\lambda < 0.3$, $4\pi L/\lambda < 3$, $\sigma_z/L < 0.3$). This implies that SPM is often applicable for low-frequency systems in the case of single-scale rough surfaces.

The use of the first-order small-perturbation model dates back to Rice (1951) [23]. A synthetic discussion and the expression derived for the scattering cross section, which is related to the roughness spectrum, is reported in [24].

3.6.2 Volume scattering

Volume scattering can be observed when the incident wave penetrates into a lower inhomogeneous half-space, such as a vegetation canopy. In this case, the wave loses part of its energy by scattering in all directions, so that part of the scattered energy is back-scattered toward the transmitter.

The intensity of volume scattering is depending on the discontinuous permittivity and on the density of the heterogeneous medium, while the scattering angle depends on the roughness of the surface separating the two media, on the average relative permittivity and on the wavelength.

Volume scattering will completely depolarize the incident signal, so that fully polarization measurements can be important for discriminating different scatterers.

Many surface features when illuminated by microwaves will act like a "volume scatterer", i.e. they scatter energy equally in all directions. Forest canopies, snow layers, are two examples. A very simplified model for the normalized RCS for unit area assumes

[21]:

$$\sigma^0_{vol_{pq}} = \alpha_{pq} \cos(\theta_i), \qquad (3.47)$$

where α_{pq} is some constant determined by the particular target properties.

For an extended volume scattering target, such as a forest, α_{pq} will remain approximately constant for all incidence angles, then, it is a more convenient measurement parameter than σ^0_{pq}.

For the computation of the normalized RCS, inhomogeneous medium can be modeled as a collection of uniformly distributed discrete scatterers, each of which deviates a fraction or the energy or any incident radiation from its original propagation direction and re-radiates towards the other scatterers. For instance, dielectric cylinders are commonly used to model branches and trunks of trees. Different statistical orientation distributions are considered for different kind of trees. Moreover, multiple interaction between the scatterers has to be considered.

For instance, in the case when the illuminated scene is represented by a vegetation layer covering a ground surface, if the incident wave is at low frequency microwaves, it has the capability to penetrate through the whole vegetation layer down to underlying ground. As the wave travels through the vegetation, it interacts with leaves, branches, tree trunks, and the terrain. Then, the scattered field contains different contributions (see Fig. 3.8), such as the ones due to (a) direct scattering by the underlying ground surface, (b) direct and indirect scattering by the crown layer, and (c) scattering by the ground-tree trunk and ground-tree crown combinations. The signal resulting from direct backscattering from woody elements within the vegetation layer is depolarized and provides the most direct information about the vertical structure of the vegetation. The interaction between the ground and the tree trunk is an example of dihedral corner reflection. It is a result of the two specular reflections of the wave onto the tree trunks and the terrain, or vice-versa. Intensity is maximal when terrain topography is flat, whereas it tends to vanish in the case of sloped topography.

3.7 Speckle

In remote sensing applications, the transmitted radar pulses illuminate, according to the radiation pattern of the transmitted antenna, an area on the ground surface which is composed of several scattering elements. The wider the illuminated area, the larger the number of scattering element is. The resulting returned echo is

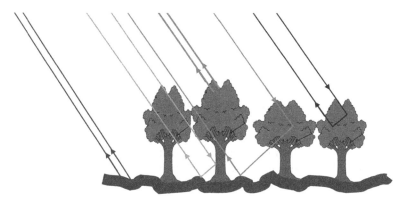

Figure 3.8. Scattering from a forest.

the coherent summation of the waves backscattered by this large number of scattering elements. These elements are randomly positioned inside the illuminated area, so that their backscattered contributions randomly combine coherently at the position where the radar antenna is located.

We can consider each single contribution in terms of the phasor associated with the backscattered wave, characterized by a given amplitude A_i and a given phase ϕ_i, and the overall backscattering signal as the summation of the different phasors, resulting in a amplitude A and a phase ϕ, as shown schematically in Fig. 3.9.

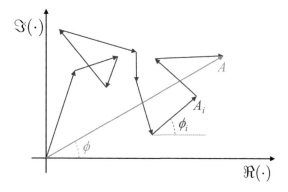

Figure 3.9. Overall backscattering signal as the summation of different phasors.

In Fig. 3.9, amplitudes and phases relevant to the different contributions differ from each other. Amplitude changes occur likely when the illuminated area is very extended, as shown in Fig. 3.10. In this case it can easily happen that parts of the illuminated area, with different reflectivity properties, reported with different colors in Fig. 3.10, can be simultaneously present in the same illuminated spot. By reducing the extension of the illuminated area, and this is the case of the resolution cell of the SAR images after the image formation process (after range and azimuth compression), it is more likely that only homogeneous reflectivity is present in the spot, as shown in Fig. 3.11, so that the phasors amplitude of Fig. 3.9 should be all almost equal. As far as the phases are concerned, their values depend on their distances between scattering elements and radar antenna. When the illuminated spot dimension is larger than the wavelength, taking also into account the random positioning of the scattering elements into the spot, the resulting phases can be considered random variables, uniformly distributed in the interval $(0, 2\pi)$.

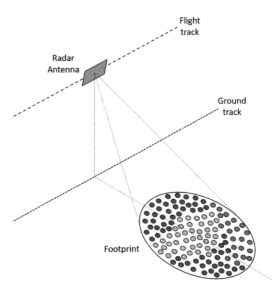

Figure 3.10. Real Antenna Footprint. Presence of different properties background in the footprint.

In order to derive a general model for the speckle we consider simplified geometries and cases. Let start considering the case of a small target dimension in terms of working wavelength, positioned at a distance r_1 from the radar. The return signal backscattered from it will be a replica of the transmitted signal (supposed to be a sinusoidal signal at frequency f and with a initial null phase), attenuated by the usual factor q and time delayed by $2r/c$. Call the amplitude and the phase of the return (received) signal by A_1 and $\varphi_1 = 4\pi r_1/\lambda$. For small incremental variation of the distance, while we can consider that the amplitude A_1 remains almost constant,

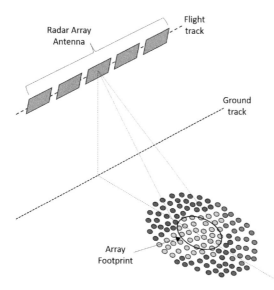

Figure 3.11. Array Antenna Footprint. Presence of homogeneous background in the footprint after the so-called range and azimuth compressions.

the phase φ_1 will change significantly over a scale of the order of the wavelength.

Consider now the case of two small targets in terms of wavelength, with similar cross sections, so that $A_1 \simeq A_2$, whose distance from the radar antenna is r_1 and r_2, respectively, with $r_1 \simeq r_2$, as shown in Fig. 3.12. The return signal backscattered from them is the superposition of the single return signals (neglecting the possible mutual coupling and the very small amplitude difference between them).

The resulting return signal will be given by:

$$V = Ae^{\varphi} = A_1e^{\varphi_1} + A_2e^{\varphi_2} = A_1e^{-j2kr_1} + A_2e^{-j2kr_2} \qquad (3.48)$$

Considering the case of a radar antenna which moves along the linear trajectory x shown in Fig. 3.12, at a distance from the targets much larger than the distance between targets ($r_0 \gg d$), we can use the following approximations:

$$r_i = \sqrt{r_0^2 + (x - x_i)^2} \simeq r_0 \left(1 + \frac{1}{2} \left(\frac{x - x_i}{r_0} \right)^2 \right), \qquad i = 1, 2 \quad (3.49)$$

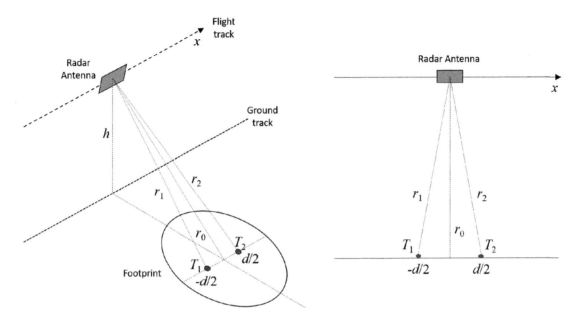

Figure 3.12. Two small targets case. 3D view (left). 2D projected view (right).

where $x_1 = -d/2$ and $x_2 = d/2$. Substituting (3.49) into (3.48), we get:

$$V \simeq 2A_1 e^{-j2k\varphi_0} e^{-j2\frac{k}{r_0}x^2} \cos\left(\frac{kd}{r_0}x\right)$$

$$|V| = 2A_1 \left| \cos\left(\frac{kd}{r_0}x\right) \right|$$

(3.50)

where φ_0 is a phase term, constant with respect to x, which depends only on geometrical parameters r_0 and d. If we plot (see Fig. 3.13) the amplitude of the backscattered signal Vs. the position x of the radar antenna, we can observe that it varies from a minimum of 0 when $x = n\lambda r_0/2d$, with n being relative numbers, which corresponds to the case when the two return signals have the same phase (but for a multiple of 2π), to a maximum of $2A_1$, when $x = \lambda r_0/4d + n\lambda r_0/2d$ and the two return signals have phases which differ by π. In this case the amplitude pattern follows a periodic behavior (absolute value of a sinusoidal signal) with a spatial frequency equal to $2d/\lambda r_0$, or equivalently with a spatial period equal to $\lambda r_0/2d$.

Fig. 3.13 shows that the same two targets are perceived in a completely different way with the movement of the radar antenna of only 15 m at a distance of 1 Km from the targets.

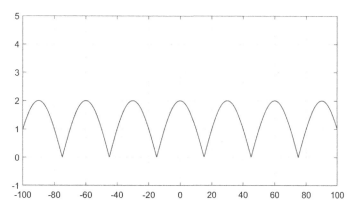

Figure 3.13. Two targets response. $A_1 = A_2 = 1, r_0 = 1000$ m, $d = 5$ m, $f = 1$ GHz. Observation interval: $x = (-100, 100)$. Spatial period: 30 m.

To complete this analysis, consider now a radar whose antenna size along x is L. As better described in the next chapter, the illuminated footprint, at 3 dB, at distance r_0 will be given by $(\lambda/L)(r_0/2)$, where the factor 2 at the denominator is due to the two-way nature of the radar. This implies that, in a footprint (i.e., a resolution cell of the radar), the signal will exhibit a number of oscillations (fluctuations from 0 to $2A_1$) equal to:

$$N_f = \frac{\lambda r_0}{2L} / \frac{\lambda r_0}{2d} = \frac{d}{L} \tag{3.51}$$

Hence, the same two targets scenario provides very different backscattered signals when their mutual position with respect to the antenna varies even of small amount. This effect, known in radar systems, and is called fading or speckle; this latter term is more in Synthetic Aperture Radar imaging.

Adding more targets sharpens this effect, in the sense that the amplitude fluctuation changes along the distance more rapidly than in the case of two targets.

Consider now a small area into the footprint, whose extension is few meters in both the direction parallel and perpendicular to the ground track, where some targets with similar cross section are randomly located (see Fig. 3.14), and consider the same radar antenna which moves along the linear trajectory x, at a distance from the targets much larger than the distances between targets. The 5 targets coordinates with respect to the center of the footprint are $T_1 \equiv (-3.73, -3.44)$, $T_2 \equiv (-1.44, 0, 25)$, $T_3 \equiv (1.24, -0.74)$, $T_4 \equiv (2.56, 1.75)$, $T_5 \equiv (3.75, 0.25)$.

Also in this case we can plot (see Fig. 3.15) the amplitude of the backscattered signal Vs. the position x of the radar antenna. The

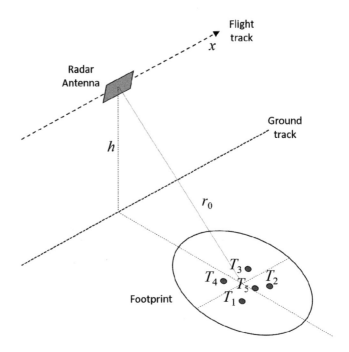

Figure 3.14. Five targets cluster case.

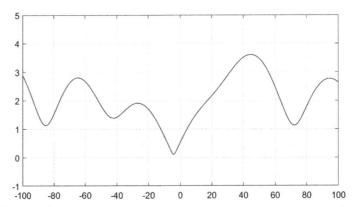

Figure 3.15. Five targets response. $A_i = 1, i = 1, .., 5, r_0 = 1000$ m, $f = 1$ GHz. Observation interval: $x = (-100, 100)$ m.

signal varies from a minimum of about 0 for $x = -3$ m, to a maximum of about 3.6 for $x = -42$ m. The same five targets cluster is seen as a stealth target or as a target whose amplitude is almost equal to the response of 4 targets at a distance of about 45 m along the flight trajectory. Few differences in the acquisition angle result in a strong difference of the return signal backscattered by the targets.

In the case of interest of SAR there is a very large number of scattering elements in a resolution cell. Moreover, the resolution cell is characterized by a natural roughness which obeys to random models. It is of course impossible to model such phenomenon with close deterministic models, unless for few cases relevant to geometric regular structures. For this reason the appropriate way to take into account the speckle consists in deriving a statistical model of the response of a surface in which there is a large number of targets that backscatters the signal toward the radar antenna. In the next subsection a statistical model of the amplitude (intensity) of the radar received signal is derived.

3.7.1 Speckle statistics

Assume that K different scatterers are randomly located in a resolution cell of the radar. The surface to which the targets belong can be rough. The signal that is received by the radar antenna, after backscattering from the different scatterers on the ground, is given in phasor form by the superposition of the elementary returns from different scatterers:

$$V = Ae^\varphi = A_R + jA_I = A\cos\varphi + jA\sin\varphi =$$

$$= \sum_{k=1}^{K} A_k e^{j\alpha_k} e^{-j\frac{4\pi}{\lambda}r_k} = \sum_{k=1}^{K} A_k e^{j\varphi_k} \tag{3.52}$$

where A_k is the amplitude of the return backscattered by the k-th target at the receiver, which depends on radar system parameters, on the target backscatter and on the attenuation of the signal due to propagation distance and possible atmospheric effects, α_k is the phase shift introduced by the backscattering (the "phase" of the target), and $\frac{4\pi}{\lambda}r_k$ is the phase delay due to the propagation distance.

Suppose that the different elementary targets are far from each other by quantities that are much smaller than their distances from the radar. This allows us to state that all the terms A_k are almost equal.

The statistical distribution of V in (3.52) is determined by several factors [25]. Here we limit our derivation to the so-called Fully Developed speckle model, also referred to by some authors as Rayleigh speckle model. At the basis of such model there are the following assumptions:
- the scatterers are very numerous and are randomly located in the resolution cell;

- the variables $A_1, \ldots, A_K$, are independent and identically distributed (iid), so that there are no dominant scatterers in a resolution cell;
- the resolution cell is far greater than the working wavelength, and it is much lower than the distance from the radar antenna; this implies that the phase terms φ_k are uniformly distributed in $(-\pi, \pi)$ iid variables, independent from $A_1, \ldots, A_K$.

Under these hypotheses, thanks to the Central Limit Theorem (CLT), the real and imaginary parts of (3.52) are both Gaussian distributed with zero mean, and same variance:

$$E\left[\Re\left(Ae^\varphi\right)\right] = E\left[A\cos\varphi\right] = \sum_{k=1}^{K} E\left[A_k\cos\varphi_k\right] =$$

$$= \sum_{k=1}^{K} E\left[A_k\right] E\left[\cos\varphi_k\right] = 0$$

$$E\left[\Im\left(Ae^\varphi\right)\right] = E\left[A\sin\varphi\right] = \sum_{k=1}^{K} E\left[A_k\sin\varphi_k\right] =$$

$$-\sum_{k=1}^{K} E\left[A_k\right] E\left[\sin\varphi_k\right] = 0$$

(3.53)

$$E\left[(A\cos\varphi)^2\right] = \sum_{k=1}^{K} E\left[(A_k\cos\varphi_k)^2\right] = \sum_{k=1}^{K} E\left[A_k^2\right] E\left[\cos^2\varphi_k\right] = \frac{\sigma^2}{2}$$

$$E\left[(A\sin\varphi)^2\right] = \sum_{k=1}^{K} E\left[(A_k\sin\varphi_k)^2\right] = \sum_{k=1}^{K} E\left[A_k^2\right] E\left[\sin^2\varphi_k\right] = \frac{\sigma^2}{2}$$

(3.54)

Variances of real and imaginary parts are equal, as $E\left[\cos^2\varphi_k\right] = E\left[\sin^2\varphi_k\right] = 1/2$. The factor $1/2$ present in the previous variances allows representing the variance of the signal in (3.52), which is the power of the signal and is related to the overall radar cross sections of the involved targets, equal to σ^2. As it can be easily shown, the real and imaginary part of (3.52) are uncorrelated random variables; being also Gaussian they are also independent thanks to the CLT.

We can eventually derive the statistical distributions of the amplitude and of the intensity signal reported in (3.52). Hence, considering the Cartesian representation of (3.52) whose joint pdf of real and imaginary parts is:

$$f_{A_R,A_I}(a_R, a_I) = \frac{1}{\pi\sigma^2} \exp\left(-\frac{a_R^2 + a_I^2}{\sigma^2}\right)$$

(3.55)

and considering the polar to Cartesian transformation formulas:

$$A = \sqrt{A_R^2 + A_I^2}$$
$$\phi = \tan^{-1}\left(\frac{A_I}{A_R}\right)$$
(3.56)

the joint pdf of the representation in polar coordinates is given by [26]:

$$f_{A,\phi}(a, \phi) = \frac{ae^{-a^2/\sigma^2}}{\pi\sigma^2}$$
(3.57)

where $a > 0$, and $\varphi \in (-\pi, \pi)$.

The pdf of the amplitude (the envelope) of the polar representation is given by:

$$f_A(a) = \int_0^{2\pi} \frac{ae^{-a^2/\sigma^2}}{\pi\sigma^2}d\varphi = \frac{2a}{\sigma^2}e^{-a^2/\sigma^2}u(a)$$
(3.58)

where $u(a)$ is the unit step function.

Eq. (3.58) is the well known Rayleigh probability distribution function. It exhibits a mean value $E[A] = \sigma\sqrt{\pi/4}$ and a variance $\sigma_A^2 = \sigma^2(1 - \pi/4)$, and represents in this case the pdf of the amplitude of the signal received by the Radar antenna whose plot is shown in Fig. 3.16.

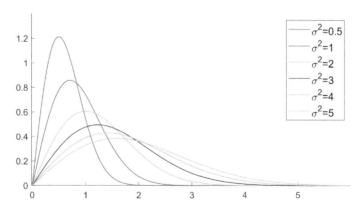

Figure 3.16. Rayleigh probability density function plotted Vs. different values of the radar cross section σ^2.

It is easy to derive from Eq. (3.58) the pdf of the intensity (the power) of the signal $R = A^2$, whose plot is shown in Fig. 3.17:

$$f_R(r) = \frac{1}{\sigma^2}e^{-r/\sigma^2}u(r)$$
(3.59)

which is the well known exponential distribution, with mean $E[R] = \sigma^2$ and variance $\sigma_R^2 = \sigma^4$.

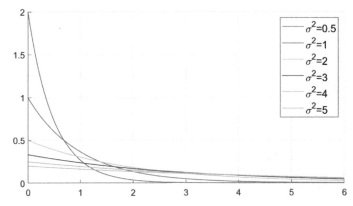

Figure 3.17. Exponential probability density function plotted Vs. different values of the radar cross section σ^2.

The information we want to retrieve from the receiver radar signal is the radar cross section. This information is carried out in the intensity signal, proportional to the NRCS according to the radar equation introduced in Chapter 1, whose pdf is given by (3.59). In particular, we can consider that the mean value $i = \sigma^2$ of this distribution represents the actual value of the intensity, proportional to the NRCS. It is easy to note that to larger values of $i = \sigma^2$ correspond distribution whose values are more spread on larger values of the received signal r (compare the green curve Vs. the dark blue one in Fig. 3.17). So, as different values of i (of RCS) lead to different distributions, we can rewrite pdf (3.59) as the following conditional pdf:

$$f_{R|I}(r|I=i) = \frac{1}{i}e^{-r/i}u(r) \qquad (3.60)$$

This reasoning leads to write the received intensity in terms of the well-known multiplicative speckle model:

$$R = I \cdot N \qquad (3.61)$$

where R is the speckle affected received noise, I is the speckle free signal which contains the information about NRCS, and N is the multiplicative speckle noise whose pdf is given by:

$$f_N(n) = e^{-n}u(n) \qquad (3.62)$$

Very often in radar and SAR applications more independent samples are incoherently averaged to reduce the effect of the speckle. In the radar context, the number M of independent samples is referred as number of looks, and the operation as multilook. In this case the pdf of the intensity signal change in the so-called M-Looks speckle distribution:

$$f_{R_M|I}(r_M|I=i) = \frac{1}{\Gamma(M)} \left(\frac{M}{i}\right)^M r_M^{M-1} e^{\frac{Mr_M}{i}} u(r_M) \qquad (3.63)$$

For $M = 1$ pdfs (3.60) and (3.63) coincide. For this reason, very often pdf (3.60) is referred as the one-look, or single-look speckle distribution. Moreover, the mean value of distribution (3.63) remains the correct one $i = \sigma^2$, while the variance becomes σ^4/M, so that it results a reduction of M times with respect to the one-look case.

A final comment of this section is related to the hypothesis summarized above. In particular, we derived the above models in the case when no dominant scatterers are present in a resolution cell. When, conversely, one or more dominant scatterers are present in a resolution cell, the pdfs of amplitude and intensity signals change. A detailed description of different speckle models in different operative condition can be found in [25].

4

Imaging radar: SAR data acquisition geometry and modes

The chapter discusses the use of radars for imaging purposes. It starts by addressing the received signal model of a Real Aperture Radar (RAR) and discussing its imaging characteristics, including the limitation affecting the azimuth resolution. It subsequently moves toward higher resolution systems by introducing the concept of aperture synthesis, or equivalently Doppler bandwidth, that leads to the improvement of the azimuth resolution of a Synthetic Aperture Radar (SAR) with respect to the RAR. The most common operational modes (stripmap, spotlight, scansar, and TOPS) are finally discussed and analyzed in the azimuth time frequency domain, thus explaining from a schematic point of view the pro and cons of each mode.

4.1 Acquisition geometry of imaging radars

Classical radar systems are conceived to detect and locate targets within a given portion of space. To this aim, the radar antenna is mounted on a fixed observation point and its beam is either mechanically or electronically steered to the scope of scanning the region of interest. The position of the detected target is thus typically referred to a spherical coordinate system, centered in the observation point: the range (distance) is measured by the radar, whereas the angular coordinates, the azimuth angle and possibly also the elevation, are provided by the pointing of the antenna illumination beam. Furthermore, radar systems provide also a measure of the back-scattering capability of the target, illuminated by the impinging radiation. When operating for surveillance, radar can be, and in practice it is, also used for the target characterization and classification.

The sensitivity of the system to the backscattering can be indeed exploited to perform the imaging of an illuminated surface. For instance, rotating radars can be employed to image surfaces of interest, such as sloping terrains (mountain flanks, etc.). The

Multi-Dimensional Imaging with Synthetic Aperture Radar
https://doi.org/10.1016/B978-0-12-821655-2.00008-5

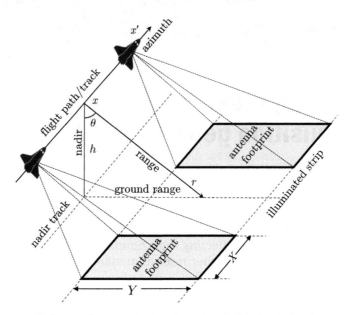

Figure 4.1. Reference Geometry for imaging radars in the classical, stripmap mode configuration.

amplitude (or its squared valued, i.e. the intensity) of the back-scattered signal provides a two-dimensional, i.e., range and azimuth angle, microwave image of the observed surface. In other words, the back-scattering measures associated with the distinguishable elements of the surface (according to the system resolution) allow exploiting the radar as a microwave imaging system.

Rotation, however, is not the only way to perform a scan of a surface of interest. In fact, imaging radars can be also installed on a moving support, such as a car [27] in order to image surface slices from the ground. In the context of active microwave remote sensing, imaging radars are widely exploited technologies for Earth Observation [28] [29]. In this case, a (usually pulsed) radar sensor is mounted on a flying platform (generally an airplane or a satellite): the area of interest, typically a strip on the Earth surface, is illuminated during the flight by exploiting a side-looking radar antenna. A common configuration is sketched in Fig. 4.1 for the case of a space platform (generally a satellite) with a side-looking antenna pointing in a fixed direction, typically broadside, i.e. orthogonal to the flight track. Such a configuration allows imaging a strip on the ground with the axis aligned to the flight direction: for this reason it is usually referred to as stripmap mode. As a first approximation, generally valid for space imaging radars, the flight path is assumed to follow a straight trajectory.

Fig. 4.1 suggests assuming a cylindrical reference system, instead of the spherical one typically considered in ground surveillance radars. With reference to Fig. 4.1, the position of a generic point P on the illuminated surface is provided by its cylindrical coordinate system (x, r, θ). The x coordinate is the axial coordinate, along the flight path (track), with the origin in a generic point: it is usually referred to as azimuth (to keep the analogy with the angular azimuth coordinate associated with the antenna rotation in classical radar systems) or along-track coordinate of the target. The r coordinate is the slant range (or simply range), sometimes also referred to as cross-track coordinate[1]: it represents the radial distance of the point P from the track. Finally, θ is the angular coordinate, usually referred to as look-angle, which allows discriminating equi-range points belonging to the plane orthogonal to the azimuth axis. The look-angle θ provides an indirect measurement of the height and hence, along with x and r, allows achieving the 3D localization of the point P.

The side-looking configuration shifts the antenna footprint, i.e. the area on the ground illuminated by the beam of the real antenna, by a fixed offset from the nadir track line. The angle between the nadir direction and the antenna beam direction, typically ranging between 20° and 50°, is called as (mid-swath) look-angle, also referred to as (mid-swath) off-nadir angle. Such a configuration allows preventing the presence of the left-right range ambiguities, i.e. the superposition of echos coming from symmetric (with respect to the nadir track) points located at the same range r that would be present for a beam illuminating the imaged surface along the vertical direction to the flight track.

The projection of the slant range (r) to the ground is referred to as ground range and hereafter is indicated as y.

A remark is now in order to differentiate the off-nadir angle from the incidence angle, i.e. the angle between the vertical and impinging radiation direction.[2] Depending on the distance between the scene and the sensor nadir, the projection from the slant to the ground may involve the necessity to account for the Earth surface curvature between the nadir track and the imaged scene. The problem is sketched in Fig. 4.2. The y axis, orthogonal to the azimuth, is taken as the projection of the range on the ground plane tangent to the ellipsoid in the scene reference point,

---

[1] The notation "cross-track" is indeed ambiguous being one of the infinite directions in the plane orthogonal to the track.

[2] Note that the incidence angle relevant to the vertical of an extended scene addressed here, typically referred to the scene center, differs from the incidence angle considered in the previous chapter, and equal to the angle between the l.o.s. and the (local) normal to the surface.

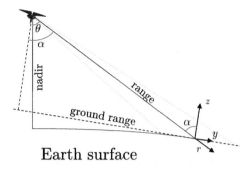

Figure 4.2. Reference geometry in the plane orthogonal to the flight path for a spaceborne system.

typically located at mid range. This choice implies that the angle α between the range direction (line of sight, or briefly "los") and the vertical direction (i.e., the direction of the z-coordinate in Fig. 4.2) orthogonal to the ground range y, referred to as incidence angle , and the look angle (θ) cannot be considered congruent. The two angles are in fact congruent only if the curvature between the nadir point and the scene center can be neglected, that is a condition generally pertinent to airborne imaging systems. Note also that (x, y, z) is typically assumed as the (scene centered) rectangular reference system, that is a reference system alternative to the (radar native) cylindrical reference system (x, r, θ). While the former is easier to be interpreted, the latter is the natural from the radar signal processing point of view and hence generally preferred choice for the system description.

Returning to Fig. 4.1, the antenna footprint is schematically represented by a rectangle, whose sizes are usually referred to as azimuth swath (parallel to the flight track) and range swath (orthogonal to the flight track) dimensions, indicated as in the previous and following chapters as X and as Y (ground range). For the sake of notation simplicity, unless differently stated, in the rest of the book we neglect the Earth curvature: the look angle should be substituted by the incidence angle in case this assumption is not valid. Letting θ_0 to be the look angle at the mid range r_0, the ground range swath S is approximately equal to:

$$Y \simeq \frac{h}{\cos^2 \theta_0} \theta_r \tag{4.1}$$

whereas the azimuth swath X is given by:

$$X =\simeq \frac{h}{\cos \theta_0} \phi_A \tag{4.2}$$

In (4.1) and (4.2), h is the sensor altitude; θ_r and ϕ_A are the angular apertures of the antenna beam in range and azimuth, respectively. In analogy to Section 2.3 for the use of capital and normal subscripts adopted for the range resolution before and after the chirp compression to increase the resolution, the capital subscript for the azimuth beam in (4.2) has been indicated to emphasize the fact that this quantity concerns the "low resolution" stage, relevant to the real aperture, i.e. before the adoption of any processing in the azimuth direction.

4.2 Resolution of a real aperture radar (RAR)

An imaging radar that delivers images acquired by simply collecting the echoes received from the scene during the acquisition interval, is referred to as Real Aperture Radar (RAR). We are now interested in evaluating the resolutions of such a simple imaging system. To this end we consider a single, ideal, scattering point located on the ground at a generic coordinate $P \equiv (x, r)$, or equivalently $P \equiv (x, y)$. Note that the height of the scatterer has been not specified: unless explicitly stated, such as in the subsequent chapters pertinent to the interferometric systems for topographic reconstruction, the role of the topography is here neglected and a flat topography surface at the reference (zero) height is assumed.

Similarly to the surveillance radar discussed in Chapter 2, radar pulses are transmitted at regular time instants t_n, separated by the Pulse Repetition Interval (PRI), i.e. the inverse of the Pulse Repetition Frequency (PRF). We will also refer to the coordinate x' associated with the along-track positions of the antenna position; moreover, we will assume the Start-Stop approximation, see Section 2.5), thus allowing to identify x' as vt_n, with v being the platform velocity. Still following the notation used in the surveillance radar, we will refer to r' as the range corresponding to the fast time t pertinent to the radar timing, so that $r' = ct/2$. We assume that the radar transmits chirp pulses $p(t)$ of form (2.31) at regular times separated by the PRI and indicates with $\tau(t_n, x, r)$ the delay of the echo by the scatterer $P(x, r)$ to the pulse transmitted at $t_n = nPRI$. After the range chirp compression, the amplitude of the signal representing the echo received by the scatterer P can be therefore expressed as:

$$|y(t, t_n)| = \Pi\left(\frac{vt_n - x}{X}\right) |f_p(t - \tau(t_n, x, r))| \qquad (4.3)$$

or equivalently for the space variable counterpart:

$$|y(r', x')| = \Pi\left(\frac{x' - x}{X}\right)|f_p(r' - R(x', x, r))| \qquad (4.4)$$

where f_p can be the transmitted pulse or its ACF depending on whether y in (4.3) and (4.4) is the received signal before or after the range compression; furthermore, $\Pi(\cdot)$ is the rectangular pulse (gate function) defined in (2.7) and:

$$R(x', x, r) = \sqrt{r^2 + (x' - x)^2} \qquad (4.5)$$

The spreading of the response of the scatterer in the image domain (x', r') provides a measurement of the azimuth and range resolutions of the imaging system, i.e. the minimum separation allowing a straightforward discrimination of two different scatterers.

Eq. (4.4) shows that the azimuth resolution ρ_A of a RAR system coincides with the azimuth antenna footprint X:

$$\rho_A = \frac{h}{\cos\theta_0}\frac{\lambda}{L_A} \qquad (4.6)$$

where we have considered that the 3 dB aperture for a planar antenna is given by (see Appendix A.91):

$$\phi_A \simeq \frac{\lambda}{L_A} \qquad (4.7)$$

with L_A being the dimension of the real (i.e. before the processing) antenna along the flight direction.

As discussed in Chapter 2, the range resolution (achieved after the matched filter for chirp pulses), is given by:

$$\rho_r = \frac{c}{2B_t}. \qquad (4.8)$$

Bandwidths of the order of a few tens of megahertz can lead to resolution of a few meters. For modern systems, the transmission of bandwidths larger than 150 MHz can easily lead to submetric range resolution. The corresponding ground range resolution

$$\rho_y = \frac{\rho_r}{\sin\alpha} = \rho_r = \frac{c}{2B_t\sin\alpha}, \qquad (4.9)$$

where α is the incidence angle with respect to the vertical, i.e. the angle between the vertical axis of the rectangular reference system located on the ground (generally at the midrange) and the direction of the impinging radiation (slant range direction), see

Fig. 4.2. Should the Earth curvature be neglected we could let $\alpha \simeq \theta$. Evidently the ground range resolution is always worse (i.e. numerically higher) than the slant range resolution, see (4.9). As for the azimuth resolution on the data collected by the radar sensor, with space-borne systems orbiting in typical Low Earth Orbits at heights of $h = 700$ km, antennas with $L_A = 10$ m and look angles of $\theta_0 = 25$ degs, would lead to $\rho_A = X$ of the order of about 4 km at C-Band ($\lambda = 5.6$ cm). It is clear the unbalancing between range and azimuth resolution for such a simple imaging radar, which simply collects the echoes from the scene and limit the processing to the implementation of the range matched filter at most.

The name RAR is related to the fact that the system resolves scatterers along the azimuth by exploiting the illumination restraint (i.e. the azimuth swath) of the real on-board radar antenna. From (4.6) it is clear that to achieve resolutions larger (i.e. numerically smaller) than ρ_A, the antenna length L_A should be

$$L_A \geqslant \frac{\lambda}{\rho_A} \frac{h}{\cos \theta_0} \tag{4.10}$$

that is, for $\rho_A = 10$ m and still $h = 700$ km and $\theta_0 = 25$ degs, L_A should be larger than 4300 m at C-Band.

The above considerations clarify that high resolution imaging is not viable with a RAR due to the difficulties in deploying large enough antennas.

4.3 SAR resolution and Doppler bandwidth

The previous section highlighted the practical impossibility of achieving large azimuth resolution by simply exploiting the amplitude characteristics of the data acquired by a real aperture radar. The Synthetic Aperture Radar (SAR) overcomes the problem by synthesizing a large aperture. The synthesis is not achieved by a real array composed of several (physical) antennas displaced along the track, but rather by exploiting the displacement of the real antenna on-board the platform moving along the azimuth track. More specifically the synthesis is implemented via a digital processing carried out by a coherent (amplitude and phase) processing of the echoes received from a target and collected by the sensor at the different positions along the illumination interval, i.e. the time corresponding to the illumination of the target by the beam of the real antenna.

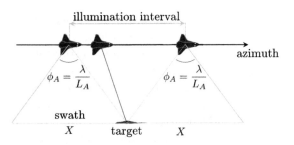

Figure 4.3. Illustration of the aperture synthesis principle to achieve a large (synthetic) aperture along the azimuth direction.

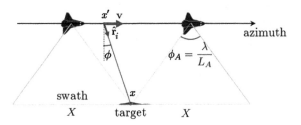

Figure 4.4. Reference geometry for the SAR Doppler Analysis.

4.3.1 Aperture synthesis approach

The aperture synthesis principle is depicted in Fig. 4.3: a point scatterer (red (mid gray in print version) dot) placed on the ground is illuminated for a large time interval. The segment covered in this interval, called illumination segment, represents the size of the synthetic aperture (L_a).

The azimuth resolution ρ_a achievable with the synthesis principle by a SAR system can be evaluated, similarly to RAR in (4.6), as:

$$\rho_a = \frac{1}{2}\frac{h}{\cos\theta_0}\frac{\lambda}{L_a} \tag{4.11}$$

where the gain factor 2 accounts for the doubling of the distances that are covered forth and back by the radiation. Such a factor leads in practice to a doubling of the propagation constant, thus changing the phase coefficient that multiplies the range from $2\pi f_0/c$ to $4\pi f_0/c = 4\pi/\lambda$ in (2.10). This scaling leads to an increase (numerical halving) of the resolution.

Simple geometrical considerations on the congruence of parallel segments depicted in Fig. 4.3 lead demonstrating that the illumination segment, i.e. the synthetic aperture length L_a, is equal to the azimuth swath X:

$$L_a = X \tag{4.12}$$

The expression of X in (4.2) highlights that a large synthetic aperture can be achieved thanks to the typical large values of the sensor height or interestingly, if not even surprising, by reducing the real antenna size L, see the expression of θ_A in (4.7).

The substitution of (4.2) and (4.12) in (4.11) leads to the following expression of the azimuth resolution achieved by a SAR system:

$$\rho_a = \frac{L_A}{2} \qquad (4.13)$$

Eq. (4.13) states that the azimuth resolution of a SAR system equals half the length of the real aperture. It also highlights that the resolution can be increased by reducing the real antenna length. In fact, the smaller the antenna, the larger the ground footprint and the larger the synthetic aperture, thus leading to a narrower synthetic beam and, therefore, to a finer resolution ρ_a. Moreover, differently from RAR imaging sensors, the resolution does not degrade with the increase of the distance. This can be easily explained by the fact that when the distance increases, the footprint on the ground increases as well, so that a larger aperture L_a can be synthesized. The increase of the synthetic aperture perfectly counterbalances the increase in distance.

4.3.2 Doppler approach

An alternative approach for the quantitative evaluation of the SAR azimuth resolution is based on the exploitation of the Doppler principle widely discussed in Section 2.5. Actually the Doppler principle inspired Carl A. Wiley who, for the first time in history, proposed in 1951 in a Goodyear Aircraft report, the idea of improving the resolution of classical real aperture imaging radars [30]. Basically, due to the change of the line of sight along the azimuth, targets sharing the same range within the antenna beam can be resolved by measuring their Doppler shift. Since the basic idea (patent "Pulsed Doppler Radar methods and Apparatus", filled in 1954 and secreted till 1964 [31]), SAR has experienced significant advancements, among the most important ones, the progress related to the development of digital processing techniques.

The SAR Doppler approach can be explained by referring to Fig. 4.4 where it is shown the geometry of a SAR system in the slant range plane, i.e. the plane defined by the azimuth and (slant) range directions. The starting and ending positions corresponding to the illumination segments are depicted in the figure. Moreover, for a generic position (x') of the synthetic antenna within the illumination segment, the platform velocity vector v and the instantaneous target line of sight $\hat{r}_i$ are also shown. The angle ϕ formed by the instantaneous line of sight direction (to the target) and the broadside direction (i.e. the direction perpendicular to the azimuth), i.e. the off-broadside angle, is referred to as aspect angle.[3] The aspect angle depends on the difference between the platform position (x') and the azimuth target position (x): $\phi = \phi(x' - x)$. The (instantaneous) Doppler can be written as:

$$f_d(x' - x) = \frac{2}{\lambda} \boldsymbol{v} \cdot \hat{r}_i = \frac{2v}{\lambda} \sin[\phi(x' - x)]. \qquad (4.14)$$

Modern spaceborne SAR systems are built to set ϕ, and therefore the Doppler span centered around zero: this condition is referred to as Zero Doppler Centroid (ZDC). ZDC can be achieved by proper beam steering, referred to as Zero Doppler Steering, aimed at compensating the effects of the Earth rotation. In the airborne case platform deviations and attitude variation may however lead to cases where the Doppler Centroid cannot be considered negligible. With the aim to evaluate the azimuth resolution of a canonical SAR system, we refer in the following to the ZDC case. We will assume therefore a symmetrical Doppler span. This implies that the maximum (absolute) value of the achieved at the beginning and at end of the illumination interval, is equal to $\lambda/(2L_A)$. Typical values of the maximum off-broadside angle are on the order of $10^{-3} \div 10^{-2}$ radians. The bandwidth spanned by the Doppler shift, B_d usually refereed to as Doppler Bandwidth is given by:

$$B_d \simeq \max_{x'-x} f_d(x' - x) - \min_{x'-x} f_d(x' - x) = 2\frac{2v}{\lambda}\frac{\lambda}{2L_A} = \frac{2v}{L_A} \qquad (4.15)$$

where again we have assumed the symmetry of the Doppler span, i.e. the symmetry of the line of sight of the target within the illumination interval around the position of closest distance (range) between the target and the sensor, characterized by $\phi = 0$, and also referred to as Zero Doppler (ZD) position.

Assuming that spectral distortions of the acquisition system are compensated with the processing described in details in the next chapter, in analogy to the range compression discussed in (2.28), the Doppler bandwidth B_d would lead to a theoretical resolution (along the time) given by the inverse of B_d. In other words the

---

[3]The definition adopted here for the aspect angle is according to the one introduced in [32] for SAR. Note that a common definition of the aspect angle, used also in the radar context, is "target centric" and generally introduced in association with the elevation angle. The aspect angle is defined as the angle between the impinging direction and the main horizontal target axis. For instance for a target corresponding to a plane for which the axis of the fuselage, the imaging of a radar looking from the front corresponds to an aspect of 180°.

(temporal) impulse response function of the system after the processing in the azimuth direction having a bandwidth B_d is equal to:

$$B_d \text{sinc}(B_d t') \longleftrightarrow \Pi\left(\frac{f_a}{B_d}\right), \qquad (4.16)$$

where t' is the time continuous variable associated with the transmission instants t_n along the azimuth x'. Finally, by considering that the association between azimuth time and azimuth coordinate is a simple scaling involving the platform velocity v, the following Doppler derived resolution expression can be achieved:

$$\rho_a = \frac{v}{B_d} \qquad (4.17)$$

which, by considering (4.17), is equivalent to the expression (4.13) of the azimuth resolution provided by the antenna synthesis viewpoint addressed in Section 4.3.1

4.4 SAR acquisition impulse response function

The previous section has investigated a specific spectral feature of the SAR signal along the azimuth direction, namely the spectral support which has been evaluated by reasoning on the Doppler frequency span. We have also evaluated the azimuth resolution, which is however achieved after the compensation of the spectral distortion affecting the signal in the useful spectral band which, similarly to the chirp compression addressed in Chapter 2, is carried out via the 2D focusing operation described in detail in the next chapter. The SAR acquisition system is, in fact, far from being distortionless. A processing of the acquired complex (amplitude and phase) data is in fact necessary to compensate the spectral distortions, mainly affecting the phase, which are intrinsic of the SAR acquisition system. As explained in details in the next chapter, the sequence of the acquisition and processing systems retrieves in such a way a distortion free response: the processing is referred to as SAR data focusing. Compared to a RAR, for which only the amplitude characteristics of the received data are of concern, see (4.4), for the case of SAR it is necessary to evaluate the whole expression of the received signal in the complex domain. To this end we refer to (4.4) and consider the phase factor related to the wave propagation in the presence of a variation of the distance between

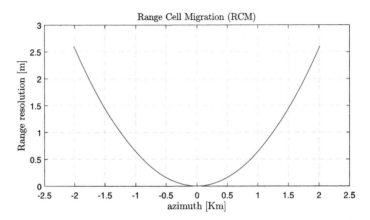

Figure 4.5. Example of Range Migration vs the azimuth within a synthetic aperture interval for a satellite system orbiting at 800 km. Parameters for the determination of the synthetic aperture interval are set with reference to a X-Band sensor with 3 m azimuth resolution.

the antenna and the target, see (2.55), thus obtaining:

$$y(r', x') = \Pi \left(\frac{x' - x}{X} \right) \exp \left\{ -J \frac{4\pi}{\lambda} R(x', x, r) \right\} f_p(r' - R(x', x, r)).$$

$$(4.18)$$

For a fixed point, i.e. a fixed (x, r) coordinate pair, the variation of R with x' given by (4.5), referred to as Range Migration (RM) or Range Cell Migration (RCM) for the discrete domain case, impacts on both the amplitude and phase of the signal.

Fig. 4.5 shows an example of range migration computed following (4.5) with the parameters set with reference to the COSMO-SkyMed (standard mode) satellite system, i.e. X-Band system with 3 m azimuth resolution: the horizontal axis represents the azimuth coordinate spanning a synthetic aperture, i.e. an interval of length X. The image shows that RM can reach values as large as the system resolution. As shown later in the book, the pixels spacing, i.e. the distance corresponding to two adjacent pixels in the image, is generally kept to the same order (generally around 95%) of the resolution for data volume reduction. In any case the pixel spacing, i.e. the image grid driven by the (azimuth and range) sampling frequency, is always lower than the resolution according to the Nyquist sampling principle of bandlimited signals. Accordingly, the range migration can cause the fact that the response of a scatterer migrates in range from one pixel to the next one during the recording of the echo in the illumination interval associated with the synthetic aperture. If not properly compensated during the aperture synthesis process this variation of the position in range of the response of the target can lead to a resolution loss with respect to the theoretical one given in (4.13).

Moving to the analysis of the phase component in (4.18) we note that the RM ($R(x', x, r)$) is multiplied by a factor proportional to the carrier frequency (i.e. inversely proportional to the wavelength): the higher the frequency, the higher the impact on the phase of the variations of the RM term. This phase variation implies phase distortions which prevents the possibility to perform, as in the case of stepped systems for the range discrimination, an azimuth focusing algorithm simply based on the application of a FT. To better analyze this factor we develop (4.5) in Taylor series under the hypothesis $r \gg X$, valid typically for satellite SAR systems, thus obtaining:

$$R(x', x, r) = \sqrt{r^2 + (x' - x)^2} \approx r + \frac{(x' - x)^2}{2r} + \mathcal{O}\left(\frac{(x' - x)^3}{r^2}\right) \quad (4.19)$$

where the approximation holds for:

$$\frac{X}{2r} \ll 1 \qquad (4.20)$$

The substitution of (4.19) in (4.18) provides:

$$y(r', x') = \Pi\left(\frac{x' - x}{X}\right) \exp\left\{-j\frac{4\pi}{\lambda}\frac{(x' - x)^2}{2r}\right\} \times$$
$$f_p(r' - R(x', x, r))e^{-j\frac{4\pi}{\lambda}r} \qquad (4.21)$$

The azimuth phase factor in (4.21) can be expressed in terms of the temporal coordinate x'/v, i.e. the time continuous variable associated with the pulse transmission times t_n:

$$\Pi\left(\frac{x' - x}{X}\right) \exp\left\{-j\frac{4\pi}{\lambda}\frac{(x' - x)^2}{2r}\right\} =$$
$$\Pi\left(\frac{x'/v - x/v}{X/v}\right) \exp\left\{-j\pi\frac{2v^2}{\lambda r}(x'/v - x/v)^2\right\} \qquad (4.22)$$

which shows that the azimuth phase response is chirp-like and therefore, similarly to the range direction, amenable to a compression step, via a matched filter processing, see Section 2.3. It is worth to note that the azimuth chirp rate depends on the range: echoes from near range are characterized by Doppler rates larger than those associated with far range targets. It is worth also noting that the extension in azimuth of the response, referred to as the

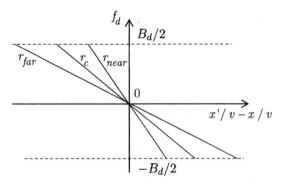

Figure 4.6. Schematic representation in the time frequency diagram of the response of a scatterer for three different range values. The slope variation decrease from near to far range is compensated by an increase of the time support (integration time).

integration time, given by:

$$T_{int} = \frac{X}{v} = \frac{\lambda r}{L_A} \tag{4.23}$$

depends on the range. The products between the chirp rate and chirp duration, that is the Doppler Bandwidth B_d in (4.14) is, on the contrary, a constant with respect to the range:

$$\frac{2v^2}{\lambda r} X/v = 2v/L_A. \tag{4.24}$$

This concept can be schematically well described in the time-frequency diagram associated with the Doppler history of the targets, i.e. the instantaneous frequency (see Appendix A.2) corresponding to the phase response of a target $P(x,r)$:

$$f_d(x'/v) = \frac{v}{2\pi} \frac{\partial \phi(x',x,r)}{\partial x'} = \frac{2v}{\lambda} \frac{\partial R(x',x,r)}{\partial x'} \tag{4.25}$$

The time-frequency diagram describing the variation of the Doppler rate for the cases of a near, mid, and far range target is shown in Fig. 4.6.

4.5 SAR data sampling and ambiguities

In the previous section the SAR signal representation has been considered with respect to the continuous azimuth x' and range r' variables, although for the former a correspondence with vt_n has been already pointed out.

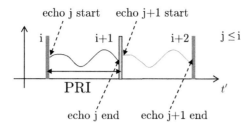

Figure 4.7. Schematic representation along the one dimensional time of the radar timing with transmitted pulses and received echoes. The echo start and end correspond to the returns of the points at the closest and farthest range, respectively.

In modern radar systems data are sampled both in azimuth and range for (on board and on ground) digital storage and processing. The sampling in the azimuth time is related to the PRI, i.e. the inverse of the PRF; along the range the signal is sampled with a sampling frequency generally indicated with F_s. We investigate in the following the limitations of SAR imaging systems operating in the stripmap mode and related to the possible choices of sampling frequencies.

The correct sampling of the range bandwidth requires a sampling according to the Nyquist criterion, i.e. $F_s \geqslant B_T$. While in range the sampling limitations are essentially related to the receiver A/D conversion apparatus and to the increase of the data rate and data volume for an increasing sampling rate, thus posing constraints on the satellite datalink, the setting of the azimuth sampling rate requires more attention. In addition to the lower limit enforced by the Nyquist criterion on the sampling of the Doppler Bandwidth, i.e.:

$$PRF \geqslant B_d = v/\rho_a, \qquad (4.26)$$

which sets a lower bound to the PRF to achieve the desired azimuth resolution ρ_a, a further limitation is related to the necessity to avoid overlaps of echoes returned from the scene. An increase in the PRF, i.e., a decrease of the interpulse interval (PRI), may lead the returns from the scene to be spread on multiple (sequential) intervals. The situation is depicted in Fig. 4.7, where three generic pulses transmitted at times t_i, t_{i+1}, t_{i+2} and the echos from the scene corresponding to the j-th pulse transmitted at $t_j \leqslant t_i$ are shown. The difference $t_i - t_j$ is related to the time needed for the pulse to travel from the transmitting sensor position to the closest target in the scene. The echoes starting and ending times correspond to the times of arrival of the return from the closest and farthest scatterers on the ground in the illuminated scene, respectively. Neglecting the pulse duration and letting the range swath be equal to S, see (4.1), the condition for the echoes from the swath required to fit the interval dedicated to the echo reception is:

$$Y_r = Y \cos \theta_0 \leqslant \frac{cPRI}{2} = \frac{c}{2PRF}, \qquad (4.27)$$

that is, the echo coming from the illuminated scene must be comprised within a time interval of length PRI. This condition sets an upper bound for the PRF for a given size of the illuminated swath size in range. By merging (4.26) and (4.27) we get

$$\frac{v}{\rho_a} \leqslant PRF \leqslant \frac{c}{2Y_r} \qquad (4.28)$$

which obviously requires

$$\frac{Y_r}{\rho_a} \leqslant \frac{c}{2v} \qquad (4.29)$$

By exploiting (4.1) and (4.13) the latter condition can be expressed as a lower bound on the real antenna area:

$$L_A L_r \geqslant \frac{4h\lambda v}{c \cos \theta_0} \qquad (4.30)$$

Eq. (4.29) highlights that the ratio between the range coverage and the azimuth resolution is bounded, that is, an increase on the resolution (numerical reduction of ρ_a) must be traded off by a reduction of the range coverage and vice-versa. As better explained in the next section, specific operational modes allow pushing the trade off in one direction or the other: a high azimuth resolution at the expense of a reduced range swath can be achieved with the spotlight (SPOT) mode; the opposite for the SCANSAR or TOPS modes.

The violation of the sampling requirement in (4.26) and (4.27) rises azimuth and range ambiguities, respectively. Ambiguities can be avoided, or equivalently the bounds can be both relaxed in order to achieve High-Resolution Wide Swath (HRWS) imaging capabilities, either via use of digital beamforming techniques that can be implemented at the receiver stage (beamforming on receive) or by means of MIMO concepts based on the use of multiple antennas and orthogonal waveform. Both approaches require, however, hardware sophistication such as multichannel operation with either a splitting of the receiver antenna into multiple panels with independent receiver capabilities or by exploiting multiple receiving satellites orbiting in formations. The description of such important solutions which probably will characterize the future developments of SAR, boosted also by the current progress on small satellites, are out of the scope of this book. The reader can refer to [28], [33] and [34] for further details.

4.6 Acquisition modes

In the following the most common SAR acquisition modes, i.e. the stripmap, spotlight and scansar modes, are described by limiting the description to the illustration of the configuration and the explanation of the respective Doppler characteristics.

4.6.1 Broadside and squinted stripmap modes

As already specified above, the acquisition mode described in Fig. 4.1 is referred to as stripmap mode. The illuminating beam is assumed to be orientated along the direction perpendicular to the azimuth track: this specific case is referred to as broadside stripmap. The feature of the broadside illumination is that the Doppler excursion during the illumination interval is centered around zero, a situation that leads to a ZDC. For this reason this particular configuration is also sometimes referred to as Zero-Doppler (ZD) mode.

Either unintentionally, as in the case of airborne sensors, or purposely, as in the case where the target should be imaged before than the sensor passes above it, the beam can be directed with an azimuth offset from the broadside direction. The configuration is referred to as squinted stripmap mode and is shown in Fig. 4.8.

The IRF associated with a squinted stripmap acquisition can be achieved by simply shifting the $\Pi(\cdot)$ function in (4.18) with respect to $x' - x = 0$. Letting x_{DC} to be the ground azimuth offset of the antenna beam, see Fig. 4.8, we have $x' = x - x_{DC}$, where $x_{DC} > 0$ and $x_{DC} < 0$ corresponds to the cases of forward and backward squint, respectively. The expression of the signal received by a generic target located at (x, r) becomes

$$y(r', x') = \Pi\left(\frac{x' - x + x_{DC}}{X}\right) \exp\left\{-j\frac{4\pi}{\lambda}\frac{(x'-x)^2}{2r}\right\} \times \\ f_p(r' - R(x', x, r))e^{-j\frac{4\pi}{\lambda}r} \tag{4.31}$$

This simple shift has significant effects on the range migration, impacting both the range location of the pulse envelope as well as the azimuth phase variation, i.e. the Doppler history of a target. In fact, by considering the expression of the target to sensor distance in the squinted geometry and letting $r_{DC} = \sqrt{r^2 + x_{DC}^2}$ and $\phi_{DC} =$

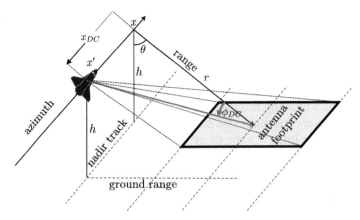

Figure 4.8. Reference geometry for imaging radars in the stripmap squinted mode configuration. In the picture the squint angle ϕ_{DC} is positive (forward looking); x_{DC} is, consequently, negative, that is discordant to the azimuth. Range and azimuth are still referred to cylindrical reference geometry pertinent the broadside direction. Alternative choices for the reference geometry, such as a conical reference geometry, as well as for the associated data processing (zero-Doppler and acquisition or native-Doppler processing) are possible, as discussed in [35].

$\arcsin(x_{DC}/r_{DC})$ we have:

$$
\begin{aligned}
R(x',x,r) &= \sqrt{r^2 + (x'-x)^2} \\
&= \sqrt{r_{DC}^2 - 2x_{DC}(x'-x+x_{DC}) + (x'-x+x_{DC})^2} \\
&= \approx r_{DC} - \sin(\phi_{DC})(x'-x+x_{DC}) + \frac{(x'-x+x_{DC})^2}{2r_{DC}}
\end{aligned}
\tag{4.32}
$$

where in the last equality we have exploited the fact that $x' - x + x_{DC}$ is small compared to r_{DC} within the duration (X) of the window function in (4.31). Eq. (4.32) highlights that $x_{DC} \neq 0$ (i.e. $\phi_{DC} \neq 0$) leads to the presence of a linear component in the range migration. The terminology used in the SAR context to identify the two x'-dependent contributions in the RCM are: the range walk for the linear term and the range curvature for the quadratic term. It is evident that $\phi_D \neq 0$ implies the presence of the range-walk, which as said, interests the location in range of the pulses as well as the Doppler history. As for the latter the presence of the linear phase component along x' leads to a shift of the azimuth (Doppler) spectrum by a quantity:

$$
f_{DC} = \frac{2v}{\lambda} \sin \phi_{DC},
\tag{4.33}
$$

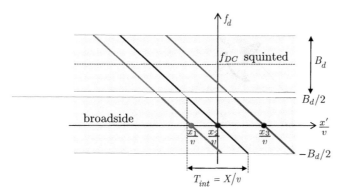

Figure 4.9. Azimuth time frequency representation in the azimuth time-frequency domain of the broadside and squinted acquisition mode. The Doppler history is represented vs the azimuth time for three targets located at positions x_1, x_2, and x_3 and identified by different colors. In the figure f_d is the Doppler Centroid corresponding to the offset angle of the beam from the broadside direction.

usually referred to as Doppler Centroid. The name explains the "DC" pedix used hereto. The time-frequency diagram for both the broadside and squinted stripmap mode is shown in Fig. 4.9 for the case of three targets located at positions x_1, x_2, and x_3.

4.6.2 Spotlight

In Section 4.5 we showed that there is a limitation in the ratio between the range swath coverage and the azimuth resolution. The spotlight configuration allows pushing the performances of SAR systems to the limit in (4.29) toward the increase (i.e. a numerical reduction) of the azimuth resolution.

As discussed in the previous paragraphs, the azimuth resolution for the stripmap case is limited by the real antenna size along the azimuth, see (4.13). Then, an increase of the resolution requires a reduction of the antenna size, which can be a problem from the implementation point of view due to the constraints on the minimum transmitted power. In fact, the smaller the antenna, the less is the efficiency for the radiation of a given power. To increase the azimuth resolution, acquisition modes different from the classical stripmap one can be adopted. The common aim is to increase the integration time, i.e. the time interval, or better the orbit segment over which the echoes from a ground target are recorded.

The spotlight configuration makes specifically use of an electronic antenna beam steering along the azimuth to illuminate a given area on the ground for a time interval longer than the one corresponding to the stripmap configuration. As already pointed

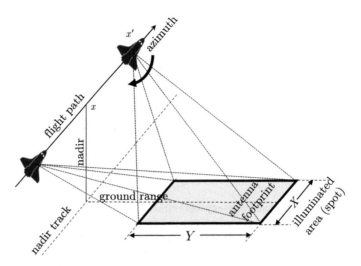

Figure 4.10. Reference Geometry for imaging radars in the very high resolution, staring spotlight configuration in which the footprint is fixed on the ground.

out, in the stripmap mode case the antenna beam illumination direction is fixed, typically at broadside: the footprint on the ground proceeds along the azimuth with a velocity equal (for the airborne case) or proportional (for the spaceborne case) to the antenna velocity (see Fig. 4.1). For the spotlight mode, the antenna steering allows, in practice, reducing (at the limit zeroing) the velocity with which the antenna beam footprint proceeds along the azimuth on ground. Fig. 4.10 shows the case where the footprint velocity is zeroed and the configuration is referred to as staring spotlight mode, whereas Fig. 4.11 shows the intermediate case also referred to as sliding spotlight mode, or hybrid mode. Fig. 4.10 shows that, in principle, it is possible to steer the antenna for a very large interval, theoretically achieving the limit of $\lambda/2$ for the azimuth resolution, corresponding to a maximum Doppler Bandwidth of $2v/\lambda$, see (4.14). In the practice the antenna steering is limited to a few degrees from the broadside to avoid the increase of the signal ambiguities. A large offset of the beam direction from the broadside configuration can, in fact, reduce the backscattered power received from the mainlobe, because of the increase of the angle with respect to the normal to the surface. Moreover it can bring the secondary lobes of the antenna beam to illuminate the ground along the broadside direction. The consequence of both actions is an SNR loss and a reduction of the Azimuth Ambiguity Signal Ratio (AASR) signal to ambiguity ratio.

The azimuth resolution increase of the spotlight mode can be further understood by referring to the azimuth time-frequency do-

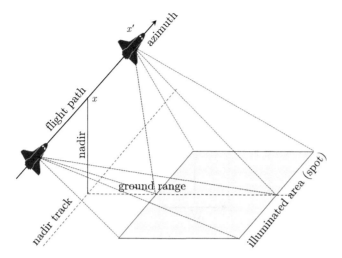

Figure 4.11. Reference Geometry for imaging radars in the sliding spotlight configuration, intermediate between the stripmap and staring spotlight modes.

main. The situation for the staring spotlight case is depicted in Fig. 4.12 again for a case of three targets. The representation of the acquisition in the time-frequency domain, i.e. the time-frequency span corresponding to the acquired frequencies along the time is highlighted in the light gray box. The comparison of Fig. 4.12 with Fig. 4.9 allows appreciating the differences between the stripmap and spotlight modes. In Fig. 4.9 the horizontal box highlight the acquisition span in the time-frequency domain for the stripmap case: it illustrates that the Doppler bandwidth of echoes coming from targets displaced along the azimuth are all aligned along the vertical axis. This agrees with the space-invariant nature of the system (assumed to fly on a rectilinear track) with respect to the azimuth. Fig. 4.12 shows on the other hand a rather different situation. First of all, due to the fact that the footprint is fixed for staring spotlight case, it can happen (as in the figure) that scatterers are not illuminated at all, see the third scatterer represented with the red (mid gray in print version) color in Fig. 4.9. Due to the steering, the light gray box is rotated clockwise with respect to the stripmap case, thus allowing to select a longer frequency segment on the illuminated scatterers. The result is a larger integration time and a larger Doppler bandwidth. The price paid for the increase of such quantities is not only a reduction of the illuminated swath, but also an inherent nonstationarity of the response along the azimuth. The illuminated scatterers, e.g. the green and blue (gray and dark gray in print version) scatterers have now Doppler frequency intervals that are displaced along the vertical axis.

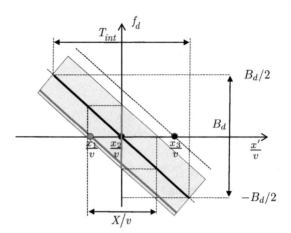

Figure 4.12. Representation in the azimuth time-frequency domain of the staring spot acquisition mode for three targets located at positions x_1, x_2, and x_3.

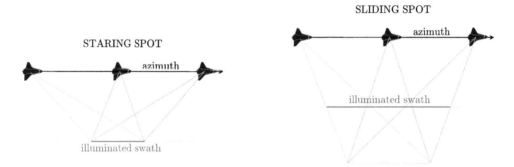

Figure 4.13. Schematic representation of the staring mode (left) and sliding spot, also known as hybrid mode (right). Differently from the right image, each target on the scene is illuminated in the left image on the whole segment.

A possibility to trade off between the azimuth resolution and scene coverage is offered by the sliding spotlight, also known as hybrid mode. In this case the footprint on the ground is not fixed but can move (i.e. slide) with a velocity controlled by the steering mechanism and lower than the one corresponding to the stripmap case. As shown in Fig. 4.13 the sliding spotlight can be achieved by steering the beam with respect to a point located under the surface: the farther the point from the surface, the more the mode moves from staring spot to stripmap. It is therefore an intermediate mode between the staring spotlight and the stripmap modes, hence the name "hybrid" mode.

As done for the stripmap and staring spotlight case it is interesting to analyze the azimuth time-frequency domain also for the hybrid mode, see Fig. 4.14. The comparison between Fig. 4.12 and

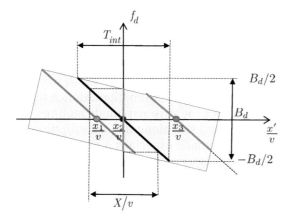

Figure 4.14. Representation in the azimuth time-frequency domain of the sliding spot, or hybrid acquisition mode for three targets located at positions x_1, x_2, and x_3.

Fig. 4.14 allows highlighting that the hybrid mode, i.e. Fig. 4.14, enlarge the coverage allowing to acquire a frequency bandwidth also for the scatterer located at azimuth x_3 (red (mid gray in print version) segment), but the integration time is reduced with respect to the staring spotlight case in Fig. 4.12.

A final comment is dedicated to the comparison of Fig. 4.9, Fig. 4.12 and Fig. 4.14. All the figures refer to a system observing the same scene, three azimuth targets, and with fixed antenna velocity (v) and antenna footprint size in azimuth (X). The ratio X/v is thus the same for all the three schemes. However, while for the stripmap case the integration time T_{int} is fixed by the ratio X/v, the operating modes corresponding to the staring and sliding spot case allow increasing T_{int} (and hence the Doppler Bandwidth, B_d). For the staring spotlight case the latter is limited only by the on- and off-switching of the instrument, whereas for the sliding case it is influenced by the steering rotation degree.

ScanSAR

With reference to the constraint in (4.29) on the ratio between the range swath coverage and the azimuth resolution, the ScanSAR mode allows pushing such ratio to the upper limit by increasing the range swath at the expense of a reduction (i.e. numerical increase) of the azimuth resolution. The reduction of the azimuth resolution decreases the Doppler bandwidth. Consequently, the PRF, can be lowered (i.e. the PRI increased) thus providing the availability of a larger interval dedicated to the echoes reception along the range with respect to the stripmap case. An increase of the range swath size can be therefore achieved.

Similarly to the spot case, there are, however, technological aspects to be accounted for achieving the desired balance between the azimuth resolution and range swath size, which can be hardly reached with the stripmap configuration. A reduction of the Doppler bandwidth for the stripmap mode requires, first of all, an increase of the antenna size along the azimuth direction: this solution is rather problematic especially for satellite case. Secondly, the increase of the range swath would require a beam reshaping in the across track direction that would unavoidably impact the antenna gain and, thus, the achieved SNR. As for the Spotlight mode case, an efficient and flexible solution can be achieved by the use of phased array, namely Active Electronically Scanned Array (AESA). The phased array can be used to implement a steering that, differently from the spot case, is operated in the across track (range) direction.

Let us assume that the SAR system operates with N_S beam positions in the cross track direction defining N_S ground subswaths. The integration time corresponding to the antenna azimuth beam size ($T_A = X/v$), that is the integration time for the classical stripmap mode, is then partitioned in a fixed number, say N_B, subintervals referred to as bursts, each with duration $T_B = T_A/N_B$. To guarantee the periodicity of the imaging scan along the subswaths the value of N_B is set to be a multiple of the number of subswaths N_S. The ratio:

$$L = \frac{N_B}{N_S} = \frac{T_A}{T_B N_S} \tag{4.34}$$

provides a number referred to as number of looks, that is the number of times that a fixed target is illuminated over the different scans. Typical values of L rarely exceed 3. Moreover, N_S is generally limited by the need to avoid illuminating very thin subswaths requiring an extremely fine shaping in range. Given a value of N_S, the limit on N_B is related to the need to avoid fast scanning that would negatively impact the illumination mechanism with the transients phases.

The illumination geometry is depicted in Fig. 4.15 for the case $N_S = 2$. The timing of the sensor is structured so that the burst duration T_B is equal to $T_A/2$, so that $N_B = 2$. In the configuration shown in Fig. 4.15 the system periodically scans the 1st and 2nd subswaths, with period $T_P = T_B N_S$. In particular, starting from the first burst, $T_B = T_A/2$ seconds (i.e. half synthetic aperture) is dedicated to the illumination of the 1st subswath the subsequent half of the synthetic aperture (time) is dedicated to the 2nd subswath,

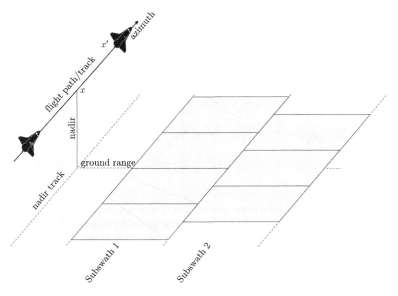

Figure 4.15. Reference Geometry for imaging radars in the scan configuration where a scan along the range is exploited to increase the swath width in the range direction. The figure refers to a scan system with two subswath and one burst for each subswath.

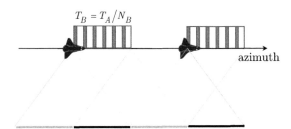

Figure 4.16. Timing and illumination of a ground subswath in the SCAN mode. Shown are the bursts of pulses illuminating a subswath. As for Fig. 4.15, the sketch refers to a scan system with two subswaths and a single look. The thick black lines on the ground identify the areas illuminated for the whole illumination time on a single burst; the thick light gray lines refer to scatterers illuminated over two adjacent bursts.

and so on for the subsequent bursts. The timing for the illumination with bursts of a subswath in the scan mode is further detailed in Fig. 4.16.

The figure also shows that the integration time of each target is halved with respect to the stripmap case. For the case of a more

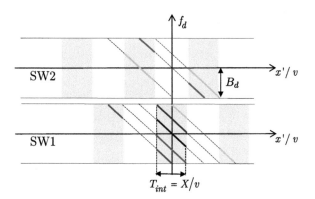

Figure 4.17. Representation in the azimuth time-frequency domain of the scan acquisition mode. The system is supposed characterized by two subswaths, one burst with an integration time equal to the half of the corresponding stripmap configuration. More targets on the ground are shown to highlight the effect of the space variant characteristics of the acquired Doppler bandwidths, x_1, x_2, and x_3.

general system of N_S subswaths the integration time is given by:

$$T_{int} = \frac{T_A}{N_S} = \frac{X}{vN_S} \tag{4.35}$$

and therefore a reduction of the resolution by a factor N_S is necessary to illuminate N_S subswaths in range.

The situation in the time-frequency domain is depicted in Fig. 4.17. Both the subswaths are depicted in the same figure. The light gray boxes correspond to the acquired domain in each burst. The reduction of the integration time and Doppler bandwidths with respect to Fig. 4.9 is evident. It is worth noting that the timing is fixed so that there are no-gaps along the azimuth, i.e. there are no target acquired for an integration time lower than T_A/N_S. However there are targets acquired in a single burst, and targets acquired in two subsequent bursts, see for instance the green and yellow (gray and mid light gray in print version) targets. From Fig. 4.17 it is evident the space-variance characteristics acquisition system along the azimuth. Even limiting to the targets illuminated in a single burst, the Doppler bandwidth shows a vertical shift. This characteristic requires special care also at the focusing stage. The detailed discussion of focusing of nonstandard modes, i.e. SPOT (sliding), ScanSAR, as well as TOPS mode discussed in the next session is outside the scope of this book. Interested readers may refer to [36,37].

A last remark is dedicated to illustrate a practical limitation of the ScanSAR mode. We start by observing that in the real case the antenna pattern in azimuth has hardly an ideal shape, i.e. described by an angular gate function. As a result the antenna gain can sensibly change over the azimuth within the mainlobe. In the stripmap case any target on the ground is swept along the whole antenna beam: the illumination starts in the fore-beam and stops in the aft-beam. From Fig. 4.17 it is evident that ground targets are illuminated in an inhomogeneous way. The position on the ground of the targets is represented by the crossing with the horizontal axis, the positions of the satellite in the active interval (i.e. in the burst) are associated with the azimuth positions in the light gray boxes. On one hand, there are targets, f.i. that represented in black, whose position is aligned along the broadside with the burst center, that is the illumination gain is centered on the broadside. On the other hand, the green and yellow (gray and mid light gray in print version) targets are located in the positions for which the system is illuminating the other subswath when the antenna is exactly in the ZD (i.e. broadside) position. As a consequence the target is illuminated in two bursts, over burst for which the antenna is in the fore and aft positions. The gain will therefore correspond to the fore-beam and aft-beam. As a consequence the illumination gain assumes a variability along the azimuth. The phenomenon is referred to as scalloping and gives rise to strips along the azimuth characterized by variation of the intensity of the received signal. The problem can be mitigated partially by increasing the number of looks. A much viable solution is however to operate a scan also along the azimuth direction, with a fore-rotation, i.e. opposite to the SPOT case, so that the beam center is swept on all the targets along the azimuth: the corresponding mode is called TOPS, i.e. the overturn of SPOT.

TOPS mode

The SCAN mode has been shown to be capable of increasing the range swath at the expense of azimuth resolution. The scanning mechanism along the range leads however to a nonstationarity along the azimuth of the antenna gain (scalloping). By referring to Fig. 4.16 is rather evident the origin of the scalloping: while the targets shown by thick black segments are illuminated by the antenna also when looking in the broadside direction, i.e. along the maximum gain. Conversely, the scatterers illuminated over subsequent bursts, i.e. the one located on the light gray segment, are never illuminated from the broadside direction, i.e. on the peak of the antenna beam. To overcome the scalloping problem, a possibility is to implement a scan also along the azimuth direction so that the peak of the illumination beam "brushes" all the scatterers along the azimuth. To do this, a scanning along the azimuth must

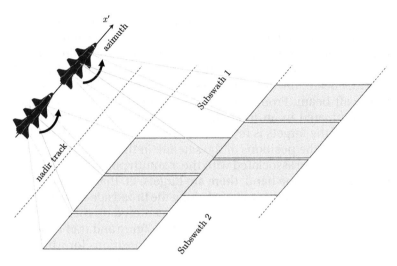

Figure 4.18. Reference Geometry for imaging radars in the TOPS configuration for the case of two subswaths.

be added to the one along the range. The sketch of the Terrain Observation with Progressive Scan (TOPS) mode is shown in Fig. 4.18. For what concerns the range direction, the system operates as in the SCAN mode by steering the beam along different subswaths. In addition, a steering of the beam is also implemented along the azimuth so that the beam peak illuminates, from burst to burst, all the point on the ground along the azimuth. This latter aspect necessarily requires that the beam is rotated, along the azimuth, in a direction of rotation opposite with respect to the SPOT (hence also the acronym TOPS) so that the footprint moves (during the illumination) faster than the satellite ground velocity. Accordingly, while in the SPOT mode the antenna footprint slows down in order to increase the illumination interval, in the TOPS mode the antenna footprint speeds up to decrease the illumination interval.

The timing for the illumination of a subswath in the tops mode is detailed in Fig. 4.19. With respect to Fig. 4.16, the beam steering along the azimuth allows illuminating along the direction of the beam peak (see the beam position at the end of the first burst and the start of the subsequent burst) even the points located at the sensor broadside in the mid point between two consecutive bursts, i.e. when the sensor is illuminating the other subswath in the mid point of the illumination interval.

The opposite behavior between SPOT and TOPS is reflected also in the time frequency diagrams, see Figs. 4.14 and 4.20, respectively. Returning to the comparison of Fig. 4.9 and Fig. 4.12, it is clear that backward rotation along the azimuth of the SPOT

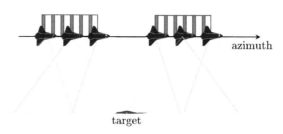

Figure 4.19. Timing and illumination of a ground subswath in the SCAN mode. Shown are the bursts of pulses illuminating a subswath. As for Fig. 4.15, the sketch refers to a TOPS system with two subswath and a single look. The rotation of the beam allows illuminating all the scatterers on the ground with similar beam spanning.

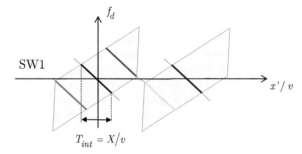

Figure 4.20. Representation in the azimuth time-frequency domain of the TOPS acquisition mode. The system is supposed characterized by two subswaths, one burst with an integration time equal to the half of the corresponding stripmap configuration. First, mid, and end points illuminated for the whole integration time of a burst are shown.

mode translates in a clockwise rotation of the "acquired time-frequency slice" in the time frequency domain. It is therefore not surprising that the TOPS mode observation, for which the rotation of the beam along the azimuth is in the forward direction, corresponds to a counterclockwise rotation (see Fig. 4.20), thus reducing the integration time.

TOPS is the background mission mode, i.e. the default mode, over the land of the ESA Sentinel-1 (S1) constellation currently operative. This mode allows for a significant increase of the range coverage, with data slices covering up to 400 Km in ground range, thus allowing a complete and regular coverage of the Earth surface.

5

2D SAR focusing

Imaging radars are designed to produce images of the observed scene in the microwave region of the electromagnetic spectrum. The Impulse Response Function introduced in Chapter 4 highlights that the acquired radar signal only returns a rough representation of the scene characterized by a poor spatial resolution, not sufficient for any practical application. However, the characteristics of the transmitted waveform and of the acquisition make the received data a wide spectrum signal in both range and azimuth directions: the intrinsic large bandwidth can be therefore exploited to increase the resolution of the image. Along the range direction, the large bandwidth is achieved by the use of sophisticated frequency modulated signals, i.e. chirps, which allows, through the processing discussed in Chapter 2, to compress the target response in an interval significantly smaller than the transmitted impulse duration. As introduced in Chapter 4, the relative motion between the sensor and the target allows achieving a large bandwidth, through the Doppler effect, also along the moving trajectory. In this chapter we discuss the processing steps required, by exploiting the Doppler bandwidth, to improve the azimuth resolution with respect to the one provided by the antenna footprint. However, differently from range compression, the azimuth processing has to account also for the RCM effect, related to the range variability of the target response in the azimuth beam during the acquisition of the radar signal. Accordingly, the processing required to produce high resolution SAR images, usually referred to as 2D SAR focusing, results to be much more challenging than that for classical surveillance radars.

5.1 Preliminary concepts

Let us consider a SAR system operating in the stripmap mode (see Section 4.6), that is illuminating a strip on the ground. In the previous chapter we have analyzed the response of a point P in a generic position (x, r), see (4.4). The illuminated scene can be represented as a superposition of points located at different az-

Multi-Dimensional Imaging with Synthetic Aperture Radar
https://doi.org/10.1016/B978-0-12-821655-2.00009-7

imuth and range coordinates, each of them scattering a fraction of the incident radiation to the sensor. Such a scattering capability is described by the radar reflectivity $\gamma(x, r)$, a complex function proportional to the ratio between the backscattered and the incident field, see Section 3.2. Letting x' and r' be the azimuth and range associated with the image pixel, i.e. the coordinates corresponding to the position of the antenna along the track and the range of the sampling time of the backscattered echo, we have:

$$\gamma(x', r') = \iint dx\, dr\, \gamma(x, r) \delta(x' - x, r' - r) \tag{5.1}$$

where the integration is carried out over the illuminated strip. A SAR system can be considered as a linear system and therefore the received signal can be expressed as a superposition of the responses by each elementary scatterer, according to the following equation:

$$y(x', r') = \iint dx\, dr\, \gamma(x, r) g(x', r'; x, r) \tag{5.2}$$

where $g(x', r', x; r)$ is the space variant system Impulse Response Function (IRF). The latter is equivalent to the signal (4.4) in Chapter 4 specialized for the transmitted pulse, i.e., before the range compression. Therefore, the conversion from the scene domain to the data domain implies the following transformation:

$$\delta\left(x' - x, r' - r\right) \to g(x', r'; x, r) \tag{5.3}$$

(x', r') and (x, r) being the output (i.e., the data domain) and the input (i.e., the domain of the unknown) points, respectively.

It is worth noting that, as long as the sensor moves along a linear trajectory, the acquisition geometry described in Chapter 4 makes the response of the system invariant with respect to the azimuth variable. Accordingly, the space variance of the IRF reduces to the range direction only:

$$g(x', r'; x, r) = g(x' - x, r'; r) = g(x' - x, r' - r; r) \tag{5.4}$$

where in the last equality, with a slight abuse of notation, the same symbol has been exploited to denote the output range axis r' (left hand side) and its centered version around the output range r (right hand side), by mean of the position $r' \to r' - r$.

Eq. (5.2) can thus be rewritten as

$$y(x', r') = \iint dx\, dr\, \gamma(x, r) g(x' - x, r' - r; r) \tag{5.5}$$

showing that the 2D signal $y(x', r')$ is an image of the function $\gamma(x, r)$ achieved through the SAR IRF in (5.4). The latter is also

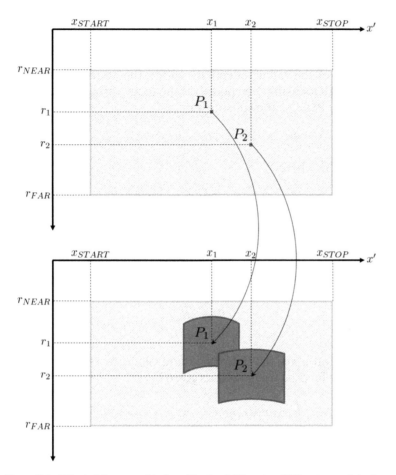

Figure 5.1. Effect of the spread induced by the SAR system PSF on two points P_1 and P_2 of the imaged surface.

referred to as the space variant (preprocessing) Point Spread Function (PSF) of the SAR system, as it spreads each point of the imaged surface within its range variant support. This spreading produces the overlap of contributions corresponding to different surface points, see Fig. 5.1, which leads to a blurring of the final image. Differently stated, the acquired signal is a defocused image of the radar reflectivity: a 2D focusing processing is thus required in order to improve the estimate of γ.

The 2D focusing should, in principle, invert the transformation (5.3) providing the acquired signal (5.5), that is:

$$g(x' - x, r' - r; r) \rightarrow \delta\left(x' - x, r' - r\right) \qquad (5.6)$$

Such inversion would allow achieving a perfect reconstruction of γ. In practice, having the SAR system a limited 2D bandwidth, the best achievable result (in term of spatial resolution) is the reconstruction of the radar reflectivity within the system bandwidth. The 2D SAR focusing then performs the range variant equalization of the SAR PSF, i.e., the substitution

$$g(x' - x, r' - r; r) \rightarrow h\left(x' - x, r' - r\right) \tag{5.7}$$

within (5.5), where:

$$h(x', r') = B_x \operatorname{sinc}\left(B_x x'\right) B_r \operatorname{sinc}\left(B_r r'\right) \tag{5.8}$$

is the space invariant (postprocessing) PSF, whose spectrum is flat within the range and azimuth system bandwidths B_r and B_x, respectively. The latter are related to the transmitted and Doppler bandwidths, introduced in Chapter 4, by the following simple scaling relations:

$$\begin{aligned} B_r &= 2B_t/c \\ B_x &= B_d/v \end{aligned} \tag{5.9}$$

where c and v are the lightspeed and platform velocity, respectively. The focused image is thus given by:

$$\hat{\gamma}(x', r') = \iint dx dr \gamma(x, r) h(x' - x, r' - r) = \gamma(x', r') \otimes h(x', r') \tag{5.10}$$

showing that each point of the imaged surface is spread, this time, within the space invariant support of the postprocessing PSF $h(x', r')$, see Fig. 5.2.

It is worth noting that the processing leading to the image $\hat{\gamma}(x', r')$ in (5.10) is aimed to recover the portion of spectrum of the radar reflectivity within the system bandwidth by compensating the distortions and the space variance intrinsically introduced by the system IRF during the acquisition. Such a processing leads to the compression of the PSF, which allows improving the image resolution.

5.2 Focusing of a single point scatterer

To get into the details of the focusing process, let us start by considering the image produced by a single point scatterer (sometimes simply referred to as a target) located on an absorbing background at fixed (known) coordinates (x_0, r_0), without lack of generality assumed to be at the scene center. In this case the radar

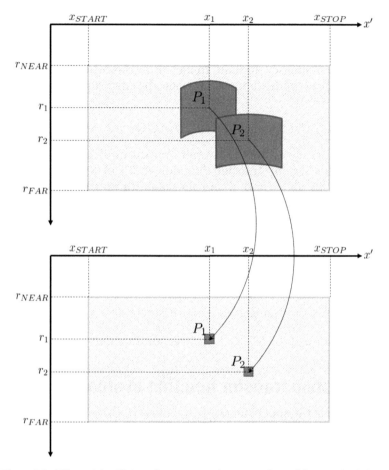

Figure 5.2. Effect of the 2D focusing on two points P_1 and P_2 of the acquired signal.

reflectivity is a Dirac delta function applied to the scatterer position and the acquired (unfocused) image is given by

$$y(x', r') = g(x' - x_0, r' - r_0; r_0) \qquad (5.11)$$

where an inessential amplitude factor has been neglected. It is worth noting that, having considered a fixed input point (and, in particular, its range r_0), the unfocused image in (5.11) can be rewritten as the convolution

$$y(x', r') = \delta(x' - x_0, r' - r_0) \otimes g_0(x', r') \qquad (5.12)$$

where, to lighten the notation, the preprocessing PSF reference to the fixed range r_0 has been denoted by the subscript "0". Eq. (5.12) shows that, in the ideal case of a single point scatterer of known position, the focusing turns out to be a classical space invariant deconvolution problem, which can be straightforwardly carried out in the frequency domain. In fact, as a consequence of the expression (5.12) of the defocused image, the spectrum of the data is given by:

$$Y(f_x, f_r) = G_0(f_x, f_r)\exp\{-j2\pi(f_x x_0 + f_r r_0)\} \tag{5.13}$$

where $G_0(f_x, f_r)$ is the 2D Fourier Transform of $g_0(x', r')$, representing the System Transfer Function (STF) of the SAR system evaluated in r_0; f_x and f_r are the azimuth and range (spatial) spectral frequencies, respectively.[1] Provided that the scatterer position (x_0, r_0) is known, the focusing can therefore be carried out by means of the inverse filter

$$F_0(f_x, f_r) = G_0^{-1}(f_x, f_r) \tag{5.14}$$

within the bandwidth of the STF, which leads to the focused image

$$\hat{\gamma}(x', r') = B_x\text{sinc}[B_x(x' - x_0)]B_r\text{sinc}[B_r(r' - r_0)] \tag{5.15}$$

Evaluation of the STF $G_0(f_x, f_r)$ is now in order.

5.3 System transfer function evaluation

To derive the spectrum $G_0(f_x, f_r)$ of $g_0(x', r')$, we start from the expression of the SAR PSF in (4.18) written with reference to the transmitted pulse (i.e., before the range compression) and with reference to the input point (x_0, r_0). It is given by:

$$g_0(x' - x_0, r' - r_0) = \Pi\left(\frac{x' - x_0}{X}\right)\exp\left(-j\frac{4\pi}{\lambda}R_0\right)p_r\left(r' - R_0\right) \tag{5.16}$$

where $p_r(\cdot)$ is the transmitted pulse in the spatial (range) domain,

$$X = \frac{\lambda r_0}{L_A} \tag{5.17}$$

is the footprint produced by the aperture L_A of the real antenna along the azimuth direction and

$$R_0 = R_0(x' - x_0) = \sqrt{r_0^2 + (x' - x_0)^2} \tag{5.18}$$

---

[1] Note that the STF concept can be introduced also for space variant systems [38] [39]. The calculus of a time-varying transfer function can be simplified for certain classes of system as, for instance, for unspread LTV, that is slowly time varying linear systems with properly limited memory (IRF support) [40].

is the sensor-to-target distance, accounting for the RCM effect described in Chapter 4.4, see the expansion in (4.19). It is worth noting that the imaged point is spread by the PSF over the domain:

$$\mathcal{D}_S(x_0, r_0) = \Pi\left(\frac{x' - x_0}{X}\right) \Pi\left(\frac{r' - R_0}{\rho_R}\right) \tag{5.19}$$

which, according to the expressions (5.17) of X and (5.18) of R_0, results to have a shape dependent on the range of the imaged point.[2] In Fig. 5.1 are qualitatively depicted in light blue (dark gray in print version) color the spreading domain $\mathcal{D}_S(x_1, r_1)$ and $\mathcal{D}_S(x_2, r_2)$, with $r_2 > r_1$: it shows that the higher the range of the point target, the higher the azimuth spread and the lower the curvature produced by R_0.

Going back to (5.16), by letting $x' - x_0 \to x'$ and $r' - r_0 \to r'$ and neglecting an irrelevant phase factor (for focusing purpose), the preprocessing PSF $g_0(x', r')$ can be written as:

$$g_0(x', r') = \Pi\left(\frac{x'}{X}\right) \exp\left(-j\frac{4\pi}{\lambda}\Delta R_0(x')\right) p_r\left(r' - \Delta R_0(x')\right) \tag{5.20}$$

where

$$\Delta R_0(x') = R_0(x') - r_0 = \sqrt{r_0^2 + x'^2} - r_0 \tag{5.21}$$

is the range deviation responsible for the RCM.

The only term dependent on the range r' in (5.16) is the transmitted pulse p_r, which is in turn delayed by the azimuth dependent range deviation ΔR_0. Accordingly, by performing the FT firstly with respect r' and subsequently with respect to x', the 2D spectrum of $g_0(x', r')$ can be factorized as

$$G_0(f_x, f_r) = G_1(f_r)G_2(f_x, f_r) \tag{5.22}$$

where

$$G_1(f_r) = \mathcal{F}_{r'}\left\{p_r(r')\right\} \tag{5.23}$$

and

$$G_2(f_x, f_r) = \mathcal{F}_{x'}\left\{p_x(x', f_r)\right\} \tag{5.24}$$

The pulse $p_x(x', f_r)$ in (5.24) is given by:

$$p_x(x', f_r) = \Pi\left(\frac{x'}{X}\right) \exp\left[-j2\pi\overline{f}_r\Delta R_0(x')\right] \tag{5.25}$$

where

$$\overline{f}_r = \frac{2}{\lambda} + f_r = \frac{2}{c}(f_0 + f_t) \tag{5.26}$$

---

[2] ρ_R in (5.19) is the range extension of the pulse $p_r(r')$, see Chapter 2.

is the range frequency $f_r = 2f_t/c$ shifted of the range carrier $2/\lambda = 2f_0/c$, f_t and f_0 being the fast time spectral frequency and the transmitted carrier frequency, respectively.

It is worth underlining that the range deviation $\Delta R_0(x')$, i.e. the RCM, leads to the presence of the range frequency dependent phase term in the azimuth pulse (5.25). This originates the coupling of the azimuth and range frequencies in the spectral domain, i.e., the impossibility of factorizing $G_2(f_x, f_r)$ in the product of two components, each one dependent on a single spectral frequency. It follows that the factorization can be carried out only if the variations of the phase term in (5.25), for x' spanning the footprint X and f_r spanning the range bandwidth B_r, can be neglected.

The spectrum $G_1(f_r)$ is the FT of the spatial (range) representation of the transmitted pulse. Accordingly, with reference to an up-chirp, the expression of $G_1(f_r)$ can be achieved via (2.44), provided that the necessary conversions from temporal to spatial domain are implemented. By neglecting an inessential amplitude term, it results

$$G_1(f_r) = \Pi\left(\frac{f_r}{B_r}\right) \exp\left(-j\frac{\pi}{\alpha_r}f_r^2\right) \tag{5.27}$$

where α_r and B_r are the spatial chirp rate and bandwidth, respectively, related to the corresponding temporal counterparts by

$$\alpha_r = \frac{4}{c^2}\alpha \tag{5.28}$$

and

$$B_r = \frac{2}{c}B_t \tag{5.29}$$

the latter defining the interval spanned by the range frequency f_r.

The spectrum $G_2(f_x, f_r)$ in (5.24) is the FT of the phase modulated pulse $p_x(x', f_r)$ in (5.25) defined over the azimuth domain, whose modulating signal depends also on the range spectral frequency f_r. According to the SPA, see Appendix A.4, it is approximated by

$$G_2(f_x, f_r) \approx p_x(x_s, f_r)e^{-j2\pi f_x x_s} \tag{5.30}$$

where x_s is the stationary phase point, achieved by equating the azimuth spectral frequency f_x and the instantaneous spatial frequency of the pulse in (5.25), see again Appendix A.4:

$$f_x = -\overline{f}_r \frac{x'}{\sqrt{r_0^2 + x'^2}} \tag{5.31}$$

Therefore, the inversion of (5.31) with respect to x' leads to the stationary phase point x_s, which is given by:

$$x_s = -r_0 \frac{f_x}{\sqrt{\overline{f}_r^2 - f_x^2}} \qquad (5.32)$$

Finally, by substituting (5.32) in (5.30) and accounting for the pulse expression in (5.25), it results

$$G_2(f_x, f_r) = \Pi\left(\frac{f_x}{B_x}\right)\exp[-j2\pi r_0 K(f_x, f_r)] \qquad (5.33)$$

where

$$K(f_x, f_r) = \sqrt{\overline{f}_r^2 - f_x^2} - \overline{f}_r \qquad (5.34)$$

and

$$B_x = \frac{2}{L_A} \qquad (5.35)$$

is the azimuth frequency bandwidth, i.e., the spatial counterpart of the Doppler (slow time) bandwidth B_d in (4.15) derived via the Doppler analysis and related to B_x as in (5.9). Note that, in deriving (5.33), the slow varying amplitude pulse in (5.25) has been evaluated in the approximated phase stationary point:

$$x_s \approx -\frac{\lambda r_0}{2} f_x \qquad (5.36)$$

achieved by stopping at the first order the Mc Laurin series expansion of the square root in (5.32) and neglecting f_t with respect f_0 in (5.26). Moreover, the larger the azimuth bandwidth (B_x) with respect to the bandwidth ($1/X$) of the amplitude of the pulse in (5.25), the better the approximation (5.33) of the spectrum of the (complex) pulse in (5.25), see Appendix A.4.

Although the expression of B_x (or, better, its temporal frequency counterpart B_d) has been already derived in Chapter 4 as the frequency interval spanned by the Doppler shift, it might be instructive to derive it again by starting from the expression (5.31) of the azimuth spectral frequency f_x. To this aim, let us rewrite (5.31) as

$$f_x = -\frac{2}{\overline{\lambda}}\sin\phi_0(x') \qquad (5.37)$$

where

$$\overline{\lambda} = \frac{c}{f_0 + f_t} \qquad (5.38)$$

is the wavelength associated with the harmonic $f_0 + f_t$ of the transmitted RF pulse and $\phi_0(x')$ is the aspect angle, i.e. the angle between the direction corresponding to the range r_0 of the target and the sensor-to-target distance (line of sight or radial direction) when the sensor is at azimuth position x'. During the sensor flight, the aspect angle spans the angular aperture of the beam radiated by the real antenna along the azimuth direction, that is $|\phi_0(x')| \leqslant \bar{\lambda}/L_A$. Such an aperture is usually very small and, therefore, (5.37) can be approximated as

$$f_x = -\frac{2}{\bar{\lambda}}\phi_0(x') \tag{5.39}$$

which allows deriving the azimuth bandwidth in (5.35) as the interval spanned by f_x when $\phi_0(x')$ spans the interval $\bar{\lambda}/L_A$.

Finally, by using Eqs. (5.17) and (5.35), the ratio between the bandwidths B_x and $1/X$ can be written as

$$\frac{2\lambda r_0}{L_A^2} = 2\lambda r_0 B_x^2 \gg 1 \tag{5.40}$$

showing that the higher the azimuth bandwidth, the better the stationary phase approximation.

It is worth noting, by the comparison between (4.14) and (5.37), that the Doppler shift frequency $f_d(x')$ observed at the azimuth position x' is a value of the slow time spectral frequency $f_a = v f_x$. In other words, for each x', the Doppler effect induced by the relative movement between the sensor and the target produces a harmonic component oscillating at a frequency f_a given by the Doppler shift $f_d(x')$. For this reason, the spectrum of the received signal associated with the slow time spectral frequency f_a (or, equivalently, with the azimuth spectral frequency f_x) and its bandwidth are also referred to as Doppler spectrum and Doppler bandwidth, respectively.

A consideration on the phase term in (5.34) is now in order. According to the Mac Laurin series expansion with respect to the range spectral frequency f_r, it can be rewritten as

$$K(f_x, f_r) = K_0(f_x) + K_H(f_x, f_r) \tag{5.41}$$

where

$$K_0(f_x) = \frac{2}{\lambda}\left[\sqrt{1 - \left(\frac{\lambda f_x}{2}\right)^2} - 1\right] \approx -\frac{\lambda}{2}f_x^2 \tag{5.42}$$

is the zero order series term, which only depends on the azimuth spectral frequency and is responsible for the defocusing of the image in the azimuth direction, whereas the residual (high order) phase term

$$K_H(f_x, f_r) = \sum_{n=1}^{+\infty} \frac{1}{n!} K_n(f_x) f_r^n \qquad (5.43)$$

couples the two spectral domains, thus being responsible for the RCM.

A final consideration is now in order. As already underlined in Chapter 4, the phase modulated pulse $p_x(x', f_r)$ in (5.25), generated by the relative movement between the transmitting sensor and the target, approximates a chirp pulse along the azimuth direction when the condition $|X/(2r_0)| = |\lambda/(2L_A)| \ll 1$ is satisfied, see Eq. (4.20). The same result is true also for its FT in (5.33)-(5.34), which is formally equivalent to (5.25). Indeed, by neglecting in (5.41) the high order term $K_H(f_x, f_r)$ associated with the RCM, the phase of (5.33) approximates a quadratic phase along the spectral frequency f_x when $|\lambda B_x/2| = |\lambda/(2L_A)| \ll 1$, which is the condition making reliable the last approximation in (5.42).

The focusing of the image produced by a single point scatterer in (x_0, r_0), pictorially shown in Fig. 5.3, is thus carried out with the inverse filter in (5.14) which, according to the phase only expression of the STF terms in (5.27) and (5.33), is also the conjugate filter. It can be written as

$$F_0(f_x, f_r) = F_1(f_r) F_2(f_x, f_r) \qquad (5.44)$$

where

$$F_1(f_r) = G_1^*(f_r) \qquad (5.45)$$

and

$$F_2(f_x, f_r) = G_2^*(f_x, f_r) \qquad (5.46)$$

The filter $F_1(f_r)$ performs the range compression, see the top-right part of Fig. 5.3, thus leading the range resolution from $\rho_R = c\tau/2$ to $\rho_r = 1/B_r$. The filter $F_2(f_r, f_x)$ compensates the RCM effect, as shown in the bottom-right part of Fig. 5.3, by means of the high order phase term $K_H(f_x, f_r)$, and performs the azimuth compression, see the bottom-left part of Fig. 5.3, by means of the phase term $K_0(f_x)$, thus leading the azimuth resolution from $\rho_A = X$ to $\rho_a = 1/B_x$.

5.4 Focusing of an extended scene

In the more realistic case of a scatterers' distribution on the illuminated ground, the received signal is expressed by (5.2) and its

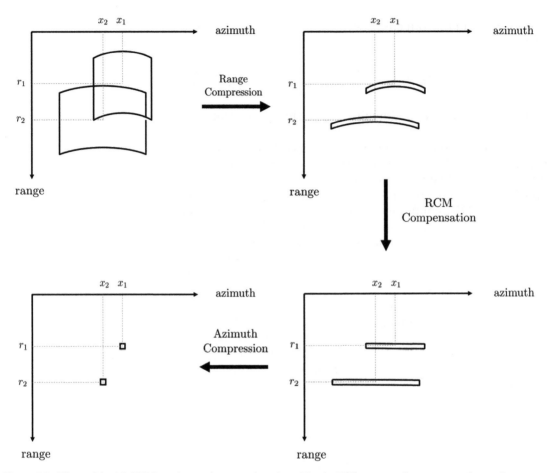

Figure 5.3. Effect of the 2D SAR focusing on the spread produced by the PSF on two point scatterers located at (x_1, r_1) and (x_2, r_2), with $r_1 < r_2$.

FT is given by

$$Y(f_x, f_r) = \iint dx\, dr\, \gamma(x, r) G(f_x, f_r; r) e^{-j2\pi(xf_x + rf_r)} \qquad (5.47)$$

where $G(f_x, f_r; r)$ is the range variant SAR STF, expressed as in Section 5.3 provided that the position $(x_0, r_0) = (x, r)$ is made, with (x_0, r_0) corresponding generally to the scene center. It is worth noting that the range variant STF can be factorized as:

$$G(f_x, f_r; r) = G(f_x, f_r; r_0) e^{-j2\pi(r - r_0)K(f_x, f_r)} \qquad (5.48)$$

where $G(f_x, f_r; r_0) = G_0(f_x, f_r)$ is the STF range invariant component, usually corresponding to the range r_0 of the scene center.

Accordingly, the received data in (5.47) can be rewritten as:

$$Y(f_x, f_r) = G_0(f_x, f_r)$$
$$\times \iint dx\,dr\,\gamma(x, r)e^{-j2\pi(r-r_0)K(f_x, f_r)}e^{-j2\pi(xf_x+rf_r)} \quad (5.49)$$

Eq. (5.49) suggests a first, very simple focusing approach, which consists on filtering the received signal with the range invariant filter $F_0(f_x, f_r) = G_0^*(f_x, f_r)$ in (5.44)-(5.46). The spectrum of the filtered image, representing an estimate of the radar reflectivity, is thus given by

$$\hat{\Gamma}_0(f_x, f_r) = \Pi\left(\frac{f_x}{B_x}\right)\Pi\left(\frac{f_r}{B_r}\right)$$
$$\times \iint dx\,dr\,\gamma(x, r)e^{-j2\pi(r-r_0)K(f_x, f_r)}e^{-j2\pi(xf_x+rf_r)} \quad (5.50)$$

where the subscript "0" refers to the range value to which the estimate is tuned on. Eq. (5.50) shows that the distortions associated with the phase term $K(f_x, f_r)$ are perfectly compensated for the input contributions at range $r = r_0$, whereas become much more significant as the contributions move away from the tuned range, thus increasing the differences $r - r_0$. Accordingly, the estimate

$$\hat{\gamma}_0(x', r') = \mathcal{F}^{-1}\left\{\hat{\Gamma}_0(f_x, f_r)\right\} \quad (5.51)$$

is almost well focused in $r' = r_0$, where it is the superposition of contributions with a low phase distortion, and becomes much more defocused as the difference $r' - r_0$ increases. For this reason, such a processing is usually referred to as narrow focusing; it is suitable for images with a small range swath, where the range differences $r - r_0$ in (5.50) can be considered small enough to limit the distortion produced by the uncompensated phase terms. The quantification of the tolerable range scene extent, although characterized by a straightforward derivation, is outside the scope of this book [41]. The block diagram of the narrow focusing is depicted in Fig. 5.4.

In order to achieve a wide focusing, i.e. a processing able to provide a well focused image also for extended range swaths, a deeper analysis of the spectrum of the received data is in order. Let us start from the narrow focused signal in (5.50). By performing the change

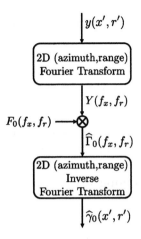

Figure 5.4. Narrow focus processing block diagram.

of variable $r - r_0 \to r$, it can be rewritten as[3]:

$$\hat{\Gamma}_0(f_x, f_r) = \Pi\left(\frac{f_x}{B_x}\right) \Pi\left(\frac{f_r}{B_r}\right)$$
$$\times \iint dx dr \gamma(x, r) e^{-j2\pi r[K(f_x, f_r) + f_r]} e^{-j2\pi x f_x} \qquad (5.52)$$

$$= S(f_x, f_r) \Gamma[f_x, K(f_x, f_r) + f_r]$$

where

$$S(f_x, f_r) = \Pi\left(\frac{f_x}{B_x}\right) \Pi\left(\frac{f_r}{B_r}\right) \qquad (5.53)$$

is the spectral support of the observed input (radar reflectivity); moreover, the phase term $\exp(j2\pi r_0 f_r)$ is supposed to be compensated and the position $\gamma(x, r + r_0) \to \gamma(x, r)$ is done for sake of notation simplicity.

Eqs. (5.52)-(5.53) show that, due to the range variant characteristic of the SAR system, the range frequency axis f_r is distorted according to the nonlinear mapping:

$$f_r \to K(f_x, f_r) + f_r \qquad (5.54)$$

when moving from the output domain (associated with $\hat{\Gamma}_0$) to the input domain (associated with Γ).

---

[3]The change of variable $r - r_0 \to r$ implies that the range swath (i.e., the range integration interval) is now centered around zero.

Such a mapping, usually referred to as Stolt mapping [42] [43], must be compensated to achieve a well focused image on the whole large range swath. The compensation is carried out by counter-deforming the output range frequency axis according to the following nonlinear mapping:

$$f_r \rightarrow \Omega(f_x, f_r) = \sqrt{\left(\frac{2}{\lambda} + f_r\right)^2 + f_x^2} - \frac{2}{\lambda} \qquad (5.55)$$

which inverts the transformation (5.54).

After this reverse mapping, the output signal is given by:

$$\hat{\Gamma}(f_x, f_r) = \hat{\Gamma}_0[f_x, \Omega(f_x, f_r)] = S_\Omega(f_x, f_r)\Gamma(f_x, f_r) \qquad (5.56)$$

where

$$S_\Omega(f_x, f_r) = \Pi\left(\frac{f_x}{B_x}\right)\Pi\left(\frac{\Omega(f_x, f_r)}{B_r}\right) \qquad (5.57)$$

is the new observed spectral support, which is now deformed with respect to the one in (5.53).

Eqs. (5.56)-(5.57) show that, by reversing the Stolt mapping, the spectrum of the radar reflectivity is perfectly recovered over a distorted support. However, the slowly varying nature of the spectral support makes negligible the distortion carried out by the nonlinear mapping, specially in the case of a satellite SAR system characterized by very high values of the range distances. In other words, with reference to (5.53) and (5.57), it is reasonable the approximation $S_\Omega(f_x, f_r) \approx S(f_x, f_r)$, which along with (5.56) allows to write the following expression for the wide focused image:

$$\hat{\gamma}(x', r') \approx \mathcal{F}_{f_x, f_r}^{-1}\left\{\hat{\Gamma}(f_x, f_r)\right\} \qquad (5.58)$$

The sequence of the steps leading to the image in (5.58) are summarizing in the block diagram in Fig. 5.5.

The compensation of the Stolt mapping is carried out by the so-called $\Omega - K$ algorithm [41], [44]. In the practical situation, being (5.57) a sampled signal, the $\Omega - K$ algorithm performs a spectral interpolation of the narrow focused signal over the grid $[f_x, \Omega(f_x, f_r)]$. This is the main drawback of such algorithm, since the spectral re-sampling is a time-consuming and approximated procedure.

A more efficient wide focusing processing, sometimes referred to as "interpolation free wide SAR focusing", can be carried out as explained in the following. The expression of the narrow focused image in (5.52) has shown that the range frequency axis f_r is distorted from the output to the input according to the Stolt mapping

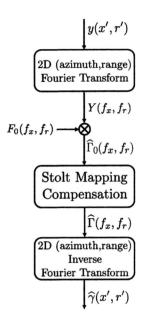

Figure 5.5. Wide focus processing block diagram.

in (5.54). Such a mapping is generally well approximated by its first order series expansion:

$$K(f_x, f_r) + f_r \approx u(f_x) + v(f_x)f_r \qquad (5.59)$$

where, with reference to (5.41)-(5.43), $u(f_x) = K_0(f_x)$ is the term accounting for the defocusing in the azimuth direction, also referred to as focus depth, and

$$v(f_x) = K_1(f_x) + 1 = \frac{1}{\sqrt{1 - \left(\frac{\lambda f_x}{2}\right)^2}} \qquad (5.60)$$

accounts for the RCM, whose effect increases when moving far from r_0, see (5.50). Obviously, although this approximation is less critical than the one exploited in the narrow focusing (where the whole Stolt mapping is neglected), the larger the difference $r - r_0$ the heavier the effects of neglecting the higher order terms in (5.59).

The narrow focused image in (5.52) can be therefore approximated by:

$$\hat{\Gamma}_0(f_x, f_r) \approx \Pi\left(\frac{f_x}{B_x}\right) \Pi\left(\frac{f_r}{B_r}\right) \Gamma[f_x, u(f_x) + v(f_x)f_r] \qquad (5.61)$$

showing that the range frequency axis of the input (i.e., the radar
reflectivity) is scaled by $v(f_x)$ and shifted by $u(f_x)$. An Inverse
Scaled Fourier Transform (ISFT) of (5.61) can be thus exploited to
achieve an estimate of the radar reflectivity in the Doppler (azimuth frequency) - range domain. To emphasize the frequency
(Doppler) - space (range) nature of the domain where this esti-
mate is defined, it will be denoted by $\hat{\Gamma}\hat{\gamma}(f_x, r')$: left and right letter
specifies the direction (azimuth and range, respectively); upper-
case and lowercase specifies the domain (frequency and space,
respectively); hat stands for estimate. According to this notation,
we have:

$$\hat{\Gamma}\hat{\gamma}(f_x, r')$$

$$= \Pi\left(\frac{f_x}{B_x}\right)$$

$$\times \int \Pi\left(\frac{f_r}{B_r}\right) \Gamma[f_x, u(f_x) + v(f_x)f_r]e^{j2\pi r'[v(f_x)f_r]}d[v(f_x)f_r] \quad (5.62)$$

$$= \Pi\left(\frac{f_x}{B_x}\right)e^{-j2\pi u(f_x)r'}\mathcal{F}_{f_r}^{-1}\left\{\Pi\left[\frac{f_r}{v(f_x)B_r}\right]\Gamma(f_x, f_r)\right\}$$

$$= \Pi\left(\frac{f_x}{B_x}\right)\Gamma\hat{\gamma}(f_x, r')e^{-j2\pi u(f_x)r'}$$

where an inessential amplitude factor has been neglected and

$$\Gamma\hat{\gamma}(f_x, r') = \int dr\, \Gamma\gamma(f_x, r)v(f_x)B_r\text{sinc}\left[v(f_x)B_r(r'-r)\right] \quad (5.63)$$

is a range estimate of the radar reflectivity $\Gamma\gamma(f_x, r)$ observed in
the azimuth frequency - range space domain. Eq. (5.62) shows that
the estimate $\hat{\Gamma}\hat{\gamma}(f_x, r')$ is defocused in the azimuth direction by the
phase term associated with the focus depth. Moreover, similarly
to the results achieved with the $\Omega - K$ algorithm, the spectrum
$\Gamma(f_x, f_r)$ is also in this case perfectly recovered on a deformed support (see the range frequency rectangular pulse in the third line of
(5.62)). Techniques as chirp scaling or chirp z-transform can be
exploited for an efficient implementation of the scaled inverse FT
[45], [8], [46]. Finally, the compensation of the focus depth phase
term, representing the azimuth focusing step, and the subsequent
Inverse FT carried out in the azimuth domain, allow achieving the
correct reconstruction of the radar reflectivity, given by

$$\hat{\gamma}(x', r') = \iint dxdr\,\gamma(x, r)e^{-j\frac{4\pi}{\lambda}r}h(x'-x, r'-r) \quad (5.64)$$

which accounts also for the phase factor neglected before and
where $h(x'-x, r'-r)$ is the postfocusing PSF. By approximating

in (5.63) the scaling factor $v(f_x)$ with its value at central frequency $f_x = 0$, that is $v(0) = 1$ (see (5.60)), within the whole Doppler bandwidth, the postfocusing PSF $h(\cdot)$ can be approximated as

$$h(x' - x, r' - r) \approx B_x \mathrm{sinc}\left[B_x(x' - x)\right] B_r \mathrm{sinc}\left[B_r(r' - r)\right] \quad (5.65)$$

thus agreeing with (5.8).

Examples of the focusing of and extended scene are provided in Figs. 5.6 and 5.7 for both high resolution ENVISAT and very high resolution COSMO-SkyMed data. Images are relevant to the same geographical area corresponding the large volcanic area of Phlegraean Fields, in the western part of the city of Naples, in the Southern part of Italy. This area is recognized to be the most widespread active volcanic system of the Mediterranean area. For the ENVISAT example in Fig. 5.6, the sequence of raw data, range compressed and fully 2D focused images show the achieved resolution improvement: these images explain on real data the steps pictorially depicted in Fig. 5.3. The image is acquired on a descending pass of the satellite: azimuth and range correspond to the vertical (from top to bottom) and horizontal (from right to left) directions, respectively. For the ENVISAT sensor, it results an azimuth resolution $\rho_a \cong 5$ m and a range resolution $\rho_r \cong 9$ m. A further example, relevant to a very high resolution COSMO-SkyMed X-Band acquisition, is provided in Fig. 5.7, where range compressed and 2D focused image are shown. Similarly to the former example, the acquisition is still carried out on a descending pass. A zoom of both ENVISAT and COSMO-SkyMed amplitude images over a residential area imaged in the upper left part of Figs. 5.6 and 5.7 is provided in Fig. 5.8. For an effective comparison, the ENVISAT image has been resampled over the COSMO-SkyMed higher sampling frequencies in both azimuth and range: The resolution improvement, being $\rho_a \cong 2.8$ m and $\rho_r \cong 1.5$ m for COSMO-SkyMed sensors, is evident.

5.5 Squinted geometry

The SAR focusing has been so far analyzed with reference to a SAR system operating in the Zero-Doppler (ZD) acquisition mode, which produces a Doppler bandwidth centered at zero frequency, see Fig. 4.9 in Chapter 4. When the acquisition geometry is characterized by the presence of a squint angle, the Doppler bandwidth turns out to be centered at a nonzero frequency usually referred to as Doppler centroid, see again Fig. 4.9 in Chapter 4, which has to be properly taken into account at focusing stage.

To see in more detail the effect of the squinted geometry on received data, let us consider the response of a SAR system acquiring

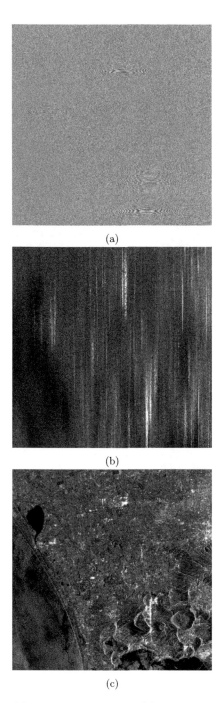

Figure 5.6. (a) Raw, (b) range compressed, and (c) fully 2D focused images for a real ENVISAT acquisition over the area of Phlegraean Fields, Naples, South Italy.

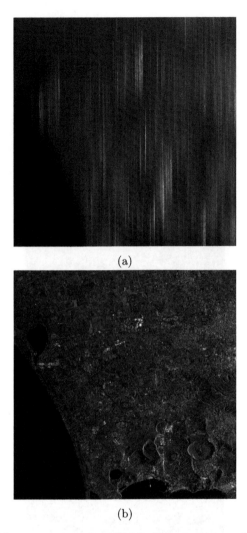

(a)

(b)

Figure 5.7. (a) Range compressed and (b) fully 2D focused images for a real very high resolution COSMO-SkyMed acquisition over the area of Phlegraean Fields, Naples, South Italy.

with a squint angle ϕ_{DC} and corresponding to a point-like target located at coordinates (x, r). According to the expression (4.31), it is given by:

$$g(x' - x, r' - r; r) = \Pi\left(\frac{x' - x + x_{DC}}{X}\right) \exp\left(-j\frac{4\pi}{\lambda}R\right) p_r\left(r' - R\right)$$

$$(5.66)$$

(a)

(b)

Figure 5.8. (a) ENVISAT and (b) COSMO-SkyMed amplitude images at full resolution over a residential area in the Northern part of the city of Naples, South Italy.

where $p_r(r')$ and $R = R(x' - x)$ are the transmitted pulse and the target-to-sensor distance, respectively; moreover:

$$X \approx \frac{\lambda r}{L_A \cos^2(\phi_{DC})} \qquad (5.67)$$

is the azimuth footprint and

$$x_{DC} = r \tan(\phi_{DC}) \qquad (5.68)$$

is the distance between the azimuth coordinates of the target and the middle of the acquisition interval, see Fig. 4.8 in Chapter 4.

The 2D FT of (5.66) can be evaluated by retracing the same steps as in Section 5.3, where the case of $\phi_{DC} = 0$ was considered with $x = x_0$ and $r = r_0$. It is given by:

$$\mathcal{F}_{x',r'}\left\{g(x'-x, x'-r; r)\right\} = G(f_x, f_r; r)e^{-j2\pi(xf_x + rf_r)} \tag{5.69}$$

where $G(f_x, f_r; r)$ is the range variant SAR STF, which can be factorized into the product of two terms $G_1(f_r)$ and $G_2(f_x, f_r; r)$, see Section 5.3. The first one is the range invariant part of the STF, given by the FT of transmitted pulse $p_r(r')$ and expressed by (5.27) in the case of an upper chirp. The second one is the range variant part of the STF, given by:

$$G_2(f_x, f_r; r) = \mathcal{F}_{x'}\left\{\Pi\left(\frac{x' + x_{DC}}{X}\right)\exp\left[-j2\pi \overline{f}_r \Delta R(x')\right]\right\} \tag{5.70}$$

where $\overline{f}_r$ is the shifted range frequency in (5.26) and $\Delta R(x')$ is the range deviation in (5.21). The FT in (5.70) can be evaluated by resorting to the SPA, see Appendix A.4, which provides:

$$G_2(f_x, f_r; r) = \Pi\left(\frac{x_s + x_{DC}}{X}\right)\exp\left[-j2\pi\left(\overline{f}_r \Delta R(x_s) + f_x x_s\right)\right] \tag{5.71}$$

where x_s is the stationary phase point given by:

$$x_s = -r\frac{f_x}{\sqrt{\overline{f}_r^2 - f_x^2}} \tag{5.72}$$

Substitution of (5.72) in (5.71), along with some mild approximations, leads to:

$$G_2(f_x, f_r; r) = \Pi\left(\frac{f_x - f_{x_{DC}}}{B_x}\right)\exp\left[-j2\pi r K(f_x, f_r)\right] \tag{5.73}$$

where

$$B_x = \frac{2}{L_A}\cos(\phi_{DC}) \tag{5.74}$$

is the azimuth bandwidth observed in the squinted geometry, $K(f_x, f_r)$ is defined in (5.34) and

$$f_{x_{DC}} = \overline{f}_r \sin(\phi_{DC}) \tag{5.75}$$

is the (range-frequency dependent) shift of the azimuth bandwidth produces by the squinted geometry, i.e. the Doppler centroid.

The wide focusing of the squinted image can be still carried out by starting from the expression of the narrow focused image. Eqs. (5.73)-(5.75) show that the presence of the squint makes the whole SAR STF $G(f_x, f_r) = G_1(f_r)G_2(f_x, f_r; r)$ different from the one in Section 5.4 only for the shift and the extension of the azimuth bandwidth. Accordingly, to account for the presence of the squint, the approximation (5.61) of the narrow focused image (which is the basis of the interpolation free wide focusing algorithm) can be slightly modified as:

$$\hat{\Gamma}_0(f_x, f_r) \approx \Pi\left(\frac{f_x - f_{DC}}{B_x}\right)\Pi\left(\frac{f_r}{B_r}\right)\Gamma[f_x, u(f_x) + v(f_x)f_r] \quad (5.76)$$

where B_x is expressed as in (5.74).

The focusing in the presence of squint can be thus carried out with similar steps leading, in absence of squint, from the narrow focused image (5.61) to the wide focused image (5.64). By means of the ISFT on the range frequency axis and after the focus depth compensation, we have:

$$\hat{\Gamma}\hat{\gamma}(f_x, r') \approx \Pi\left(\frac{f_x - f_{DC}}{B_x}\right)\int dr\,\Gamma\gamma(f_x, r)v_{DC}B_r\mathrm{sinc}[v_{DC}B_r(r' - r)] \quad (5.77)$$

where the approximation $v(f_x) = v(f_{DC}) = v_{DC}$ for each $f_x \in [f_{DC} - B_x/2, f_{DC} + B_x/2]$ has been done. Finally, after the IFT on the azimuth frequency axis, the focused image is:

$$\hat{\gamma}(x', r') = \iint dx\,dr\,\gamma(x, r)e^{-j\frac{4\pi}{\lambda}r}h(x' - x, r' - r)e^{j2\pi f_{DC}(x'-x)} \quad (5.78)$$

which, again, accounts also the phase factor neglected before and where

$$h(x' - x, r' - r) = B_x\mathrm{sinc}[B_x(x' - x)]v_{DC}B_r\mathrm{sinc}[v_{DC}B_r(r' - r)] \quad (5.79)$$

is the postfocusing PSF.

5.6 MATLAB® examples

5.6.1 Raw data simulation

In this example, the raw data from isolated targets acquired in a stripmap configuration is simulated. Acquisitions parameters relevant to the ESA ENVISAT mission are reported in Table 5.1.

Table 5.1 Simulation parameters.

Parameter	Value
Wavelength	$\lambda = 5.62$ cm
Transmitted chirp Bandwidth	$Bt = 16$ MHz
Transmitted chirp Duration	$T = 26.656$ µs
Doppler Centroid	$f_{DC} = 305.13$ Hz
Chirp Sampling Frequency	$f_s = 19.2$ MHz
Azimuth antenna length	$L_A = 10$ m
Pulse Repetition Frequency	$PRF = 1709.5$ Hz
Range at scene center	$r_0 = 814773.864$ m
Incidence angle at scene center	$\theta = 23°$
Platform velocity	$v = 7120$ m/s
Height of the sensor	$H = 750$ km

The code starts with the loading of system and acquisition parameters contained in the file `parameters_ENVISAT.mat`, definition of some constant.

```
load parameters_ENVISAT.mat

%% Physical Constants
c = physconst('LightSpeed');                %Speed of light
```

The code proceeds with the definition of the Azimuth Antenna Pattern (AAP) to weight echoes during the synthetic aperture according to the antenna radiation pattern. The AAP is modelled as $aap = c_1 sinc^4\left(\frac{L_A}{2v}f\right) + c_2$: parameters c_1 and c_2 have been estimated on a real ENVISAT data patch 3960 records long in the azimuth direction, resulting in $c_1 = 0.59065$ and $c_2 = 0.40935$. The estimated AAP coefficients are provided in Fig. 5.9.

```
%% Azimuth Antenna Pattern
naap = 3960;
faap = ([0:naap-1]'/(naap-1)-0.5)*PRF;
aap = 0.59065*((sinc(L_A/2/v*faap)).^4)+0.40935;
figure('Name', 'Azimuth Antenna Pattern', 'NumberTitle', ...
    'off');
plot(faap, aap), grid on
xlabel('{\it azimuth frequencies} ...
    [Hz]','Interpreter','latex')
```

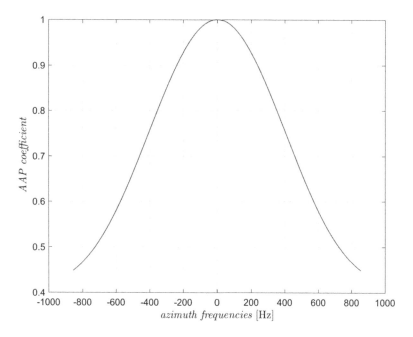

Figure 5.9. Azimuth Antenna Pattern coefficients.

The simulation accounts for the presence of a squint angle during the acquisition. The squint determines a ground azimuth offset x_{DC} of the antenna beam, resulting in an azimuth frequency shift f_{DC}, see (4.31) and (4.33). In this simulation, the Doppler centroid is set as a fraction of the PRF:

```
%% centroid Doppler parameters
fact = +0.2;                  %can be either >1 or <1;
f_dc = PRF*fact;
phi_dc = asin(lambda*f_dc/(2*v));
x_dc = -tan(phi_dc)*r0;
```

Based on the system parameters and exploited squint angle, the evaluation of the synthetic aperture length at scene center r_0 is carried out.

```
%% Synthetic Aperture (at scene center)
X = (lambda/L_A)*r0/((cos(phi_dc))^2.);
```

It results $X \approx 4582$ m at scene center.

The position of arbitrary Nt ideal targets is therefore defined: In the example, a single target at the scene center, i.e. $x' = 0$, $r' = r_0$, is simulated:

```
%% Definition of Targets
Nt = 1;                    %number of targets
xtv = zeros(Nt, 1);
ytv = zeros(Nt, 1);
ztv = zeros(Nt, 1);
rtv = zeros(Nt, 1);

%% position of the first target
xtv(1) = 0;
ztv(1) = 0;
rtv(1) = r0;
ytv(1) = rtv(1)*sin(theta*pi/180);

% %% (Optional) position of the 2nd target
% xtv(2) = 1.1*X ;
% ztv(2) = 0 ;
% rtv(2) = r0 ;
% ytv(2) = rtv(2)*sin(theta*pi/180);
```

The position of the sensor moving along the orbit is also simulated. Prior, the output image azimuth dimension is automatically set according to the given squint angle, target position and synthetic aperture length:

```
%% Azimuth Image dimension setting
naz = 2*floor((max(abs(x_dc+xtv))+X)*(PRF/v));
xx = v/PRF*[-naz/2:naz/2-1]';     %xx is the vector ...
    sampling the azimuth direction

%% position of the sensor
H = r0*cos(theta*pi/180);
xs = xx;
ys = 0;
zs = H;
```

Once targets and sensor positions are defined, range image dimension can be also automatically set and raw data matrix is initialized:

```
%% Range Image dimension setting
Rmax = sqrt((ys-ytv).^2 + (zs-ztv).^2 + ...
    ((x_dc-X/2-xtv).^2).*(x_dc <= xtv) + ...
    ((x_dc+X/2-xtv).^2).*(x_dc > xtv));
Rmin = sqrt((ys-ytv).^2 + (zs-ztv).^2 + ...
    (((x_dc+X/2-xtv).^2).*(x_dc <= xtv) + ...
    ((x_dc-X/2-xtv).^2).*(x_dc > xtv)).*(abs(x_dc-xtv) > ...
    X/4)+ (((x_dc-xtv).^2)).*(abs(x_dc-xtv) <= X/4));
```

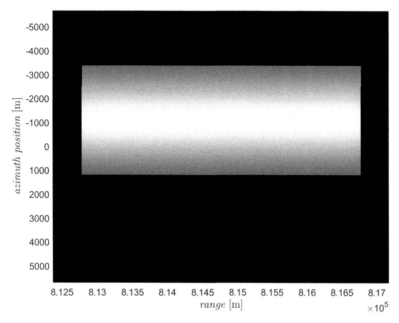

Figure 5.10. Amplitude of the simulated raw data associated with a single point target in presence of squint.

```
nrg = floor(1.2*(max(Rmax)—min(Rmin)+c*T/2)*(2*fs/c));

%% raw data coordinates
rnear = r0—nrg/2*(c/2/fs);          %range of the first ...
    image sample
rr = rnear+[0:nrg—1]'*(c/2/fs);     %rr is the vector ...
    sampling the range direction

%% raw data inizialitazion
raw  = complex(zeros(naz, nrg), zeros(naz, nrg));
```

Raw data associated with each target is finally generated. Azimuth antenna pattern is estimated and applied by interpolating the original function at the actual Doppler frequency. See Figs. 5.10, 5.11 and 5.12.

```
%% transmitted range chirp
alpha = Bt/T;

%% raw data simulation
for pp=1:Nt
    display(['Target ', num2str(pp)])
    xt = xtv(pp);
    yt = ytv(pp);
```

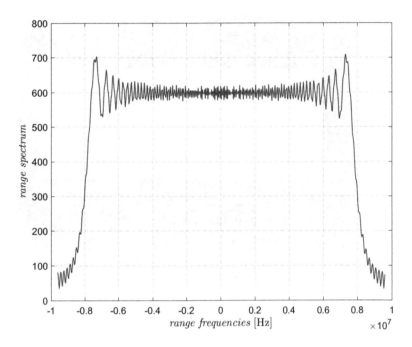

Figure 5.11. Range spectrum of the simulated raw data.

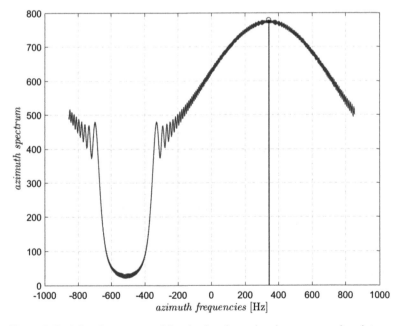

Figure 5.12. Azimuth spectrum of the simulated raw data in presence of squint. Red (mid gray in print version) line pinpoints the simulated Doppler centroid.

```
    zt = ztv(pp);

    for kk=1:naz
        if (abs(xs(kk) - xt - x_dc) ≤ X/2)
            R = sqrt( (xs(kk)-xt)^2 + (ys-yt)^2 + ...
                (zs-zt)^2 );

            %% antenna pattern
            fd = (2*v/lambda)*((xs(kk)-xt)/R);
            caap = interp1(faap-f_dc, aap, fd);
            %%

            s1 = caap*exp(complex(0, ...
                4*pi*alpha/c^2*((rr-R).^2))).*( abs(rr-R) ...
                < c*T/4);
            s2 = exp(complex(0, -4*pi/lambda*R));

            raw(kk, :) = raw(kk, :) + s1.'*s2;
        end
    end
end

figure, imagesc(rr, xx, abs(raw)), colormap gray
xlabel('{\it range} [m]','Interpreter','latex')
ylabel('{\it azimuth position} [m]','Interpreter','latex')

rawf = abs(fftshift(fft2(raw)));
fx = ([0:naz-1]/(naz-1)-0.5)*PRF;
fr = ([0:nrg-1]/(nrg-1)-0.5)*fs;

figure('Name', 'Range Spectrum', 'NumberTitle', 'off');
plot(fr, sum(rawf/naz, 1)), grid on
xlabel('{\it range frequencies} [Hz]','Interpreter','latex')
ylabel('{\it range spectrum}','Interpreter','latex')


figure('Name', 'Azimuth Spectrum', 'NumberTitle', 'off');
plot(fx, sum(rawf/nrg, 2)), grid on
hold on; stem(mod(f_dc-PRF/2, PRF)-PRF/2, ...
    max(sum(rawf/nrg, 2)), 'r')
xlabel('{\it azimuth frequencies} ...
    [Hz]','Interpreter','latex')
ylabel('{\it azimuth spectrum}','Interpreter','latex')
```

5.6.2 Narrow focusing processing

The following code implements the narrow focus processing described in Section 5.3. It operates both on simulated data generated in the previous example or on real data. Real data are relevant to the ENVISAT acquisition over the area surrounding the city of Naples, South Italy, provided in Fig. 5.6.

```
load parameters_ENVISAT.mat

load raw_20030311_ENV_6200x1700.mat        %to load real ...
    data
load raw_squint_simulated.mat              %to load ...
    simulated data

raw = raw - mean(raw(:));

%% Physical Constants
c = physconst('LightSpeed');               %Speed of light

%% Zero padding
rawo = raw;
[nazo, nrgo] = size(rawo);
phi_dc = asin(lambda*f_dc/(2*v));
X = (lambda/L_A)*r0/((cos(phi_dc))^2.);
azp = ceil(X*PRF/v);
rgp = ceil(T*fs);

raw = zeros(nazo+azp, nrgo+rgp);
raw(ceil(azp/2)+1:ceil(azp/2)+nazo, ...
    ceil(rgp/2)+1:ceil(rgp/2)+nrgo)=rawo;

%% Nominal Image dimensions
[naz, nrg] = size(raw);

%% Image axis definition
rr = r0+([0:nrg-1]-round(nrg/2))'*(c/2/fs);         ...
    %range axis
xx = v/PRF*([0:naz-1]-round(naz/2))';               ...
    %azimuth axis

%% Raw data visualization
%to visualize simulated raw data
figure, imagesc(rr, xx, abs(raw)), truesize, colormap gray
%to visualize real raw data
figure, imagesc(rr, xx, imresize(abs(raw), [naz/6, nrg]), ...
    [0 2.5*mean(abs(raw(:)))]), truesize, colormap('gray')
set ( gca, 'xdir', 'reverse' )
xlabel('{\it range} [m]','Interpreter','latex')
ylabel('{\it azimuth} [m]','Interpreter','latex')
```

Picture of real ENVISAT raw data is provided in Fig. 5.13: in the visualization, vertical and horizontal directions represent azimuth and range, respectively. Specifically, axes are oriented to resemble the acquisition on a descending orbit, i.e. their direction is inverted.

The spectrum of the real data in Figs. 5.14 and 5.15 show a behavior similar to those provided for simulated raw data in Figs. 5.11 and 5.12, but for a slight distortion of the range chirp spectrum.

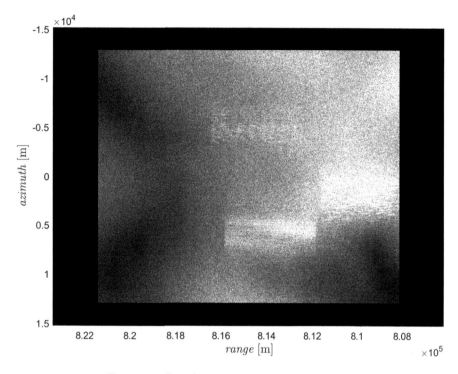

Figure 5.13. Raw data for a real ENVISAT acquisition.

```
%% Raw data spectrum
rawf = fftshift(fft2(raw));
fx = ([0:naz-1]/(naz-1)-0.5)*PRF;
fr = ([0:nrg-1]/(nrg-1)-0.5)*fs;

figure('Name', 'Range Spectrum', 'NumberTitle', 'off');
plot(fr, sum(abs(rawf)/naz, 1)), grid on
xlabel('{\it range frequencies} [Hz]','Interpreter','latex')
ylabel('{\it range spectrum}','Interpreter','latex')

figure('Name', 'Azimuth Spectrum', 'NumberTitle', 'off');
plot(fx, sum(abs(rawf)/nrg, 2)), grid on
hold on; stem(mod(f_dc-PRF/2, PRF)-PRF/2, ...
    max(sum(abs(rawf)/nrg, 2)), 'r')
xlabel('{\it azimuth frequencies} ...
    [Hz]','Interpreter','latex')
ylabel('{\it azimuth spectrum}','Interpreter','latex')
```

The SAR system transfer function is therefore evaluated accounting for the presence of a possible acquisition squint angle as described in Section 5.5.

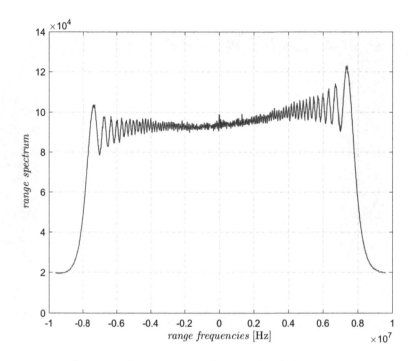

Figure 5.14. Range spectrum of the real ENVISAT raw data.

```
%% Evaluation of SAR System Transfer Function
f_r = ([0:nrg-1])*((fs*2/c)/nrg);
f_r = f_r - (fs*2/c).*(f_r ≥ fs/c);
B_r = (2/c)*Bt;
alpha_r = (4/c^2)*Bt/T;
G1 = exp(complex(0., -(pi/alpha_r).*f_r.^2)).*(abs(f_r) ≤ ...
    B_r/2 );
G1 = repmat(G1, naz, 1);
F1 = conj(G1);

f_x = ([0:naz-1])*(PRF/naz)-f_dc;
f_x = (f_x - (PRF).*(f_x ≥ PRF/2)+f_dc)/v;
phi_dc = asin(lambda*f_dc/(2*v));
B_x = (2/L_A)*cos(phi_dc);

f_r1 = (2/lambda)+f_r;
[f_r1_mesh, f_x_mesh] = meshgrid(f_r1, f_x);
f_dc1_mesh = f_r1_mesh*sin(phi_dc);

K = sqrt(f_r1_mesh.^2 + f_x_mesh.^2)-f_r1_mesh;
G2 = exp(complex(0., ...
    2*pi*r0*K)).*(abs(f_x_mesh-f_dc1_mesh) < B_x/2);
F2 = conj(G2);
F = F1.*F2;
```

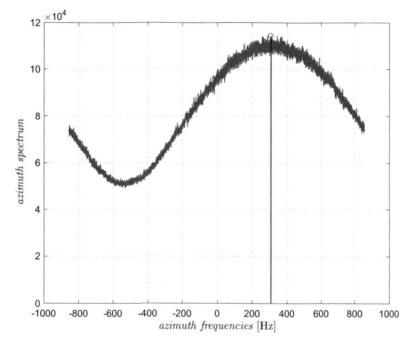

Figure 5.15. Azimuth spectrum of the real ENVISAT raw data. Red (mid gray in print version) line pinpoints the acquisition Doppler centroid.

It is interesting at this point to focus on the effect of the first term, i.e. F_1 that implement the chirp range compression: Figs. 5.16 and 5.17 provides images after range compression of both simulated and real data.

```
%% Range Compression
rawf = fft2(raw);
yc = ifft2(rawf.*F1);

%visualization of range—compressed simulated data
figure('Name', 'Range—compressed data', 'NumberTitle', ...
    'off');
imagesc(rr—r0, xx, abs(yc)), colormap gray
set ( gca, 'xdir', 'reverse' )
xlabel('{\it r—$r_0$} [m]','Interpreter','latex')
ylabel('{\it azimuth} [m]','Interpreter','latex')
axis([—1000 +1000 —4000 2000])

%visualization of range—compressed real data
figure('Name', 'Range—compressed data', 'NumberTitle', ...
    'off');
imagesc(rr, xx, imresize(abs(yc), [naz/6, nrg]), [0., ...
    2.5*mean(abs(yc(:)))]), truesize, colormap gray
```

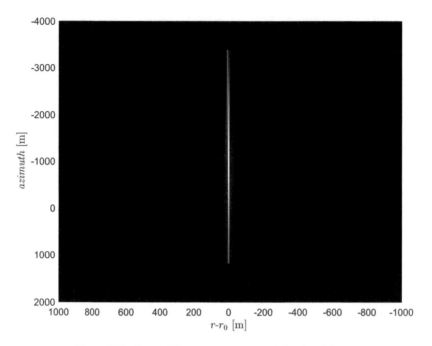

Figure 5.16. Zoom of the range compressed simulated data.

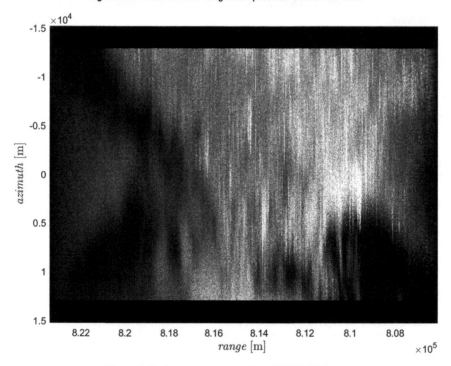

Figure 5.17. Range compressed real ENVISAT data.

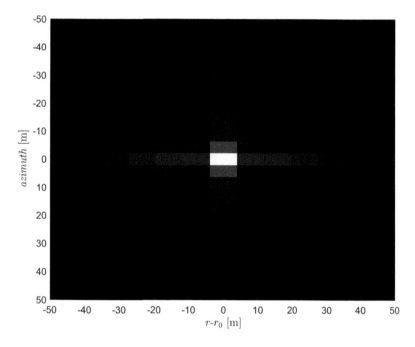

Figure 5.18. Zoom of the 2D focused image achieved by simulated data.

```
set ( gca, 'xdir', 'reverse' )
xlabel('{\it range} [m]','Interpreter','latex')
ylabel('{\it azimuth} [m]','Interpreter','latex')
```

Finally, the implementation of the complete filter $F = F_1 \cdot F_2$ provides the 2D focused image. Final focused images on both simulated and real data are provided in Figs. 5.18 and 5.19

```
yf = ifft2(rawf.*(F));

%visualization of 2D focused simulated data
figure('Name', '2D focused data', 'NumberTitle', 'off');
imagesc(rr-r0, xx, abs(yf)), colormap gray
set ( gca, 'xdir', 'reverse' )
xlabel('{\it r-$r_0$} [m]','Interpreter','latex')
ylabel('{\it azimuth} [m]','Interpreter','latex')
axis([-50 +50 -50 50])

%visualization of 2D focused real data
figure('Name', '2D focused data', 'NumberTitle', 'off');
imagesc(rr, xx, imresize(abs(yf), [naz/6, nrg]), [0, ...
    2.5*mean(abs(yf(:)))]), truesize, colormap gray
set ( gca, 'xdir', 'reverse' )
xlabel('{\it r} [m]','Interpreter','latex')
ylabel('{\it azimuth} [m]','Interpreter','latex')
```

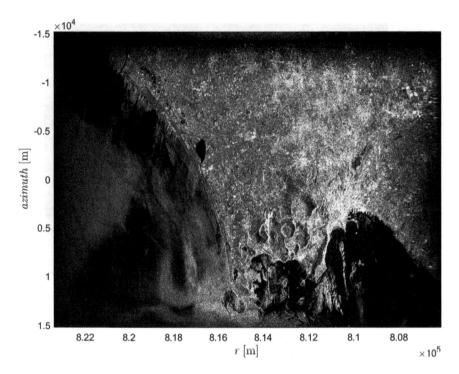

Figure 5.19. 2D focused real ENVISAT data.

In case of simulated data, comparison of the output focused signal associated with the single point target with respect to the expected model is also provided. An oversampling of the image is carried out. In the oversampling operation, azimuth spectral offset associated with the presence of a possible Doppler centroid has to be properly addressed. See Figs. 5.20 and 5.21.

```
%oversampling
ovs = 10;                     %oversampling factor
rr_ovs = r0+([0:ovs*nrg-1]-ovs*round(nrg/2))'*(c/2/fs/ovs);
xx_ovs = v/PRF/ovs*([0:ovs*naz-1]-ovs*round(naz/2))';

%range oversampling
app = fft(yf, [], 2);
app1 = [app(:, 1:floor(nrg/2)-1), zeros(naz, ...
    (ovs-1)*nrg), app(:, floor(nrg/2):end)];
yfor = ifft(app1, [], 2);

%azimuth oversampling
cx = exp(complex(0, -2*pi*[0:naz-1]'*f_dc/PRF));
cx = repmat(cx, 1, ovs*nrg);
app = fft(yfor.*cx, [], 1);
```

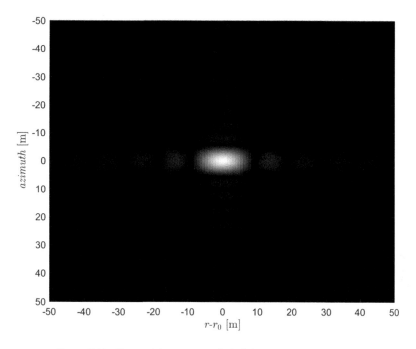

Figure 5.20. Zoom of the oversampled 2D focused simulated data.

```
app1 = [app(1:floor(naz/2)-1, :); zeros((ovs-1)*naz, ...
    ovs*nrg); app(floor(naz/2):end, :)];
yfo = ifft(app1, [], 1);

cxo = exp(complex(0, +2*pi*[0:ovs*naz-1]'*f_dc/PRF/ovs));
cxo = repmat(cxo, 1, ovs*nrg);
yfo = yfo.*cxo;

figure('Name', '2D focused data', 'NumberTitle', 'off');
imagesc(rr_ovs-r0, xx_ovs, abs(yfo)), colormap gray
xlabel('{\it r-$r_0$} [m]','Interpreter','latex')
ylabel('{\it azimuth} [m]','Interpreter','latex')
axis([-50 +50 -50 50])

[iio, jjo] = find(abs(yfo) == max(abs(yfo(:))));
%Range plot
yf_rg_ovs = abs(yfo(iio, :));
yf_rg_ovs = yf_rg_ovs/max(abs(yf_rg_ovs));
yf_rg_sim_ovs = ...
    sinc(2*Bt/c*(rr_ovs-r0))./sinc(2*Bt/c/nrg/ovs*(rr_ovs-r0));
figure('Name', 'range cut', 'NumberTitle', 'off');
plot(rr_ovs-r0, abs(yf_rg_sim_ovs)), grid on
hold on, plot(rr_ovs-r0, yf_rg_ovs)
xlim([-10*c/2/Bt 10*c/2/Bt])
xlabel('{\it r-$r_0$} [m]','Interpreter','latex')
ylabel('{\it normalized amplitude}','Interpreter','latex')
```

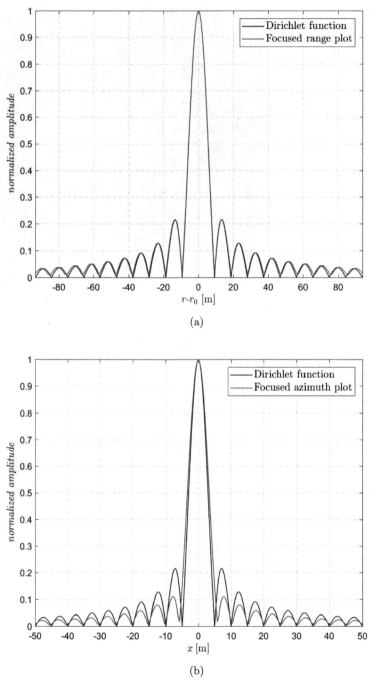

(a)

(b)

Figure 5.21. (a) Range and (b) azimuth oversampled plots corresponding to a single point target compared to the expected Dirichlet functions.

```
leg1=legend('Dirichlet function', 'Focused range plot');
set(leg1, 'Interpreter','latex')

%Azimuth plot
yf_az_ovs = abs(yfo(:, jjo));
yf_az_ovs = yf_az_ovs/max(abs(yf_az_ovs));
yf_az_sim_ovs = ...
    sinc(2/L_A*xx_ovs)./sinc(2/L_A/naz/ovs*xx_ovs);
figure('Name', 'Azimuth cut', 'NumberTitle', 'off');
plot(xx_ovs, abs(yf_az_sim_ovs)), grid on
hold on, plot(xx_ovs, yf_az_ovs)
xlim([-10*L_A/2 10*L_A/2])
xlabel('{\it x} [m]','Interpreter','latex')
ylabel('{\it normalized amplitude}','Interpreter','latex')
leg1=legend('Dirichlet function', 'Focused azimuth plot');
set(leg1, 'Interpreter','latex')
```

6

SAR interferometry

Interferometric SAR (InSAR) is a technique able to provide the 3D localization of the ground targets present in the observed scene. This is achieved by properly combining the data acquired from two different antenna phase centers in the plane orthogonal to the azimuth direction, either by exploiting repeated passes of a single antenna system or by using of another antenna on a single platform. The separated phase centers allow imaging the scene from slightly different incidence angles (angular diversity) and provide a range parallax that can be exploited for the measurement of the height distribution of the ground scatterers over the imaged scene.

By exploiting a specific InSAR configuration characterized by repeated passes with limited, in principle absence of angular diversity, it is also possible to limit (nullify in the ideal case) the topographic contribution and emphasize the phase contribution due to possible surface displacements occurring in the time lag between the two passes. The accuracy on the displacement measurement reaches the wavelength fraction, i.e. centimeter. The technique is known as Differential Interferometry SAR (DInSAR) and has opened many applications of SAR in the area of natural hazards. DInSAR is widely used for the analysis of large deformations such as, for example, those generated by major earthquakes or eruptive activities in volcanic areas. InSAR and DInSAR are discussed in this chapter by addressing their fundamentals from a geometrical point of view, together with the specific noise sources and the steps necessary to extract the useful information.

6.1 InSAR basic principles for topographic applications

Fig. 6.1 shows the acquisition geometry of a SAR system in the plane orthogonal to the flight direction in a fixed azimuth coordinate x. For sake of simplicity, the Earth curvature of the reference surface (f.i. a universal ellipsoid) has been neglected. The point **S** represents the position of the sensor and **P** the position of a scat-

Multi-Dimensional Imaging with Synthetic Aperture Radar
https://doi.org/10.1016/B978-0-12-821655-2.00010-3

terer at range r and height z: $\mathbf{P}_R$ is the position of the reference point, that is the point at the same range located on the reference surface represented, thanks to the curvature absence assumption, by the horizontal dashed line in Fig. 6.1.

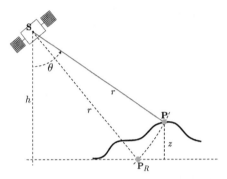

Figure 6.1. Acquisition geometry of a satellite SAR system in the plane orthogonal to the flight direction.

The 3D localization of the scatterer is provided by the knowledge of the three coordinates azimuth x, range r, and look angle θ. However, for each azimuth and range coordinates pair, the acquired 2D SAR image is an estimate of the superposition of the radar reflectivity contributions coming from the illuminated scatterer distribution with the azimuth and resolution cell. Even in the case of infinite resolution, the latter distribution includes also the scatterers located along the equirange arc of circle (dashed arc in Fig. 6.1), typically well approximated by the elevation segment, i.e. the portion of line tangent the arc of circle in the reference point. As a consequence, as also pointed out in Chapter 1, even in the case of an infinite resolution, the mere measurement of the range of a scatterer (**P** in Fig. 6.1) does not allow one to discriminate it from the other scatterers located on the (equirange) arc of circle centered on **S**, for example from that located on the reference surface ($\mathbf{P}_R$).

The measurement of the look angle, and hence the full (in association with the azimuth and range) localization of the scatterers, can be achieved by exploiting an additional SAR image relevant to the same, or better unchanged scene and acquired with a slight different imaging angle. This situation is pictorially shown in Fig. 6.2 where the geometry of an InSAR system is sketched again with reference to the plane orthogonal to the azimuth axis (the flight direction axis).

$\mathbf{S}_1$ and $\mathbf{S}_2$ in Fig. 6.2 are the sensor positions from which the first (master) and second (slave) image are acquired, respectively. We assume for the moment that the images are acquired at the

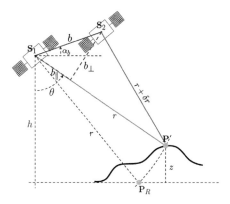

Figure 6.2. Acquisition geometry of a satellite InSAR system in the plane orthogonal to the flight direction.

same time: this assumption will be subsequently relaxed and the problems related to the presence of a time lag between the two acquisitions will be discussed in detail.

The two sensors are assumed to be separated by a baseline vector b, which is determined by the baseline length b (baseline modulus) and the baseline tilt angle α_b evaluated with respect to the direction perpendicular to the nadir of the master sensor. The angular imaging diversity characterizing the two SAR acquisitions leads the scatterer located in P to have two different ranges in the master and slave images, hereafter indicated with r and $r + \delta r$ respectively. As discussed in Chapter 1, the intersection of the two circles in the plane orthogonal to the azimuth and centered on S_1 and S_2 with radii r and $r + \delta r$ provides the localization of the target: the range (travelled path) variation from one image to the other (range parallax δr) is thus the key quantity that allows measuring the look angle θ.

From a mathematical point of view, the relationship between the path difference δr and the look angle θ, for a fixed range r, is achieved by applying the "law of cosines" (Al-Kashi's theorem) [47] to the triangle $S_1 P S_2$:

$$r + \delta r = \sqrt{r^2 + b^2 - 2rb \sin(\theta - \alpha_b)} \qquad (6.1)$$

which, by developing the square root in the hypothesis of $r \gg \delta r$ and $r \gg b$, leads to

$$\delta r \simeq -b \sin(\theta - \alpha_b) = b_\parallel \qquad (6.2)$$

It is worth noting that being, $\theta = \theta(x, r)$, i.e., a function of the azimuth and range related to the topography, δr is in turn a function variable over the image. The explicit dependence on the azimuth and range has been omitted for sake of notation simplicity. Such a simplification will be frequently carried out in the book also for other quantities variable over the image, e.g., $z(x, r)$ to avoid weighting down the notation. In other cases it will be necessary to refer to a single number for the whole scene for quantities variable with x and r: in this case it will be implicitly assumed the reference to a fixed point, generally the scene center.

Eq. (6.2) shows that the traveled path difference (range parallax) is given by the *parallel baseline* component along the (master) line of sight of the illuminated scatterer. The latter in turn depends on the topography z through the look angle $\theta = \theta(x, r)$. Accordingly, by measuring δr and knowing the baseline vector, it is possible to invert (6.2) to evaluate θ.

Finally, the measurement of the height is provided by:

$$z = h - r \cos \theta \qquad (6.3)$$

where h is the height of the master sensor with respect to the reference surface.

Eqs. (6.2) and (6.3) provide the basic (pure geometric) relationships between the topographic path difference δr and z.

6.1.1 SAR stereometry

A possibility for measuring δr is given by radargrammetry, a technique that exploits the backscattering coefficient measured by the image amplitude. Similarly to classical photogrammetry, which exploits optical stereometry to retrieve the 3D shape of object from (at least) two optical images, the amplitude, or intensity information retrieved from the focused SAR images, acquired with angular diversity as in Fig. 6.2, can be exploited to determine the topography of the illuminated scene. More specifically, SAR systems measure the backscattering scene properties as a function of the range. As a consequence of the imaging diversity provided by the stereoscopic view, the backscattering from a given area on the surface will be located, in the two images, at two different ranges. Assuming that the backscattering does not change from one image to the other, the determination of the location of the same area in the two images provides the measurement of the two ranges, whose difference represents an estimate of the range parallax, i.e. δr.

The localization of the same area in the two images can be digitally provided by image correlation techniques. The accuracy in the measurement of the range difference can be in this case a fraction (typically up to 1/16) of the range resolution of the system, generally from one to a few meters. The associated accuracy on the estimation of the topography can be achieved by evaluating the sensitivity of the radargrammetric system.

The sensitivity of a radargrammetric measurement of the range parallax is related to the minimum value of z (say σ_z) that allows one to "appreciate" a variation of δr, say $\sigma_{\delta r}$. More precisely the sensitivity is the ratio between $\sigma_{\delta r}$ and σ_z. A high sensitive interferometric system provides a high variation $\sigma_{\delta r}$ for a low value of σ_z. The sensitivity of SAR stereometric system can be determined by (6.2) and (6.3) as follows:

$$\frac{d\delta r}{dz} = \frac{d\delta r}{d\theta}\frac{d\theta}{dz} = \frac{b_\perp}{r\sin\theta} \tag{6.4}$$

which provides;

$$\sigma_z = \sigma_{\delta r}\left|\frac{d\delta r}{dz}\right|^{-1} \tag{6.5}$$

Eq. (6.4) shows that the sensitivity of the system is related to the orthogonal baseline component. It is in fact this component that regulates the angular diversity of the two "radar sights"; the greater the orthogonal baseline, the greater the diversity of the images and therefore the greater the sensitivity of the stereometric system. It should be however considered that SAR is a coherent system. As discussed later in this section, the fact that SAR provides an estimate of the backscattering in a given pixel by (coherently) integrating the reflectivity over the resolution cell may lead, for large baseline, to a significant change of the value of the backscattering achieved from the two SAR images. For spaceborne systems, operating from hundreds of kilometers (typically around 800 Km) far from the Earth surface, the orthogonal baseline is generally bounded up to hundreds of meters to limit the variation of the backscattering estimations in the two images. Accordingly, the sensitivity of a SAR stereometric system is intrinsically very low, due to the ratio $b_\perp/r$, which is typically of the order of 10^{-3}, or less, for the satellite case. As a consequence of (6.5), the use of a technique, such as radargrammetry (SAR stereometry), able to provide accuracy of fractions of meters on the estimate of δr, has definitely limited applicability.

6.1.2 SAR interferometry

To significantly improve (numerically decrease) the accuracy on the δr measurement, SAR Interferometry exploits precisely the coherent nature of SAR systems, in particular takes benefit of the possibility to access the phase of the focused image.

More specifically, InSAR is the technique that allows the interferometric (phase based) determination of δr starting from two focused SAR images. We refer to (5.64) and suppose, initially, that the bandwidth is large enough to consider the postfocusing PSF in (5.65) as approximating an (ideal) Dirac delta function. In this case we have that the image is a faithful reconstruction of the backscattering with a phase factor proportional to the range. The interferometric phase difference (phase beating) in each pixel of two images acquired under the angular diversity shown in Fig. 6.2 is, in the ideal case of absence of noise, thus proportional to the range parallax δr:

$$\varphi(x,r) = \frac{4\pi}{\lambda}\delta r(x,r) \qquad (6.6)$$

leading to the possibility of measuring the range variation with an accuracy of a fraction of the wavelength λ. By noting that the latter is centimetric at microwaves, it is thus clear the improvement on the accuracy in the measurement of δr achieved by SAR Interferometry with respect to SAR stereometry (radargrammetry). As a matter of fact, SAR Interferometry is a widely used methodology to map the Earth topography on the world-scale, see the Shuttle Radar Topography Mission [48] and Tandem-X mission [49].

We discuss in the following the effect of relaxing the infinite bandwidth assumption. To this end we start from the expressions of the focused master and slave SAR images. We still assume both images acquired simultaneously and expressed with reference to the cylindrical coordinates system relevant to the master sensor. We also assume the absence of any Doppler Centroid difference between the two images. Following (5.64), the master and slave images are given by:

$$\widehat{\gamma}_1(x',r') = \iint dx dr \gamma(x,r) e^{-j\frac{4\pi}{\lambda}r} h(x'-x,r'-r) \qquad (6.7)$$

$$\widehat{\gamma}_2(x',r') = \iint dx dr \gamma(x,r) e^{-j\frac{4\pi}{\lambda}(r+\delta r)} h(x'-x,r'-r-\delta r) \qquad (6.8)$$

respectively, where $h(x'-x,r'-r)$ is the postfocusing PSF in (5.65). In the real case of a finite bandwidth it is evident from (6.7) and (6.8) that the images value in a generic pixel (x',r') are the result of integration of the reflectivity function over a finite (i.e. not infinitesimal) area with (slightly) different complex weighting functions.

More in detail, let $\mathbf{P}(x',r')$ be the generic imaged point on the ground corresponding to the pixel (x',r') in the master image, located at range distance r' from the master antenna and $r'+\delta r'$ from the slave antenna ($\delta r' = \delta r(x=x',r=r')$). The aim of the

InSAR technique is the measurement of the range parallax $\delta r'$ for point $\mathbf{P}(x', r')$ from the two estimates of its radar reflectivity γ provided by the master and slave images in (6.7) and (6.8). It should be noted that the objective of measuring $\delta r'$, i.e. the range parallax of the point $\mathbf{P}(x', r')$ located at the center of the pixel, is a theoretical schematization, strictly valid only if the scattering in the resolution cell is uniform or, conversely, associated with a single (dominant) scatterer at the pixel center. In all the realistic cases, the most "reasonable" aim of SAR Interferometry is to estimate the phase difference between the result of the integrals in (6.7) and (6.8), which involves a weighting of the phase factors by the amplitude of the reflectivity function. In other words, if a single (dominant) scatterer is present (even displaced from the pixel center) within the resolution cell at coordinate (x_0, r_0), i.e. $\gamma(x, r) = \gamma_0 \delta(x - x_0) \delta(r - r_0)$ we have:

$$\widehat{\gamma}_1(x', r') = \gamma_0 e^{-j\frac{4\pi}{\lambda}r_0} h(x' - x_0, r' - r_0) \tag{6.9}$$

$$\widehat{\gamma}_2(x', r') = \gamma_0 e^{-j\frac{4\pi}{\lambda}(r_0 + \delta r(x_0, r_0))} h(x' - x_0, r' - r_0 - \delta r_0) \tag{6.10}$$

so that the phase difference between the master and slave image in (x', r') would be sensitive to the range parallax $\delta r(x_0, r_0)$ of the dominant scatterer (i.e., not of the pixel center).

In real cases, i.e. nonimpulsive $\gamma(x, r)$, due to the angular view parallax the samples of the images in the same pixel (x', r') in (6.7) and (6.8) provide the estimates of the radar reflectivity associated with two different areas on the ground, both centered at the azimuth x' and distant r' from the position of the respective acquiring sensor.

According to the acquisition geometry of the InSAR system, the coordinates of the sample of the slave image corresponding on the ground to the point $\mathbf{P}(x', r')$, imaged by the master at (x', r'), are $(x', r' + \delta r')$. This effect, which is in fact exploited in SAR stereometry to measure the range parallax, is evident by looking at (6.7) and (6.8), and even more at their specification for a single dominant point in (6.9) and (6.10). As a consequence, to extract the phase difference from the same point $\mathbf{P}(x', r')$ on the ground, it is required a remapping of the slave domain mathematically expressed by the following substitution:

$$\widehat{\gamma}_2(x', r') \rightarrow \widehat{\gamma}_S(x', r') = \widehat{\gamma}_2(x', r' + \delta r') \tag{6.11}$$

where the subscript "S" has been exploited to emphasize the "slave" role of the image. This remapping, called *coregistration* and discussed in more details in a subsequent section in this chapter, involves generally a space variant interpolation to bring the slave image into the acquisition geometry of the master image. It is worth underlining that the dependence of $\delta r'$ on x' and r' makes the coregistration procedure a deformation of the slave domain rather than a simple rigid shift. More specifically, the different positions from which the master and slave images are acquired lead to a deformation of the slave domain with respect to the one of the master: the coregistration performs the counter-deformation that compensates the deformation introduced by the system geometry. An apparent nonsense seems at this point to raise up: to estimate $\delta r'$ via SAR Interferometry, one needs to priorly know $\delta r'$ in order to remap the slave image onto the master image. The nonsense is solved by noting that an accuracy of the pixel fraction on $\delta r'$ is sufficient for the registration process to obtain, through the phase difference of the aligned images, a measurement of $\delta r'$ accurate to the level of the fraction of the wavelength. As explained in a following section dedicated to this issue, the remap for registration can be for instance inferred by radargrammetry or by an external (even low resolution) DEM.

After the coregistration step, the two SAR images can be expressed as:

$$\widehat{\gamma}_M(x',r') = \iint dx dr \gamma(x,r) e^{-j\frac{4\pi}{\lambda}r} h(x'-x,r'-r) \qquad (6.12)$$

$$\widehat{\gamma}_S(x',r') = \iint dx dr \gamma(x,r) e^{-j\frac{4\pi}{\lambda}(r+\delta r)} h(x'-x,r'-r+\delta r'-\delta r) \qquad (6.13)$$

where in (6.12) the subscript "M" has been introduced to emphasize the "master" role of the image.

To understand the subsequent steps of the InSAR processing, let us rewrite (6.12) and (6.13) as:

$$\widehat{\gamma}_M(x',r') = A_M(x',r') e^{-j\frac{4\pi}{\lambda}r'} \qquad (6.14)$$

$$\widehat{\gamma}_S(x',r') = A_S(x',r') e^{-j\frac{4\pi}{\lambda}(r'+\delta r')} \qquad (6.15)$$

where:

$$A_M(x',r') = \iint dx dr \gamma(x,r) e^{j\frac{4\pi}{\lambda}(r'-r)} h(x'-x,r'-r) \qquad (6.16)$$

and

$$A_S(x',r') = \iint dx dr \gamma(x,r) e^{j\frac{4\pi}{\lambda}(r'-r+\delta r'-\delta r)} h(x'-x,r'-r+\delta r'-\delta r) \qquad (6.17)$$

We now note that $A_M \simeq A_S$: in the limiting case of $\delta r' = \delta r$, that is a path difference for all scatterers in the resolution cell equal to that associated with the pixel (e.g. pixel center) we have $A_M = A_S$. Eqs. (6.14) and (6.15) show that the term of interest $\delta r'$ is coded within the difference between the phases of the master and slave images, hereafter referred to as *interferometric phase*. However, with respect to (6.6) a disturbance is present due to the difference between A_s and A_M. We have thus:

$$\varphi(x', r') = \frac{4\pi}{\lambda}\delta r' + \varphi_w(x', r') \qquad (6.18)$$

In (6.18), $\varphi_w(x', r')$ plays the role of a noise phase term and is given by the difference between the phase values of $A_M(x', r)$ and $A_S(x', r')$ which, in turn, can be considered as complex multiplicative noise terms affecting the useful signal in the master and slave images, respectively.

Unfortunately, the interferometric phase in (6.18) is not directly accessible and, thus, two additional steps are required to retrieve the path difference $\delta r'$.

The first one is a very simple step consisting in the extraction of the principal argument (denoted by the operator Arg[·]) of the beating of the two images, that is

$$\phi(x', r') = \text{Arg}\left[\widehat{\gamma}_M(x', r')\widehat{\gamma}_S^*(x', r')\right] = \langle\varphi(x', r')\rangle_{2\pi} \qquad (6.19)$$

providing the wrapped value $\langle\cdot\rangle_{2\pi}$ (i.e., restricted to the $(-\pi, \pi]$ interval) of the interferometric (full) phase. The image corresponding to the wrapped phase in (6.19) is usually referred to as *interferogram* and shows sharp variations (fringes) where the full phase in (6.18) is a multiple value of the 2π, see Fig. 6.3.

The second step is the nontrivial *phase unwrapping* (PhU) transformation

$$\varphi(x', r') = \text{Ph}\left[\phi(x', r')\right] \qquad (6.20)$$

denoted by the operator Ph[·] and leading from the wrapped phase (6.19) to the full phase (6.18) (not restricted to the $(-\pi, \pi]$ interval). It is aimed to invert, pixel by pixel, the nonlinear and surjective wrapping operator in (6.19) and usually is a very difficult task. Several PhU algorithms have been proposed in the literature [50–52]. The solution of the problem implies the determination (and compensation) of the 2π jumps present in the measured wrapped phase (difference). More insights on this problem are provided in Section 6.7. However, the higher the rate of the fringes in the interferogram, the more difficult the PhU step. It is, thus, useful evaluating the height variation that generates a topographic phase change of 2π, usually referred to as *ambiguity height*

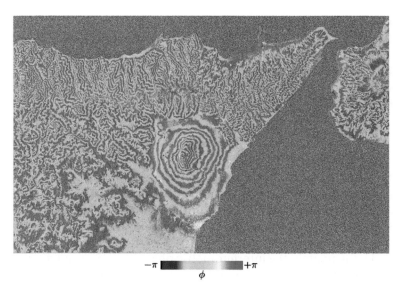

Figure 6.3. Topographic fringes generated from a Sentinel-1 interferometric SAR pair acquired by two close acquisitions on 4th and 10th January 2020 over the area of the Etna volcano, South Italy. The image shows how the interferometric fringes follow the ground topography.

or *phase cycle* [53,54]. By resorting (6.2), (6.3) and (6.20), it results:

$$\frac{d\varphi}{dz} = \frac{4\pi}{\lambda} \frac{d\delta r}{d\theta} \frac{d\theta}{dz} = \frac{4\pi}{\lambda} \frac{b_\perp}{r \sin\theta} \tag{6.21}$$

The ambiguity height $z_{2\pi}$ is thus achieved by solving (6.21) with respect dz and equating $d\varphi$ to 2π:

$$z_{2\pi} = \frac{\lambda r}{2} \left| \frac{\sin\theta}{b_\perp} \right| \tag{6.22}$$

where:

$$b_\perp = b\cos(\theta - \alpha_b) \tag{6.23}$$

is the *orthogonal baseline*, that is the baseline component orthogonal to the (master) line of sight. According to (6.23), for the Sentinel-1 interferogram reported in Fig. 6.3, each fringe corresponds to a height variation of $z_{2\pi} \approx 275$ m, being $b_\perp \approx 52$ m, $\lambda = 0.056$ cm, $\theta \approx 37°$, and $r \approx 853$ km. By counting the number of fringes, the estimated height of the Etna volcano in the center of the image is about 3.300 m.

A high ambiguity height, which leads to a lower fringe rate, is thus desirable for limiting the PhU errors and can be achieved by decreasing the orthogonal baseline. However, a large baseline, that is a numerically low value of $z_{2\pi}$ is desirable to provide a high sensitivity to the topography.

The sensitivity of the interferometric system, which is a simple extension of that evaluated for SAR stereometry, easily follows from (6.4), (6.5) and (6.18):

$$\frac{\sigma_\varphi}{\sigma_z} = \left|\frac{d\varphi}{dz}\right| = \frac{4\pi}{\lambda}\left|\frac{d\delta r}{d\theta}\frac{d\theta}{dz}\right| = \frac{4\pi}{\lambda r}\left|\frac{b_\perp}{\sin\theta}\right| = \frac{2\pi}{z_{2\pi}} \tag{6.24}$$

The target height estimate accuracy is thus:

$$\sigma_z = \sigma_\varphi\left|\frac{d\varphi}{dz}\right|^{-1} = \sigma_\varphi\frac{z_{2\pi}}{2\pi} \tag{6.25}$$

and considering that generally σ_φ is a small fraction of 2π, the accuracy in the estimation of the height by SAR Interferometry is a small fraction of $z_{2\pi}$.

Eq. (6.24) shows the key role played also in the interferometric system by the orthogonal baseline in the estimation of the topography. Similarly to the radargrammetric system, the greater the orthogonal baseline, the greater the sensitivity of the interferometric system (or equivalently the lower the ambiguity height), and therefore the greater (numerically lower) the height estimation accuracy. It should be however pointed out that, as mentioned for the stereometric case, an increase in the orthogonal baseline introduce changes in the backscattering estimates at the two images, which are even more critical for the phase with respect to the amplitude involved in SAR radargrammetry. Essentially an increase of the orthogonal baseline leads to an amplification of the term $\delta r' - \delta r(x, r)$ in the exponential phase factor in (6.17), with (x, r) spanning the resolution cell (support of the h function). Consequently, the phase noise contribution accounted for by φ_w in (6.18), associated with the interferometric beating of A_M and A_S, will be more significant, thus leading to an increase of σ_φ. In other word $\sigma_\varphi = \sigma_\varphi(b_\perp)$ is a growing function of $b_\perp$. This effect is known as geometrical decorrelation and will be addressed in more detail, for a canonical case corresponding to a sloping terrain, in the subsequent sections. The product $\sigma_\varphi z_{2\pi}$ involves therefore a growing function of $b_\perp$ (σ_φ) and a decreasing function of $b_\perp$ ($z_{2\pi}$): once $\sigma_\varphi = \sigma_\varphi(b_\perp)$ is functionally expressed, an optimal orthogonal baseline, i.e. the orthogonal baseline that minimizes σ_z can be defined [54].

6.1.3 Canonical topography case

The interferometric phase φ is linked to the topography of the illuminated scene by means of the topographic path difference δr

as in (6.18): its relationship with the topographic profiles can be thus determined by analyzing the variation of δr with the range for a given value of the azimuth. To this aim, let us consider the slant reference system (r, s) depicted in green (gray in print version) color in Fig. 6.4, which represents the best Cartesian approximation of the polar system describing the radar acquisition geometry in a given plane orthogonal to the azimuth direction: such a reference system has origin on $\mathbf{S}_1$ and slant range axis r directed as the segment $\overline{\mathbf{S}_1 \mathbf{P}_0}$, which crosses the slant height axis s, sometimes also referred to as "elevation", in the reference topographic point $\mathbf{P}_0$ identified by the slant range r_0 and the look angle θ_0. Note that the approximation of the polar reference system with the Cartesian one is valid locally around P_0 and therefore when r_0 is large. It is in fact, the larger r_0, the smaller the curvature of the polar reference system and hence the larger the angular sector in which the approximation can be considered valid. Under this hypothesis, the projection of the point $\mathbf{P}$ on the axis r of the Cartesian reference systems becomes equal to the range of the polar reference system, i.e. the distance of $\mathbf{P}$ from $\mathbf{S}_1$.

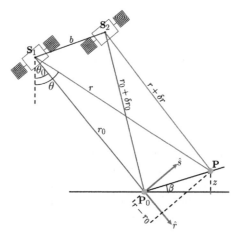

Figure 6.4. Acquisition geometry of a satellite InSAR system in the plane orthogonal to the flight direction for a canonical planar terrain with a slope β.

For now we assume a generic topographic profile described by the function, $s = s(r)$, thus excluding singular cases which will be commented later in the section as a limiting case of a linear topography. The generic topographic point $\mathbf{P}$, seen by the sensor $\mathbf{S}_1$ at the slant range r with the look angle $\theta(r) = \theta_0 + \Delta\theta(r)$, in the slant reference system is referred by the coordinates r and $s(r)$, where:

$$s(r) = r \sin[\Delta\theta(r)] \tag{6.26}$$

The topographic path difference $\delta r(r)$ corresponding to the point **P** in the interferometric system constituted by the pair of sensors $\mathbf{S}_1$ and $\mathbf{S}_2$, which is related to the topography and system parameters by means of (6.2), is thus given by:

$$\begin{aligned}\delta r(r) &\simeq -b\sin[\theta_0 + \Delta\theta(r) - \alpha_b]\\ &= b_\parallel\cos[\Delta\theta(r)] - b_\perp\sin[\Delta\theta(r)]\end{aligned} \tag{6.27}$$

where $b_\perp$ and $b_\parallel$ are the orthogonal and parallel baseline components, respectively, along the line of sight of the reference topographic point $\mathbf{P}_0$. Note that we have neglected the subscript "0" highlighting the dependence of the baseline components on $\mathbf{P}_0$, according to previous considerations clarifying that, when dependence is omitted, the involved quantities refer to the reference point. By comparing (6.26) and (6.27), the path difference can be rewritten as:

$$\delta r(r) \simeq b_\parallel\cos[\Delta\theta(r)] - b_\perp\frac{s(r)}{r} \tag{6.28}$$

which around $r = r_0$ can be approximated as:

$$\delta r(r) \simeq b_\parallel - \frac{b_\perp}{r_0}s(r) \tag{6.29}$$

which can be further developed locally by introducing the terrain slope $\dot{s}(r)$ in the slant Cartesian reference system, with $\dot{s}(r)$ being the first derivative of $s(r)$:

$$\delta r(r) \simeq b_\parallel - \frac{b_\perp}{r_0}\dot{s}(r_0)(r - r_0) \tag{6.30}$$

Eq. (6.30) shows that the range parallax can be locally approximated as a linear function of the range, where the angular coefficient increases with the ratio $b_\perp/r_0$, that is the angular parallax, and with the terrain slope.

Finally, by considering that $b_\parallel = \delta r(r_0)$, (6.29) can be rewritten as:

$$\delta r(r) - \delta r(r_0) \simeq -\frac{b_\perp}{r_0}[s(r) - s(r_0)] \tag{6.31}$$

Eq. (6.31) shows that, by moving from the reference point, the topographic path difference variation, i.e. $\delta r(r) - \delta r(r_0)$ (providing the interferometric phase variation from $\mathbf{P}_0$ to $\mathbf{P}$), is sensitive to the slant height variation through the orthogonal baseline component relevant to the reference point. Accordingly, the larger the orthogonal baseline component, the higher the phase variation and therefore the more sensitive the interferometric phase to the topography. It is worth to notice that, similarly to the visual system, the component of the baseline that induces a variation of the signal depending on the height profile is the one orthogonal to the "view" direction, i.e. to the line of sight. Consequently, when subscript to b is omitted, it is implicitly assumed the reference to the orthogonal baseline component.

A canonical topographic profile, useful to understand some peculiar behaviors of the topographic path difference, is the linear profile with slope β with respect to the ground, see Fig. 6.4. In this case the topographic point **P** with slant range r has slant height given by:

$$s(r) = (r - r_0)\cot(\theta_0 - \beta) \tag{6.32}$$

which leads (6.31) to:

$$\delta r(r) - \delta r(r_0) \simeq -\frac{b}{r_0}(r - r_0)\cot(\theta_0 - \beta) \tag{6.33}$$

Eq. (6.32) shows that the topographic path difference $\delta r(r)$ is a linear function of the range r, whose angular coefficient depends on terrain slope β. Moreover, it shows that even in the case of flat Earth ($\beta = 0$) the path difference changes linearly with the range. The angular coefficient of $\delta r(r)$ increases for β approaching θ_0. The condition $\beta = \theta_0$ (layover limit) provides an infinite angular coefficient: this situation corresponds to a planar terrain whose profile is orthogonal to the l.o.s. direction. The whole topography profile would be in this case imaged at range r_0 and the path difference would be therefore characterized by an infinite derivative in $r = r_0$.

6.2 Differential SAR interferometry

InSAR is strictly associated with the possibility to measure a path difference δr with an accuracy which reaches a small fraction of the transmitted wavelength. For 3D localization, InSAR exploits an acquisition spatial diversity to make the path difference δr sensitive to the height z of the ground target through the perpendicular baseline b. Conversely, Differential SAR Interferometry (DInSAR) exploits the temporal diversity (time separation) of the acquisitions obtained by repeated passes over the same area at different epochs, ideally from exactly the same orbit to zero the spatial diversity. The aim is in this case to make δr sensitive to path variations induced by possible deformation of the ground surface.

Consider the general case of the acquisition geometry depicted in Fig. 6.5 relevant to a repeated pass configuration in which the point $\mathbf{P}(t_1)$ is sensed at the time t_1 by the sensor at position $\mathbf{S}_1$ at a distance r. Assume also that at a subsequent acquisition time $t_2 > t_1$, subject to a ground deformation, the point moves to the new position $\mathbf{P}(t_2)$ and that it is now sensed by the sensor at position $\mathbf{S}_2$ at a different range $r + \delta r$. In this case the path variation δr includes both the target height contribution induced by perpendicular baseline and the relative motion of the observed target with respect to the sensor.

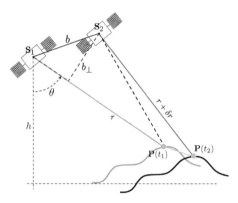

Figure 6.5. Acquisition geometry in a repeat pass configuration.

Specifically, in presence of repeated pass acquisitions, the model for the interferometric phase specializes in:

$$\varphi = \frac{4\pi}{\lambda}\delta r = \frac{4\pi}{\lambda}\delta r_T + \frac{4\pi}{\lambda}\delta r_d + \varphi_a + \varphi_w \qquad (6.34)$$

where δr_T is the contribution to the range variation induced by the spatial baseline related to the target height as in (6.2) and (6.3), δr_d is the component along the radar LOS of the ground displacement. In (6.34) φ_w is, similarly to the InSAR case, the term that account for the phase noise and decorrelation, i.e. the phase variation of the estimated backscattering contribution in the two images, i.e. A_M and A_S in (6.16) and (6.17). It should be however pointed out that for the case under analysis of acquisitions repeated at different epochs, we have to account for also possible changes in the scene backscattering coefficient (function γ in the integrals) between the master and slave image. More precisely assuming that deformations occur on the scale of the resolution cell, (6.16) and (6.17) should be substituted by:

$$A_M(x', r') = \iint dx\,dr\,\gamma(x, r, t_1) e^{j\frac{4\pi}{\lambda}(r'-r)} h(x' - x, r' - r) \qquad (6.35)$$

and

$$A_S(x', r') = \iint dx dr \gamma(x, r, t_2) e^{j\frac{4\pi}{\lambda}(r'-r+\delta r'_T - \delta r_T)} \times$$
$$h(x' - x, r' - r + \delta r' - \delta r) \tag{6.36}$$

Finally, φ_a in (6.34) represents the additional stochastic phase term induced by the variation of wave propagation delay through the atmosphere during the repeated passes which shows spatial correlation lengths generally much larger the resolution cell size. The sensitivity to deformation of Differential Interferometry is very high: LOS surface displacements of half of the wavelength reflect into a phase change of 2π, i.e. an interferometric fringe:

$$\delta r_d = \frac{\lambda}{2} \rightarrow \frac{4\pi}{\lambda} \delta r_d = 2\pi \tag{6.37}$$

The accuracy on the deformation measurement is:

$$\sigma_{\delta r} = \sigma_\varphi \frac{\lambda}{4\pi} \tag{6.38}$$

and therefore not scaled negatively by the (large) ratio between r/b as for the height accuracy, see (6.24) and (6.25).

Other phase sources in (6.34) act as disturbance terms for the determination of the deformation. Specifically, the phase model shows that a single interferogram contains, in origin, both the topographic and deformation components: the first term is negligible only in the case of a spatial baseline close to zero, i.e. when the two data takes are acquired with passes repeated 'almost' over the same orbit, which is however typically not the case of satellite real acquisitions. In practice, this component can be estimated and compensated through the knowledge of an external DEM as well as of precise orbit information and acquisition parameters. Specifically, acquisitions are typically provided with ancillary information reporting orbital state vectors, i.e. vectors of position and velocity that together with their time (epoch) uniquely determine the trajectory of the orbiting SAR satellite in space. By combining the knowledge of the orbits together with acquisition times, i.e. range delay and azimuth Zero Doppler time, and including a prior knowledge of the 3D profile of the imaged area provided by the external DEM, it is possible to estimate on a pixel basis the expected synthetic range r, i.e. the distance of the ground points localized by DEM coordinates and the zero Doppler satellite positions (see Chapter 4). Accordingly, an estimate of the path difference $\widehat{\delta r}_T$ induced by the spatial baseline can be also derived and compensated from the measured interferogram: this operation is referred to as *Zero Baseline Steering* as it is equivalent to "turn" the actual acquisition geometry into a virtual one characterized by a spatial baseline equal to zero. However, the uncertainty in the knowledge of the DEM, principally related to the limited accuracy and spatial resolution of the available DEM, results in a residual topographic contribution $\delta r_{eT} = \delta r_T - \widehat{\delta r}_T$, depending on the amplitude of the DEM error and/or the actual baseline length. This residual contribution could affect the possibility to isolate the deformation component. Accordingly, a first condition to achieve the measurement of the displacement δr_d is that the residual topography δr_{eT}, i.e. the DEM error, must be limited, typically a fraction or at most equal to the ambiguity height. Large deformation components, as f.i. those associated with main earthquakes, typically involve displacements of several tens of centimeters (i.e. several times $\lambda/2$ and therefore leading to phase variations much larger than 2π). These displacements are normally clearly distinguishable from a residual topography components generated from DEMs with an accuracy higher (i.e. numerically lower) than the ambiguity height (the latter leading to a variation of the phase of 2π). Earthquakes (coseismic) displacements maps can be therefore easily measured with the DInSAR technology. In fact, DInSAR has provided to geologists information about displacements associated with most of the largest recent tectonic ruptures with a spatial density (i.e. coverage) that cannot be achieved by any other classical technique (levelling, GPS, etc.). An example of the application of DInSAR for earthquake monitoring is provided in Fig. 6.6.

The effect of the DEM inaccuracy is particularly relevant in urban areas imaged by very high resolution SAR system: available global DEMs typically represent the ground surface and do not reproduce ground structures like buildings. The residual topographic error associated with the difference between the height of the building and the DEM surface can easily compare or even overcome overcome the height of ambiguity, resulting in significant topographic phase texture in differential interferograms.

An example of the effect of the residual topographic phase contribution in differential interferogram is provided in Fig. 6.7. In this figure, the buildings pictured in the optical image of a district of the urban area of Naples, South Italy, are well recognizable as phase fringes in the differential interferogram generated from a pair of COSMO-SkyMed acquisitions characterized by a perpendicular baseline of about $b \simeq 214$ m, corresponding to a height of ambiguity $z_{2\pi} \simeq 30$ m.

In presence of repeated passes, the interferometric phase model in (6.34) includes a further phase term φ_a generated by the different phase delays encountered by the propagation through

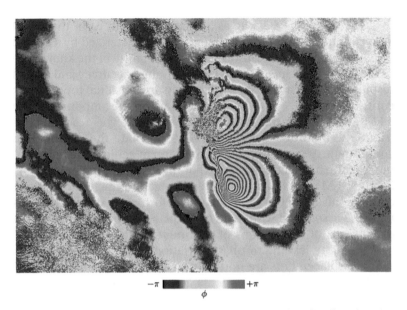

Figure 6.6. Example of the Differential Interferogram of the Bam (Iran) earthquake in 2003 generated from Envisat data acquired on December 03, 2003 and January 07, 2004.

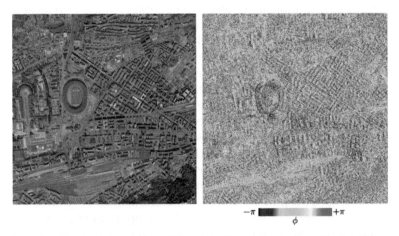

Figure 6.7. Example of the effect of the residual topographic phase in a COSMO-SkyMed differential interferogram in urban area.

the atmospheric layer of the electromagnetic waves and intrinsically related to the fact that acquisitions are carried out at different times. During the acquisition, the propagation velocity of electromagnetic waves through the atmosphere is reduced with respect that in the vacuum according to the refraction index of the layer. As a consequence, by assuming the propagation through the vacuum, an additional phase delay will be present due to the radiation slowering.

The two major layers of the atmosphere (troposphere and ionosphere) are mainly responsible for this delay: specifically, main sources are associated with pressure, temperature and partial pressure of water vapor present in the troposphere and the Total Electron Content (TEC) of the ionosphere. The total zenith delay ZD can be expressed in terms of the refractivity $N = 10^6(n - 1)$ measuring, in parts per million, the excess of the actual refractivity index n of the medium with respect to the vacuum [55]:

$$ZD = 10^{-6} \int_z^H N(t)dt \qquad (6.39)$$

where the integral spans from the height z of the point under investigation, which could be in case of DInSAR the height of the SAR image pixel, and the height of the atmospheric layer H. The difference ΔZD of the zenith delays at the different acquisition epochs impacts the DInSAR phase measurements through the *atmospheric phase delay* (APD) φ_a:

$$\varphi_a = \frac{4\pi}{\lambda \cos\theta} \Delta ZD \qquad (6.40)$$

where θ is as usual the look (off nadir) angle. Differently from the thermal and decorrelation noise, which generally significantly change from pixel to pixel, the APD shows a large spatial correlation and therefore immune to local filtering. APD cannot be thus mitigated as decorrelation noise at the expense of a spatial resolution loss by simple local averaging (multilook), but rather requires the knowledge of the volume distribution of $N(z)$, which is composed by different contributions referred as hydrostatic N_{hydr}, wet N_{wet}, and ionospheric N_{iono} components, depending on physical parameters as temperature T, pressure P, water vapor partial pressure e, electronic density of the ionosphere n_e, liquid cloud water content W_{cloud}, and on the radar frequency f [56]. For C- and X-Band SAR systems, the ionospheric contribution typically shows weak spatial variations and introduces mainly slowly varying phase components which can be easily compensated. Its variation is indeed more significant with L-Band SAR systems.

A first possibility for the compensation of APD is provided by the use of meteorological models aimed at estimating the 3D distribution of temperature, atmospheric pressure and water vapor fraction to compute the expected hydrostatic and wet components of the refractivity. Unfortunately, such numerical models have shown limitations, mainly because of their spatial resolution, which is too poor to adequately model the local tropospheric phenomena on the SAR scale [57]. A further solution for the compensation of APD is provided by the use of tropospheric delays estimated from GPS data: similarly to SAR microwaves, GPS signals are delayed during the propagation in atmosphere as well. Such effect is the main error source in the positioning accuracy of GPS. The processing of GPS data includes the estimation of APD in the zenith direction, which can be mapped into the SAR line-of-sight direction. This solution, however, relies on the availability of a sufficiently dense GPS network [58,59].

Differently from approaches based on the use of external information, APD estimation can be also handled by directly exploiting the characterization of the interferometric phase along the height in the presence of a stratified atmosphere. As a first approximation, the atmospheric phase delay can be directly related to the portion of the atmospheric layer travelled by the radiation, which is in turn related to the topography of the observed scene. A simple linear regression among the interferometric phase and the height is usually effective in reducing the APD disturbance, particularly in presence of large height variations. However, it should be underlined that any signal correlated with the topography would be in this case filtered out, including that possibly associated with the deformation, as for instance the deformation signals correlated with the topography in volcanic areas [60].

The APD contribution is strictly related to the temporal separation of the acquisitions: It vanishes in presence of simultaneous or quasisimultaneous acquisitions. In these configurations, the temporal separation is sufficiently short to allow neglecting changes in the atmospheric layer: electromagnetic waves that contribute to form the interferometric acquisition pair are corrupted by the same APD, which is therefore compensated in the phase difference operation. This configuration is usually exploited for the generation of DEMs. In the last years, the NASA SRTM and the German TanDEM-X mission exploited a bistatic configuration, i.e. a configuration composed by a transmitting/receiving antenna and a second antenna only operating in receiving mode. The resulting interferograms are automatically compensated for the APD and representative of the topographic contribution only.

A last application is the along-track Interferometry (AT-InSAR) in which the two SAR antennas are displaced along the radar flight track: the SAR data relevant to the same target are consequently acquired by the different antennas with a very short time lag (few ms). Accordingly, for each illuminated target, the interferogram is sensitive to the changes (if any) of the sensor-to-target distance occurred during this time lag. AT-InSAR allows monitoring fast-varying phenomena occurring in the observed area, such as those related to currents and waves in marine scenarios [61] or to the vehicle fluxes in marine as well as land scenarios [62].

6.3 Phase statistic

We start first by referring to an InSAR system with simultaneous acquisition: the extension to the DInSAR case, which follows rather straightforwardly, is summarized later in this section. Eqs. (6.14) and (6.15) along with (6.16) and (6.17) express the coregistered master and slave images, which allow achieving the interferometric phase in (6.18), here rewritten for convenience:

$$\varphi(x', r') = \frac{4\pi}{\lambda} \delta r' + \varphi_w(x', r') \qquad (6.41)$$

where, as usual, $\delta r' = \delta r(x', r')$ is the range parallax of the pixel (x', r'). In (6.41), the interferometric phase noise $\varphi_w(x', r')$ is first of all due to the difference between the phase values of the complex multiplicative noises $A_M(x', r')$ in (6.16) and $A_S(x', r')$ in (6.17) produced by the path difference function $\delta r(x, r)$ and, thus, by the angular diversity under which each point on the ground is imaged, intrinsic of the interferometric acquisition geometry. The interferometric phase noise is zeroed when $\angle A_M(x', r') = \angle A_S(x', r')$. Such a condition, can be however verified only in ideal or liming cases for which $\delta r(x, r) = \delta r'$ for each point on the ground involved in the formation of the samples of the (coregistered) images in (x', r'). In fact, this can happen either in the ideal case of a system with an infinite (numerically infinitesimal) range resolution, or in the presence of a specific terrain profile parallel to the range, i.e. with $\beta = \theta_0 - \pi/2$, see (6.32) and (6.31). The latter condition leads however to the limiting case of shadow, for which the scattering by the scene is blinded by the topography. Note that, the case corresponding to an ideal point scatterer for which the complex reflectivity can be described by an impulsive function, leads to a deterministic phase difference which cannot be considered as a noise contribution.

As a consequence, in realistic cases, the phase noise in (6.41) produced by the system angular diversity is an unavoidable contribution to the interferometric phase. Moreover, being associated to a multiplicative contribution, it cannot be reduced by increasing the transmitted power.

Additional phenomena can contribute to the composition of the interferometric phase noise in (6.41). Among them, it is worth mentioning the possible errors in the coregistration step and, in case of repeated pass InSAR system (i.e. DInSAR), the changes in the radar reflectivity of the imaged scene that can occur in the time interval between the two acquisitions: both of them, indeed, contribute to the phases of the multiplicative noise terms $A_M(x', r')$ in (6.16) and $A_S(x', r')$ in (6.17). Furthermore, the unavoidable thermal noise present at receiver stage and superimposed to the acquired signal leads the master image in (6.14) and the slave image in (6.15) to be more realistically expressed as:

$$\widehat{\gamma}_M(x'r') = A_M(x', r')e^{-j\frac{4\pi}{\lambda}r'} + w_M(x', r') \tag{6.42}$$

$$\widehat{\gamma}_S(x', r') = A_S(x', r')e^{-j\frac{4\pi}{\lambda}(r'+\delta r')} + w_S(x', r') \tag{6.43}$$

where $w_M(x', r')$ and $w_S(x', r')$ are complex additive noise terms corresponding to the thermal noise after the focusing and the coregistration step. By omitting the dependence on the coordinates (x', r') for sake of simplicity, the beating between the master and slave image, sometimes also referred to as *complex interferogram*, is:

$$\widehat{\gamma}_M\widehat{\gamma}_S^* = A_M A_S^* e^{j\frac{4\pi}{\lambda}\delta r'} + A_M w_S^* e^{-j\frac{4\pi}{\lambda}r'} + w_M A_S^* e^{j\frac{4\pi}{\lambda}(r'+\delta r')} + w_M w_S^* \tag{6.44}$$

whose principal argument $\phi = \text{Arg}\left[\widehat{\gamma}_M\widehat{\gamma}_S^*\right]$ is an estimate of the wrapped value ϕ_0 of the useful phase term

$$\varphi_0 = \frac{4\pi}{\lambda}\delta r' \tag{6.45}$$

impaired by the presence of both the additive and the multiplicative noise terms. Moreover, by modeling the multiplicative and the additive noise terms as properly distributed random variables, it is possible achieving the following equality

$$\text{Arg}\left[E\left(\widehat{\gamma}_M\widehat{\gamma}_S^*\right)\right] = \langle\varphi_0\rangle_{2\pi} = \phi_0 \tag{6.46}$$

showing that the principal argument of the statistical expectation of the complex interferogram in (6.44) is the noise free wrapped value ϕ_0: in this sense, with a slight abuse of language, we can say that ϕ is an unbiased estimator of ϕ_0.

To demonstrate this result, let us model the radar reflectivity γ associated with the imaged scene as a zero-mean complex circular Gaussian white random process. Furthermore, let us assume the same statistical model also for the two contributions of thermal noise superimposed to the received signals. Finally, let us assume the noise terms to be uncorrelated each other and uncorrelated with the radar reflectivity. Under these hypotheses we have:

$$E\left(\widehat{\gamma}_M\widehat{\gamma}_S^*\right) = E\left(A_M A_S^*\right)e^{j\varphi_0} = E\left(A_M A_S^*\right)e^{j\phi_0} \tag{6.47}$$

where the correlation $E\left(A_M A_S^*\right)$ between the two multiplicative noise terms results to be a real and nonnegative number thus leading to the equality (6.46). The demonstration that $E\left(A_M A_S^*\right)$ is a real number can be found in [53]; intuitively, the fact that the correlation between the A_M and A_S is a real and positive number is the consequence of the assumption that φ_0 in (6.45) is the expected phase of the complex interferogram in (6.47).

The statistically averaged complex interferogram in (6.47) is related to the correlation coefficient between the two samples of the master and slave images [54]:

$$\chi = \frac{E\left(\widehat{\gamma}_M \widehat{\gamma}_S^*\right)}{\sqrt{E\left(|\widehat{\gamma}_M|^2\right)}\sqrt{E\left(|\widehat{\gamma}_S|^2\right)}} = \rho \exp(j\phi_0) \qquad (6.48)$$

where:

$$\rho = \frac{|E\left(\widehat{\gamma}_M \widehat{\gamma}_S^*\right)|}{\sqrt{E\left(|\widehat{\gamma}_M|^2\right)}\sqrt{E\left(|\widehat{\gamma}_S|^2\right)}} = \frac{|\langle \widehat{\gamma}_M, \widehat{\gamma}_S \rangle_E|}{\|\widehat{\gamma}_M\|_E \|\widehat{\gamma}_S\|_E} \qquad (6.49)$$

is the amplitude, usually referred to as *coherence* of the interferogram. In (6.49), $\langle \cdot, \cdot \rangle_E$ and $\|\cdot\|_E$ are the inner product and the norm, respectively, defined on the linear space of random variables; the condition $\rho \in [0, 1]$ is straightforwardly achieved by applying the Schwartz inequality to the ratio in (6.49). The amplitude ρ and the (wrapped) phase ϕ_0 of χ characterize the probability density function (pdf) of the wrapped interferometric phase ϕ. Such a pdf has been derived in [63] and is given by the restriction to the interval $(-\pi, \pi]$ of the 2π periodic function:

$$p_\varphi(x) = \frac{1 - \rho^2}{2\pi} \frac{1}{1 - \rho^2 \cos^2(x - \varphi_0)} \cdot \\ \left\{ 1 + \frac{\rho \cos(x - \varphi_0) \arccos\left[-\rho \cos(x - \varphi_0)\right]}{\left[1 - \rho^2 \cos^2(x - \varphi_0)\right]^{1/2}} \right\} \qquad (6.50)$$

Note that the restriction to the $(-\pi, \pi]$ of the function in (6.50) is symmetric around φ_0 only if the restriction is centered in φ_0. Accordingly, in the general case of φ_0 different from 0 or a 2π (integer) multiple, the restriction of $p_\varphi(\varphi)$ to the $(-\pi, \pi]$ interval is not symmetric with respect to ϕ_0, which therefore cannot be considered the statistical mean of the wrapped phase.

The plot of the pdf of the wrapped phase is shown in Fig. 6.8 for several values of the coherence ρ and for φ_0 equal to a integer multiple of 2π (and, thus, for $\phi_0 = 0$): the interferogram is uniformly distributed for $\rho = 0$ (thus it has no information), becomes more concentrated around its mean ϕ_0 when ρ increases, and degenerates to a Dirac delta function in the limit case of $\rho = 1$. The coherence ρ is, thus, a measure of the interferogram reliability.

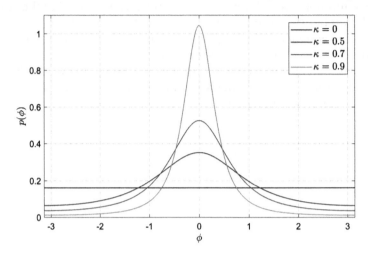

Figure 6.8. Interferogram pdf for different values of ρ.

An increase of the interferometric noise produces a reduction of coherence and, thus, a reduction of correlation between the master and slave images, also referred to as *decorrelation effect*: each of the sources of the interferometric noise among those listed before is responsible of a specific decorrelation effect, which we are going to analyze.

6.4 Decorrelation effects

Let us properly rewrite the terms composing the correlation coefficient (6.48). By resorting to the expressions (6.42) and (6.43) of the master and slave images, respectively, it results:

$$E\left(\widehat{\gamma}_M \widehat{\gamma}_S^*\right) = E\left(A_M A_S^*\right) e^{j\frac{4\pi}{\lambda}\delta r'} \tag{6.51}$$

$$E\left(|\widehat{\gamma}_M|^2\right) = E\left(|A_M|^2\right) + E\left(|w_M|^2\right) = E\left(|A_M|^2\right)\left(1 + \frac{1}{\text{SNR}_M}\right) \tag{6.52}$$

$$E\left(|\widehat{\gamma}_S|^2\right) = E\left(|A_S|^2\right) + E\left(|w_S|^2\right) = E\left(|A_S|^2\right)\left(1 + \frac{1}{\text{SNR}_S}\right) \tag{6.53}$$

where SNR_M (SNR_S) is the signal to thermal noise ratio on the master (slave) image. The correlation coefficient in (6.48) can be thus rewritten as:

$$\chi = \chi_a \chi_n \tag{6.54}$$

where

$$\chi_a = \frac{E\left(A_M^* A_S\right)}{\sqrt{E\left(|A_M|^2\right) E\left(|A_S|^2\right)}} e^{j\frac{4\pi}{\lambda}\delta r'} \tag{6.55}$$

and:

$$\chi_n = \frac{1}{\sqrt{\left(1 + \mathrm{SNR}_M^{-1}\right)\left(1 + \mathrm{SNR}_S^{-1}\right)}} \tag{6.56}$$

are the contributions related to the multiplicative and the additive (thermal) noises, respectively.

The contribution χ_a in (6.55) accounts for several (all but the noise) sources of decorrelation: the *geometrical decorrelation*, produced by the angular diversity and, thus, intrinsic of the InSAR system; the *registration decorrelation*, which reduces the coherence when errors of registration occur; the *Doppler centroid decorrelation*, produced by the presence of two different Doppler centroid in the interferometric pair; the *temporal decorrelation*, experienced in the case of repeated pass InSAR system and accounting for changes in the radar reflectivity of the imaged scene.

The contribution (6.56), describes the *thermal noise decorrelation* and, as expected, is an increasing function of the Signal to Noise Ration (SNR) on both the images. Areas with low backscattering provide lower values of the SNR and therefore lower values of χ_n, i.e. higher decorrelation.

The correlation coefficient χ_a is now further developed. We will initially assume that the acquisitions are simultaneous, i.e. acquired with a single pass or "nearly" single pass (Tandem) InSAR system, thus neglecting the temporal decorrelation contribution. In this case, according to the exploited statistical model, the autocorrelation function of the radar reflectivity is:

$$E[\gamma(x_1, r_1)\gamma^*(x_2, r_2)] = \sigma_\gamma^2 \delta(x_1 - x_2)\delta(r_1 - r_2) \tag{6.57}$$

where σ_γ^2 is the constant power spectral density of $\gamma(x, y)$, and the terms composing χ_a can be evaluated by resorting to the expression (6.16) of A_M and (6.17) of A_S:

$$E(A_M A_S^*) = \sigma_\gamma^2 \iint dx dr h(x' - x, r' - r) \times \\ h(x' - x, r' - r + \delta r' - \delta r)e^{-j\frac{4\pi}{\lambda}(\delta r' - \delta r)} \tag{6.58}$$

$$E(|A_M|^2) = \sigma_\gamma^2 \iint dxdr h^2(x'-x, r'-r) \qquad (6.59)$$

$$E(|A_S|^2) = \sigma_\gamma^2 \iint dxdr h^2(x'-x, r'-r+\delta r'-\delta r) \qquad (6.60)$$

A remark is now in order: in practical cases the variation of the term $\delta r' - \delta r$ is comparable to the transmitted wavelength but negligible with respect to the range resolution associated to the postfocusing PSF h. As a consequence, by accounting for the expression (5.65) of the PSF, (6.58)-(6.60) can be simplified as:

$$E(A_M A_S^*) = \sigma_\gamma^2 B_x^2 B_r^2 \iint dxdr \, \text{sinc}[B_x(x'-x)]\text{sinc}[B_x(x'-x+\varepsilon_x')] \times$$
$$\text{sinc}[B_r(r'-r)]\text{sinc}[B_r(r'-r+\varepsilon_r')] \times$$
$$e^{-j2\pi \Delta f_{DC}(x'-x)} e^{-j\frac{4\pi}{\lambda}(\delta r'-\delta r)}$$

$$(6.61)$$

$$E(|A_M|^2) = E(|A_S|^2) = \sigma_\gamma^2 B_x B_r \qquad (6.62)$$

where the constants ε_x' and ε_r' (with respect to the integration variable) have been now introduced to account for a possible coregistration error (in both azimuth and range directions) and the frequency:

$$\Delta f_{DC} = f_{DC}^S - f_{DC}^M \qquad (6.63)$$

to account for different Doppler centroid of the master (f_{DC}^M) and slave (f_{DC}^S) acquisitions.

The further evaluation of the integral in (6.61) requires the knowledge of the topographic profile coded in the phase term. We consider the canonical linear profile in (6.33), with slope β and referred to the imaged point at coordinates (x', r') on the ground which can be considered locally valid for most of the regular terrain profiles of interest in practical cases:

$$\delta r' - \delta r = -\frac{b}{r'}(r'-r)\cot(\theta-\beta) = -k_\beta(r'-r) \qquad (6.64)$$

where the orthogonal baseline component b and the look angle θ correspond to the line of sight of the ground pixel, and $k_\beta = b\cot(\theta-\beta)/r'$. Note that in principle although referring to the generic pixel center, to lighten the notation we neglected the use of primed coordinates for all quantities, but for the range r', leaving their meaning understood. Substitution of (6.64) in (6.61), along with (6.62), leads to the following factorization of the correlation coefficient:

$$\chi_a = \chi_r \chi_x e^{j\frac{4\pi}{\lambda}\delta r'} \qquad (6.65)$$

where

$$\chi_r = B_r \int dr \, \text{sinc} \, (B_r r) \, \text{sinc} \left[B_r (r + \varepsilon_r') \right] e^{-j 2\pi \tilde{f}_r r} \qquad (6.66)$$

and

$$\chi_x = B_x \int dx \, \text{sinc} \, (B_x x) \, \text{sinc} \left[B_x (x + \varepsilon_x') \right] e^{-j 2\pi \Delta f_{DC} x} \qquad (6.67)$$

are the contribution of χ_a coming from the integration along the range and azimuth directions: hereto, for sake of simplicity, we will refer to them as range and azimuth correlation coefficient, respectively.

The range correlation coefficient (6.66) has been achieved by letting $\tilde{f}_r = 2 k_\beta / \lambda$ and by means of the change of variable $r' - r \to r$. Similarly, in the azimuth correlation coefficient (6.67), the difference Δf_{DC} in (6.63) has been considered and the change of variable $x' - x \to x$ has been performed. It is worth noting that, but for the name of the involved variables, (6.66) and (6.67) are expressed by the same integral. Accordingly, evaluation of one of them, specifically along the range, is now addressed: the result will be *mutatis mutandis* extended to the other case.

Eq. (6.66) is the value in $\tilde{f}_r$ of the Fourier transform of the product of two cardinal sine functions in the spatial (range) domain, which can be rewritten as a convolution in the spectral domain:

$$\chi_r = \frac{1}{B_r} \int \Pi \left(\frac{\xi}{B_r} \right) \Pi \left(\frac{f - \xi}{B_r} \right) e^{j 2\pi \varepsilon_r' \xi} d\xi \qquad (6.68)$$

Finally, by resorting to the factorization property in Appendix A.3, (6.68) can be rewritten as:

$$\chi_r = \Lambda \left(\frac{2 k_\beta}{\lambda B_r} \right) \text{sinc} \left[\varepsilon_r' \left(B_r - \frac{2}{\lambda} |k_\beta| \right) \right] e^{j \frac{4\pi}{\lambda} \varepsilon_r' k_\beta r'} \qquad (6.69)$$

Similarly, the azimuth correlation coefficient in (6.67) is:

$$\chi_x = \Lambda \left(\frac{\Delta f_{DC}}{B_x} \right) \text{sinc} \left[\varepsilon_x' \left(B_x - |\Delta f_{DC}| \right) \right] e^{j 2\pi \varepsilon_x' \Delta f_{DC}} \qquad (6.70)$$

The range correlation coefficient in (6.69) accounts for range misregistration and spatial decorrelation effects; the azimuth correlation coefficient (6.70) accounts for azimuth misregistration and Doppler centroid decorrelation effects: all of them produce a decrease of the coherence (the amplitude of χ_a), whose analysis is now in order.

As for the misregistration decorrelation, by supposing $k_\beta = 0$ and $\Delta f_{DC} = 0$, (6.69) and (6.70) shows that the maximum degree of decorrelation, that is a coherence equal to zero, is achieved when $\varepsilon'_r = 1/B_r$ or $\varepsilon'_x = 1/B_x$, corresponding to a range and azimuth coregistration error equal to the range and azimuth resolutions, respectively.

As for the spatial decorrelation, by supposing $\varepsilon'_r = 0$, (6.69) reduces to:

$$\chi_r = \Lambda \left(b \frac{2\cot(\theta - \beta)}{\lambda r' B_r} \right) e^{j\frac{4\pi}{\lambda}\delta r'} \tag{6.71}$$

showing that high values of orthogonal baseline reduce the coherence of the interferometric phase, up to make definitively $\chi_a = 0$ when b assumes the critical value:

$$b_{\perp c} = \frac{\lambda r' B_r}{2\cot(\theta - \beta)} \tag{6.72}$$

usually referred to as *critical baseline*. In this case, according to Fig. 6.8, the interferogram is uniformly distributed within the interval $(-\pi, \pi]$ and, therefore, the information on the topography carried by the interferometric phase is totally lost in the noise.

Similar consideration holds for the Doppler centroid decorrelation, which leads to a total loss of coherence, that is $\chi_a = 0$, when $|\Delta f_{DC}| = B_x$.

In the considered case of linear topographic profile, the spatial decorrelation has an interesting interpretation in the range frequency domain. To address this aspect, let us rewrite the coregistered master image in (6.12) and slave image in (6.13):

$$\widehat{\gamma}_M(x', r') = \iint dx\, dr\, \gamma(x, r) e^{-j\frac{4\pi}{\lambda}r} h(x' - x, r' - r) \tag{6.73}$$

$$\widehat{\gamma}_S(x', r') = \iint dx\, dr\, \gamma(x, r) e^{-j\frac{4\pi}{\lambda}(r + \delta r)} h(x' - x, r' - r + \delta r' - \delta r) \tag{6.74}$$

By substituting (6.64) in (6.74) and neglecting as usual $\delta r' - \delta r$ within the PSF, the master and slave images can be written as:

$$\widehat{\gamma}_M(x', r') = \iint dx\, dr\, \tilde{\gamma}(x, r) h(x' - x, r' - r) \tag{6.75}$$

$$\widehat{\gamma}_S(x', r') = \iint dx\, dr\, \tilde{\gamma}(x, r) e^{j2\pi \tilde{f}_r r} h(x' - x, r' - r) \tag{6.76}$$

where $\tilde{\gamma}(x, r) = \gamma(x, r) \exp(-4\pi r/\lambda)$ and $\tilde{f}_r = 2k_\beta/\lambda$ is the same spatial frequency as in (6.65); furthermore, an inessential (for this discussion) exponential term outside the integral in (6.76) has been neglected. Eqs. (6.75) and (6.76) clearly shown that, being the PSF the same for both the master and slave images, the portion of spectrum of $\tilde{\gamma}(x, r)$ observed in the slave image is shifted along the range frequency domain of $\tilde{f}_r$ with respect to the one observed in the master image. In Fig. 6.9, this shift is shown for a fixed azimuth frequency (the azimuth frequency dependency is then omitted to simplify the notation): the larger the shift, the smaller the common range bandwidth observed by both the images, the higher the decorrelation. As a consequence, the two images become totally decorrelated when the range spectral shift $\tilde{f}_r$ equals the range bandwidth B_r, so that no common range frequencies are present in the pair of images. Such a condition is verified when:

$$\tilde{f}_r = \frac{2k_\beta}{\lambda} = \frac{2b\cot(\theta - \beta)}{\lambda r'} = B_r \qquad (6.77)$$

which, solved with respect to b, leads to the same expression of the critical baseline in (6.72). As a final remark, it is worth underlining that an increase of coherence can be achieved in the considered case of linear topography by filtering out the frequency outside the common range bandwidth, see again Fig. 6.9: this approach is usually referred to as common band (CB) filtering [64]. In case of nonlinear topography, the CB filtering is no longer effective and a possible solution consists in approximating the topography as being locally linear within the SAR PSF length. With this assumption, a space-variant, locally adaptive, CB filtering can be designed in either the spectral or time domain [65–67].

Similar interpretation can be done for the Doppler centroid decorrelation. As already shown in Chapters 4 and (5), the presence of a Doppler centroid f_{DC} in a SAR image produces a shift of the imaged spectrum around the frequency f_{DC}. Accordingly, the presence of two different Doppler centroid f_{DC}^M and f_{DC}^S produces a partial overlap of the Doppler spectrum of the master and slave images. Such an overlap decreases by increasing the difference $\Delta f_{DC} = f_{DC}^S - f_{DC}^M$, thus leading to a total absence of common Doppler bandwidth when $|\Delta f_{DC}| = B_x$, which is the same condition discussed previously in this section for zeroing the azimuth correlation coefficient in (6.70).

The critical baseline expressed in (6.72) is amenable also of an interesting geometrical interpretation. To this end we refer to Fig. 6.10 where it is shown the reference geometry as usual in the plane range plane, i.e. in the plane orthogonal to the azimuth direction. We have again assumed a sloped terrain (red (mid gray in print version) line) with a slope β. From the figure it is rather evident that, for a given range resolution ρ_r, the extension of the cell

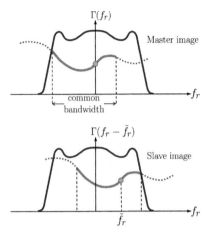

Figure 6.9. Range spectral shift induced by the perpendicular baseline on the imaged reflectivity function. The spectrum of the PSF, which is the same for both images, is in red (mid gray in print version) color. The spectrum of the reflectivity function is in green (gray in print version) color except for part common to the master and slave images, which is in azure color.

along the elevation direction is equal to $\rho_r \cot\theta - \beta$. We consider now the resolution cell as a uniform radiating antenna, which radiates back the illuminating e.m. radiation. Following Appendix A.7, we have that the 3 dB main lobe aperture is approximately given by the ratio $\lambda/(2\rho_r \cot(\theta - \beta))$, where the factor 2 as usual accounts for the doubling of the range for the forth and back path, thus doubling the phase shifts. The critical baseline is therefore just the length of the segment (again orthogonal to the range) illuminated by the main lobe at a distance r, that is maximum separation orthogonal to the range that leaves two antennas within the main-lobe of the radiation emitted by the antenna corresponding to a range resolution. It results therefore that the higher (numerically smaller) the range resolution, the larger the mainlobe and therefore the critical baseline.

Before concluding the discussion on the radar echo decorrelation in interferometric SAR, a remark is dedicated to the discussion on the peculiarities of DInSAR applications, which require the use of a repeated pass interferometric system. In this case the time dependence of the radar reflectivity function has to be accounted for and therefore (6.57) has be rewritten as:

$$E[\gamma(x_1, r_1; t)\gamma^*(x_2, r_2; t + \tau)] = \sigma_\gamma^2(\tau)\delta(x_1 - x_2)\delta(r_1 - r_2) \quad (6.78)$$

where $\sigma_\gamma^2(\tau)$ is the power density of the radar reflectivity $\gamma(x, r; t)$ (assumed to be also time stationary) in the spectral domain cor-

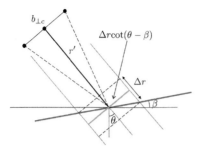

Figure 6.10. Schematic representation of the baseline decorrelation.

responding to the spatial domain (x, r), which depends on the separation τ between the instants of acquisition of the two images: although the modeling of that dependence if very difficult and varying with the characteristics of the imaged environment, it is reasonable assuming that σ_γ^2 is a decreasing function of τ. By accounting for the temporal dependence of γ described above, the correlation coefficient χ_a becomes:

$$\chi_a(\tau) = \frac{\sigma_\gamma^2(\tau)}{\sigma_\gamma^2(0)} \chi_a(0) \tag{6.79}$$

where $\chi_a(0)$ is the correlation coefficient expressed in (6.65)-(6.67) corresponding to simultaneous acquisitions. Eq. (6.79) shows that the temporal separation between the two images produces a reduction of $\chi_a(0)$ according to the scaling factor $\sigma_\gamma^2(\tau)/\sigma_\gamma^2(0)$, which varies with τ inside the interval $[0, 1]$. An example of the impact of the temporal decorrelation is provided in Fig. 6.11 where the interferometric phase and associated coherence for different pairs characterized by increasing temporal separation is shown: the estimation of the coherence from images acquired by operative systems is discussed in the next section. Images are acquired by the COSMO-SkyMed constellation over the Madrid, Spain, urban area and the examples are relevant to a suburban zone including a residential district, road network as well as natural scenarios. By increasing the temporal separation, it can be easily noticed that coherence keeps high values in correspondence for built-up areas while it rapidly degrades over natural surfaces more prone to variation of the vegetation, where the interferometric phase also tends to assume a noisier behavior.

A last remark is now in order. The discussion on the decorrelation effects has been carried out by assuming a scattering mechanism distributed over the imaged scene and described by the radar reflectivity function $\gamma(x, r)$. In case of a scattering mechanism con-

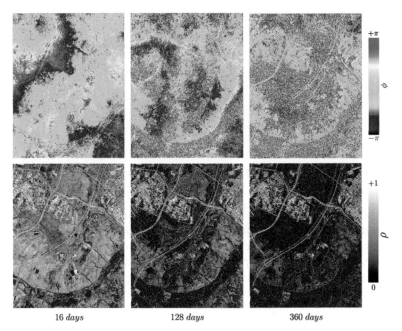

16 *days*   128 *days*   360 *days*

Figure 6.11. Interferometric phase ad coherence for different temporal separation relevant to a suburban area of Madrid, Spain, on COSMO-SkyMed data.

centrated in a given point (x_0, r_0) on the ground, the radar reflectivity function can be expressed as:

$$\gamma(x, r) = \gamma_0 \delta(x - x_0)\delta(r - r_0) \qquad (6.80)$$

where γ_0 is the backscattering coefficient associated with the concentrated scatterer. By substituting (6.80) in the expressions (6.16) and (6.17) of the master and slave images, respectively, it is possible to (easily) analyze the changes in the expression of the different terms contributing to the correlation coefficient. Interestingly, in this case the spatial decorrelation results to be absent, thus making the interferometric phase associated with a concentrated scatterer a very noiseless measure, at least for topographic (single pass) applications.

6.5 Effect of multilook on SAR interferograms

As formerly discussed, interferograms are affected by phase noise contributions, i.e. system noise as thermal noise, geometrical and temporal decorrelation as well as coregistration errors. All these factors impact the possibility to extract reliable information: The presence of noise limits the possibility to achieve a correct phase unwrapping, seriously degrading the final interferometric results. The correlation coefficient in (6.48) between the master and slave samples forming the interferogram and specifically its amplitude, i.e. the coherence, see (6.49), represents a measurement of the phase dispersion due to noise of the interferogram (see Eq. (6.50) for the circular complex Gaussian data case). On the other side, the phase of the correlation coefficient represents the expected, in statistical sense, interferometric phase. The evaluation of the correlation coefficient becomes therefore crucial to achieve, at the same time, an effective estimation of the interferometric phase as well as an indication of the reliability of this estimation. However, to achieve this task, statistical expectations should be carried out: by invoking the ergodicity property, a sufficiently large collection of random samples can be exploited to represent the average statistical properties of the entire process. In practical implementation of interferometry, those samples are usually drawn from a spatial window $\Omega(p)$ representing the neighborhood of the current pixel p (for the sake of simplicity a single spatial index is used) and statistical average of the (generic) random variable x is approximated by the arithmetic (sample) mean:

$$E[x(p)] \rightarrow \frac{1}{L} \sum_{l \in \Omega(p)} x(l) \tag{6.81}$$

with $L = |\Omega(p)|$ being the cardinality of the set $\Omega(p)$. According to (6.81), the correlation coefficient can be operatively estimated as:

$$\widehat{\chi}(p) = \frac{\sum_{l \in \Omega(p)} \left(\widehat{\gamma}_M(l) \widehat{\gamma}_S^*(l) \right)}{\sqrt{\sum_{l \in \Omega(p)} \left(|\widehat{\gamma}_M(l)|^2 \right)} \sqrt{\sum_{l \in \Omega(p)} \left(|\widehat{\gamma}_S(l)|^2 \right)}} = \widehat{\rho} \exp(j\widehat{\phi}_0) \tag{6.82}$$

The estimated wrapped interferometric phase $\widehat{\phi}_0$, which only depends on the numerator in (6.82), is thus obtained from the average of values $\widehat{\gamma}_M(l) \widehat{\gamma}_S^*(l)$ observed in neighboring pixels: This operation is usually referred to as *multilooking*.

Multilook interferogram is therefore opposed to the *single-look* interferograms obtained by the point-by-point complex beating of the interferometric signal pair on a pixel basis. Under the Gaussian assumption, it can be demonstrated that the multilook operation, performed on complex data after the interferometric beating, allows performing a ML estimation of the interferometric phase [68]. Under the same assumptions of Section 6.3, the pdf function in

(6.50) can be extended in the multilook case as [69]:

$$p_\varphi(x) = \frac{\Gamma(L+1/2)(1-\rho^2)^L \rho \cos x - \varphi_0)}{2\sqrt{\pi}\Gamma(L)\left[1-\rho^2\cos^2(x-\varphi_0)\right]^{L+1/2}} \cdot$$
$$\frac{(1-\rho^2)^L}{2\pi}{}_2F_1\left(L,1;\frac{1}{2};\rho^2\cos^2(x-\varphi_0)\right)$$

(6.83)

with $\Gamma(\cdot)$ and ${}_2F_1(\cdot)$ being the gamma and the Gaussian hypergeometric functions, respectively, and L is the number of exploited looks. The behavior of the pdf in (6.83) for different values of L, for a given value of coherence ρ, is provided in Fig. 6.12.

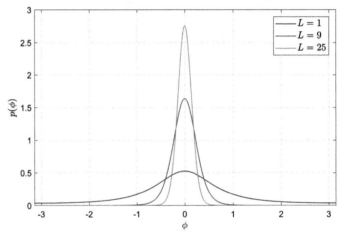

Figure 6.12. Multilook Interferometric phase pdf for different values of L and $\rho = 0.7$.

In its simplest form, the spatial average in (6.82) is implemented through a moving average window in which the window $\Omega(p)$ reduces, generally, to a small rectangular box surrounding the pixel p under investigation. An example of the application of a 5x5 moving average filter to an interferometric pair acquired by the COSMO-SkyMed sensors is provided in Fig. 6.13. The area frames a railway station and railroad tracks: The figure highlights how the spatial averaging allows filtering out the phase noise present in the single-look interferogram and puts in evidence the higher reliability degree of the interferometric phase corresponding to higher values of the estimated coherence. As evidenced by the amplitude image, those large coherence values correspond to man-made features, less affected by temporal decorrelation. Differently, the surrounding vegetated and cultivated field, for which the coherence is much lower, phase noise is evident even after the multilook operation.

single − look interferogram   *amplitude image*

5x5 moving average interferogram   *5x5 moving average coherence*

Figure 6.13. Effect of filtering effect on the interferometric phase induced by a 5x5 moving average window compared to the single-look interferogram.

The assumption behind the use of the moving average window is that the pixels closest to the current pixel are likely to share the same underlying signal: essentially, homogeneous scene reflectivity in the local window is assumed and selected samples follow an independent and identical (statistical) distribution (i.i.d.). However, this basic assumption holds only where the signal shows limited spatial variability, i.e. a low-pass characteristic, and not in the presence of textures, man-made structures or in boundaries between different regions: due to the high-pass content of the complex reflectivity, like the moving average window would produce an unacceptable loss of spatial detail. In such situations, the interferometric phase tends to exhibit nonstationary and non-homogeneous characteristics, conflicting with the i.i.d. assumption. Needless to say, multilooking reduces in any case the effective spatial resolution, but this can be an acceptable loss for many interferometric applications where only signals with small spatial variability are of interest, e.g. spread deformation signal. In this case multilook allows at the same time a significant advantage in terms of computational complexity: after the multilook operation the interferogram can be also downsampled to reduce the size.

In general, to ensure the i.i.d. assumption, filtering should be carried out without mixing samples from areas characterized by different backscattering: Adaptive multilooking procedures are based on a specific setting of the averaging window Ω consisting only of statistically homogeneous pixels (SHP), that is samples supposed to be originated from homogeneous areas from the scattering and topography point of view. Different methods have been proposed to improve the selection of these pixels, the simplest ones are based on directional masks [70] or, in the case of availability of time series of acquisitions, on the similarity of samples amplitude. The latter case frames however in the topic of the next chapter.

Non-Local techniques extend the potentialities of adaptive multilooking. They are based on the evaluation of patch similarity over the interferometric image pair: A generic pixel is assumed to share the same statistical distribution of the current pixel if the patches that surround the two pixels are similar. The Non-Local characteristic comes from the fact that patch similarity search is not restricted in a small search window and therefore pixel values somewhat far apart can be averaged together. The measure of the patch-based similarity allows implementing a weighted combination of those pixel values far apart to carry out a weighted ML estimation of the interferometric phase and of the coherence. Non-Local techniques typically outperform standard moving average multilook in terms of both noise reduction and contour preservation [71,72].

6.6 Coregistration of SAR acquisitions

For practical implementation, a mandatory requirement for the generation of interferogram is that the involved master and slave acquisitions should be precisely overlapped, that means the same sensed object on the ground should be aligned with a subpixel accuracy in the two images. This assumption is, in origin, not satisfied because images are generally acquired from different positions, typically even with nonparallel orbits. The same ground target is therefore commonly imaged at different positions in the two images due to the different imaging geometry. Even more, images can be acquired with different (azimuth-range) sampling grid. Coregistration of the images, mathematically described in (6.11), is therefore always required prior to the pixel-to-pixel complex beating to avoid the coherence losses previously described in Section 6.3.

Although the discussion can be addressed in the realistic case of finite bandwidth, to simplify the description we consider the ideal case of infinite bandwidth for which the PSF becomes an ideal Dirac delta function, see Section 6.1.2. We note that in this case stereometry would already achieve an accurate measurement of the range parallax δr. The hypothesis is however carried out on one hand to avoid the use of primed coordinates, thus lightening the notation, and, on the other hand, to simplify the integral equations in (6.7) and (6.8) as follows:

$$\widehat{\gamma}_1(x,r) = \gamma(x,r)e^{-j\frac{4\pi}{\lambda}r} \tag{6.84}$$

$$\widehat{\gamma}_2(x,r) = \gamma(x-\delta x, r-\delta r)e^{-j\frac{4\pi}{\lambda}r} \tag{6.85}$$

where an additional azimuth misalignment δx has been also considered for the reasons explained above. Eqs. (6.84) and (6.85) show that the same image pixel, located at (x,r), in the master and slave images could sample the ground reflectivity at different locations, i.e. shifted by δx and δr.

As explained in Section 6.1.2, the orbit parallax makes δr to be dependent on the pixel, i.e. azimuth and range. In the general case of repeated passes, especially in presence of crossing orbits causing a relative rotation and scaling of the slave image with respect to the master, it results that both δx and δr are in general dependent on the azimuth and range coordinate, i.e.:

$$\begin{aligned} \delta x &= \delta x(x,r) \\ \delta r &= \delta r(x,r) \end{aligned} \tag{6.86}$$

The dependence of the image shifts on the position within the image does not allow reaching a sufficiently precise alignment through a simple image shift, but rather requires a distortion, sometimes referred to as warping, of the slave image.

The functions in (6.86), which describe for each pixel of the master image the location in the slave image as shown in Fig. 6.14, are commonly referred to as *warp functions*. The warp functions could be not known a-priorly: in this case they must be estimated from the SAR images. Consequently, the image registration procedure can be divided into two main steps: *a)* warp functions evaluation and *b)* slave image resampling by means of proper interpolation.

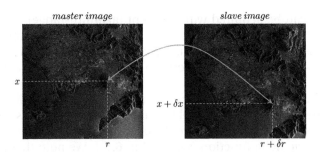

Figure 6.14. Scheme of the image coregistration problem.

One possibility for the estimation of the warp function is based the use of models involving a parametric description. A frequent choice relies on the use of bivariate polynomial transformation of say PQ degree:

$$\delta x(x,r) \simeq \sum_{p=0}^{P-1}\sum_{q=0}^{Q-1} a_{x_{p,q}} x^p r^q$$

$$\delta r(x,r) \simeq \sum_{p=0}^{P-1}\sum_{q=0}^{Q-1} a_{r_{p,q}} x^p r^q$$

(6.87)

The two set of PQ coefficients representing the parameters of the expansion can be estimated by knowing say $K \geqslant PQ$ measurements pairs

$$\delta x_1 = \delta x(x_1, r_1), ..., \delta x_K = \delta x(x_K, r_K)$$

and

$$\delta r_1 = \delta r(x_1, r_1), ..., \delta r_K = \delta r(x_K, r_K)$$

and by solving the following linear problem:

$$\delta x_k \simeq \sum_{p=0}^{P-1}\sum_{q=0}^{Q-1} a_{x_{p,q}} x_k^p r_k^q$$

$$\delta r_k \simeq \sum_{p=0}^{P-1}\sum_{q=0}^{Q-1} a_{r_{p,q}} x_k^p r_k^q$$

(6.88)

$$k = 1, ..., K$$

which can be treated as two separate linear systems of K equations in PQ unknowns, i.e. the coefficients $a_{x_{p,q}}$ and $a_{r_{p,q}}$, that can be solved in the least square sense. In this way, the warp function estimation translates in the problem to achieve the K measurement pairs $(\delta x_k, \delta r_k), k = 1, \ldots K$.

A common solution in image registration is to identify the position in the two images of *ground control points*, that is features on the ground precisely identifiable in both the master and slave images. This choice, widely adopted in the registration of optical images, is however generally impractical in the case of SAR because of the presence of the speckle noise and of the required subpixel registration accuracy. For SAR image registration, the processing strategy is generally based on the identification of suitable *tie points* around which image patches are extracted from the master and slave images. The patches are assumed to be small enough to guarantee that the warp functions in the patch can be assumed to be constant (i.e. they simplify in a relative rigid shift between the patches). At the same time they should be wide enough to allow a robust estimation of the shift by limiting the effect of noise, particularly speckle. The rigid shifts in the tie points are those that maximize the cross-correlation coefficients between the master patch and shifted versions of the slave patch. The patches need to upsampled to allow a subpixel shift estimation. Due to the bandlimited nature of the (complex) SAR images, the oversampling is implemented by means of the classical, well known, Whittaker–Kotelnikov-Shannon theorem based on the use of the cardinal sine function (Dirichlet function in the discrete case). After the patch upsampling two different choices can be considered: amplitude (i.e. incoherent) or complex (i.e. coherent) cross-correlation. Coherent cross-correlation maximization on the full complex (amplitude and phase) data can be more accurate than incoherent (amplitude-only) approach, but in general less robust, especially in presence of patches covering decorrelated areas (e.g., vegetated areas).

An alternative to the parametric warp function estimation is based on a purely geometrical approach [73]. In this case, the knowledge of an external DEM as well as of precise satellite orbit ephemerides, acquisitions timing and sampling allows the geometric determination of the position of a ground target in both the master and the slave images. More in detail, the geometrical approach to the computation of the warp functions is based on determination of the azimuth-range coordinates of a given point on the ground in the SAR geometries in both master and slave images. The procedure relies on the fact that SAR images are processed with respect to a reference Doppler geometry, almost always Zero Doppler geometry, so that points are located in the image at the closest ranges (zero-Doppler positions). Let $\mathbf{P} = (x, y, z)$ be the vector collecting the coordinates of a given point in a Cartesian reference system, which for simplicity here we assume having the x axis coincident with the azimuth of the reference image. The azimuth and range coordinates (x, r) in any image can be determined by the solving for (t', r) the so-called Range-(zero) Doppler equations:

$$\begin{cases} r = |\mathbf{P} - \mathbf{S}(t')| \\ \mathbf{v}(t') \cdot (\mathbf{P} - \mathbf{S}(t')) = 0 \end{cases} \tag{6.89}$$

where $\mathbf{S}(t')$ is the vector function describing the sensor trajectory and $\mathbf{v}(t')$ is the vector describing the satellite velocity.

The Range-Doppler system of equations allows converting the Cartesian coordinates of the points into the SAR image coordinates. This is achieved by determining the distance vector between the ground point and the sensor at zero-Doppler position (zero-Doppler focusing geometry). The first equation is simply the distance of the satellite position from the point. The left hand side of the second equation evaluates instead the scalar product between the velocity vector and the vector connecting the ground point $\mathbf{P}$ and the sensor along the orbit at the generic position $\mathbf{S}(t')$: In a zero-Doppler geometry, these two vectors must be orthogonal.

The system of equation in (6.89) can be used for both the master and slave image to determine the position of the same pixel in the two images as follows. For each pixel of the master image, i.e. $(t'_1 = t, r_1 = r)$, t'_1 allows evaluating the master satellite position $S_1(t'_1)$. Given r_1, the two equations are subsequently solved for (x, y, z) located on the external (rough) DEM, describing $z = z(x, y)$. Once $\mathbf{P}$ is evaluated, the system of equation in (6.89) is again considered, now for the slave image characterized by the trajectory $S_2(t'_2)$. The second equation in (6.89) only depends on the azimuth-time coordinate t'_2 and can be solved iteratively to estimate t'_2, which can be converted to the azimuth pixel, and therefore to the azimuth coordinate, through the Pulse Repetition Frequency (PRF). Once t'_2 has been estimated, the quantity $(\mathbf{P} - \mathbf{S}_2(t'_2))$ identifies the vector connecting the satellite position (at zero-Doppler) and the point on the ground; its modulus provides therefore the range to the slave $r_2 = r + \delta r$ which can be converted into the pixel position in range through the sampling frequency and the (known) time lag corresponding to the delay between the pulse transmission and the sampling of the first echo from the scene. Eq. (6.86) can be finally specialized in:

$$\delta x(x, r) = \delta x_g(x, r) + a_x$$
$$\delta r(x, r) = \delta r_g(x, r) + a_r \tag{6.90}$$

where $\delta x_g(x, r)$ and $\delta r_g(x, r)$ are the geometrically derived warp function and two offsets, a_x and a_r, have been considered to account for possible offset errors in the knowledge of the parameters required for the geometrical determination of the warp functions. These offsets can be evaluated following the same lines of the previously discussed parametric expansion of the warp function, with the advantage that only two parameters are now needed.

The second step of image registration, i.e., the image resampling process conceptually involves a 2D interpolation of the slave image [74] depending on the estimated warp functions: separable interpolation, acting separately on each dimension, can be generally exploited to limit the computational effort.

A last remark is dedicated to the assumption that the DEM must be known in the case of geometrical evaluation of the warp functions. It is clear that the DEM should be known with an error that guarantees the required subpixel accuracy on the determination of δr, which can be shown to be more sensitive than δx on the topography. Generally such a level of accuracy can be reached by using even a rough DEM. To this end we resort the sensitivity of the stereometric system, which has shown that, due to the presence of the ratio r/b, a subpixel accuracy on δr relates to an error on the DEM of several tens/hundreds of meters. Therefore, freely available global DEM, e.g. the NASA SRTM DEM [49], can well describe the geometrical warp function for the orthogonal baselines typically adopted in InSAR/DInSAR. In the case of InSAR for DEM reconstruction, even a rough DEM, or in its absence, the Earth ellipsoid, can provide a preliminary estimation of the geometric warp functions. After registration and the other steps involved in SAR Interferometry, a refined, i.e. more accurate DEM is then provided as a product of the processing chain. Conversely, for DInSAR a DEM with a good accuracy is in any case needed beside the warp function estimation, to limit the effect on the phase signal of the topographic error during the zero-baseline steering step.

6.7 Phase unwrapping

In both the InSAR and DInSAR contexts, the desired physical information of interest, i.e. the topography and/or the displacement of the observed scene, can be directly measured from the interferometric phase through the measurement of the path difference, see (6.33) and (6.34). Unfortunately, being extracted from complex signals, only the principal argument of the phase, i.e. its restriction to the $(-\pi, \pi]$ interval, called *wrapped phase*, can be directly accessed. The wrapping operation, denoted as in (6.19), is schematically depicted in Fig. 6.15 for a one-dimensional case.

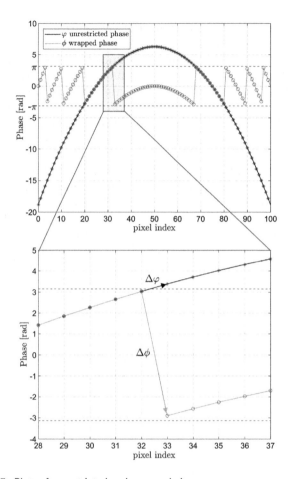

Figure 6.15. Plots of unrestricted and wrapped phase.

Measured interferometric phase observations are therefore related to the unrestricted phase, hereafter referred to sometimes as *"absolute"* or *"true"* phase, by a nonlinear operator that forces the phase signal to belong to the principal interval. However, to access either the topography or displacement information, the unrestricted phase must be known. The Phase Unwrapping (PhU) operation essentially consists of inverting the nonlinear mapping operator, thus recovering the absolute, unrestricted, phase signal from the observed wrapped measurements. The highly nonlinear nature of the problem makes PhU to be one of most critical step within any interferometric processing workflow. PhU is, in fact, frequently the source of errors that randomly bias the retrieved products, that is DEM or deformation signal.

The aim of the phase unwrapping is to recover, for each image pixel, the integer number of cycles k to be added to the wrapped phase ϕ in order to access the unrestricted phase value φ:

$$\varphi(x, r) = \phi(x, r) + 2\pi k(x, r) \tag{6.91}$$

As frequently carried out also in the previous sections, to simplify the notation we have interchanged the dashed and nondashed azimuth and range coordinates. In many of the following parts the explicit dependence on the azimuth and range will be even let be understood. According to the definition of the problem, wrapping the solution $\varphi(x, r)$ must obviously provide the starting signal $\phi(x, r)$.

If no a-priori information about φ is available, or equivalently no constraint on the solution is given, phase unwrapping results to be an ill-posed problem; an infinite number of solutions, once wrapped, are in fact compatible with the input phase signal. The wrapped phase itself, although being unrealistic for describing the topography or deformation of a scene, is a possible solution. A realistic solution among all the possible ones can be found by introducing constraints on the variation of the phase from pixel to pixel. The samples of the true phase pattern are usually modelled as coming from a "continuous" signal, which is sampled at a sufficiently high sampling rate so that the variations of the phase (phase differences) between neighboring pixels are supposed to be limited, namely lower than π. This phase difference constraint is also referred to as the Itoh condition [75].

If the Itoh condition is verified, as discussed in the following the absolute phase can be simply recovered by estimating in each pixel its variation with respect to the adjacent pixel. In fact, according to (6.91), the true phase difference $\Delta\varphi$ between two adjacent pixels, as depicted in the right plot in Fig. 6.15, differs from the difference $\Delta\phi$ of the wrapped phase difference $\Delta\phi$ for integer multiples of 2π: obviously they became equal if their restriction to the $(-\pi, \pi]$ interval is considered:

$$\langle \Delta\varphi \rangle = \langle \Delta\phi + 2\pi \Delta k \rangle = \langle \Delta\phi \rangle \tag{6.92}$$

where $\langle \cdot \rangle$ is the wrapping operator, i.e. the operator which restricts the absolute phase to $(-\pi, \pi]$ (the pixel dependency has been neglected for simplicity). At this point, if the variation of the absolute phase is limited, precisely if $\Delta\varphi \in (-\pi, \pi)$ according to the Itoh condition, then the wrapping operator on the left hand side of (6.92) has no effect on its argument. Under this hypothesis it results that:

$$\Delta\varphi = \langle \Delta\phi \rangle \tag{6.93}$$

To summarize, while (6.92) is always valid, (6.93) holds only when the Itoh condition is verified. In latter case the phase difference of the absolute (wanted) phase at the left hand side of (6.93), can be thus directly retrieved from the wrapped measurements, more specifically by restricting to the $(-\pi, \pi]$ the variation of the wrapped phase, i.e. the right hand side of (6.93). This concept is schematically represented in the right image in Fig. 6.15 obtained by zooming the plot in the left image close to a wrapping jump (from π to $-\pi$) abscissa. The first order variation of the wrapped phase between pixels crossing the jump, $\Delta\phi$, indicated by the green (gray in print version) vector is lower than $-\pi$ (i.e. in modulus larger than π); the corresponding variation of the unrestricted phase $\Delta\varphi$ (black vector) can be achieved by adding 2π to $\Delta\phi$, or equivalently by wrapping $\Delta\phi$ thus explaining the rationale of (6.93).

Although relevant to a one-dimensional case, the concept described in Fig. 6.15 can be extended to the 2D or even multidimensional case. Should the hypothesis that the variation of the phase between adjacent pixel be (in modulus) lower than π in every pixel of an interferogram (Itoh condition valid everywhere), by restricting all the variations of the measured wrapped phase, one would directly get a correct estimation of all the variations of the absolute (i.e. unrestricted) phase. An integration of these phase variations starting from a reference point with known elevation/motion along any path connecting all the pixels of the image, easily solves the PhU problem.

In real SAR interferograms, however, either due to the unavoidable noise or to the presence of fast variations or even discontinuities of the phase signal, for instance induced by the presence of steep topography or discontinuities of the deformation signal, the Itoh condition is frequently violated. Inconsistencies in the estimated absolute phase variations appear in this case and make the retrieved unwrapped phase to be highly dependent on the integration path. This result is a direct consequence of the fact that the estimated phase differences may show nonzero circulation over closed circuits, defined for instance over squares or triangles, thus exhibiting a rotational component.

To better explain this point, let us consider the elementary triangle depicted in Fig. 6.16 formed by the oriented edges connecting three generic pixels (vertices or nodes) labeled as $(1, 2, 3)$. It results that, differently from the circulation of the unrestricted phase gradients, the circulation Q_c of the restriction of the gradients of the wrapped phase may not be always 0 and therefore give rise to

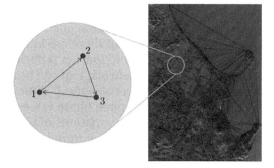

Figure 6.16. Grid of triangles over a sparse map of coherent pixels.

a residue ($Q_c \neq 0$):

$$
\begin{cases}
\Delta\varphi_{21} = \varphi_2 - \varphi_1 \\
\Delta\varphi_{32} = \varphi_3 - \varphi_2 \\
\Delta\varphi_{13} = \varphi_1 - \varphi_3
\end{cases}
\rightarrow \Delta\varphi_{21} + \Delta\varphi_{32} + \Delta\varphi_{13} = 0 \qquad (6.94)
$$

but

$$
\begin{cases}
\Delta\phi_{21} = \langle\phi_2 - \phi_1\rangle \\
\Delta\phi_{32} = \langle\phi_3 - \phi_2\rangle \\
\Delta\phi_{13} = \langle\phi_1 - \phi_3\rangle
\end{cases}
\rightarrow \Delta\phi_{21} + \Delta\phi_{32} + \Delta\phi_{13} = Q_c \in (-3\pi, 3\pi]
$$

$$(6.95)$$

where $Q_c \in (-3\pi, 3\pi]$ results from the fact that Q_c is the sum of three terms belonging to $(-\pi, \pi]$. Moreover, being the restricted variations in (6.95) achieved by the true phase variations in (6.94) by summing 2π (integer) multiples, see (6.91), we have:

$$
Q_c = \Delta\phi_{21} + \Delta\phi_{32} + \Delta\phi_{13} = 2\pi(k_{21} + k_{32} + k_{13}) = 2\pi k_c \qquad (6.96)
$$

so that the circulations can be only multiples of 2π. This condition, together with the previous one $Q_c \in (-3\pi, 3\pi]$, further restricts the residue to be -2π or 2π, i.e., $Q_c \in \{-2\pi; 0; +2\pi\}$ or equivalently $k_c \in \{-1; 0; +1\}$.

   To avoid the propagation of the error, the estimated phase variations associated with violations of the Itoh condition need either to be excluded from the integration path or to be properly corrected. This point is further detailed in the following. To keep the highest level of generality, we suppose to handle the phase unwrapping problem on a sparse grid of say P pixels.[1] On such a grid of P vertices (nodes) we define a network of triangles, generally

---

[1]The case of a full (nonsparse) rectangular grid is a specific configuration under this general assumption.

built up by using a Delaunay triangulation which is characterized by the property to maximize the minimum angle of the triangles, i.e. to avoid "sliver" triangles [76] and therefore long edges more prone to the violation of the Itoh condition. Let Q be the number of resulting edges (branches) of the network, that can be represented as a graph of order P and size Q. Let also the P-dimensional vector φ to collect the (wanted) unwrapped phase values and $\Delta\varphi$ the Q-dimensional vector collecting the variations of the unwrapped phase values over all the edges. We now introduce at this point the *topology matrix*, i.e. the matrix $\mathbf{D}$ containing only the elements $\{-1, 0, 1\}$ describing the oriented network (graph). The generic q-th row of this matrix describes the q-th edge leaving the node l and reaching the node p: such a row has all zero elements but for 1 and -1 for the p-th and l-th columns, respectively.[2]

$$\mathbf{D} = \begin{bmatrix} . & . & . & . & . & . & . & \cdots & . \\ 0 & . & 0 & -1 & 0 & . & 1 & 0 & \cdots & 0 \\ . & . & . & . & . & . & . & . & \cdots & . \end{bmatrix} \tag{6.97}$$

so that

$$\mathbf{D}\varphi = \Delta\varphi \tag{6.98}$$

describes the linear system of equation relating φ and $\Delta\varphi$. The problem in (6.98) is overdetermined because $Q > P$ but has only $P - 1$ independent equations so that there are infinitely many (∞^1) solutions, thus requiring the setting of an initial condition, i.e. the assumption to know one of the P phase values, i.e. one of the elements of φ. The vertex in which the phase is known (usually assumed to be equal to zero) is, as defined also earlier in this section, the reference node: without lack of generality we assume the last node to be the reference node with zero value (*ground* node). Introducing the additional equation associated with the reference node, or equivalently reducing the number of unknowns to $P - 1$, allows the possibility to invert the problem, thus reconstructing φ from $\Delta\varphi$. In the real case we have however only an estimate of $\Delta\varphi$, say $\Delta\hat{\varphi}$ evaluated from the wrapped phase values, i.e., achieved by restricting (wrapping) the variations of the wrapped phase signal ϕ as discussed so far. Assuming the Itoh condition not verified everywhere leads to the presence of inconsistencies in the measured phase variations so that the overdetermined system of equations:

$$\begin{aligned} \mathbf{D}\varphi &= \Delta\hat{\varphi} \\ \varphi_P &= 0 \end{aligned} \tag{6.99}$$

has now generally no solution.

[2] In network theory the topology matrix is defined as the transpose of the matrix defined here.

Having the problem in (6.98) $P - 1$ unknowns (being $\varphi_P = 0$) and $Q > P$ equations one could think to select the $P - 1$ linearly independent equation by eliminating $Q - P + 1$ linearly dependent equations and use the remaining equations in (6.99). The rationale is to eliminate equations that are "likely" to have phase variation estimation errors. The strategy evidently corresponds to choosing among all the integration paths defined on the network, the one that avoid measurement inconsistencies. The problem is how to identify the inconsistencies and therefore the equations to be eliminated. One of the first strategies was based on the so-called *branch cut* phase unwrapping method [51]. This method was based on the observation that each edge violating the Itoh condition generates a pair of $(-2\pi, 2\pi)$ residues in the two (oriented) circuits (circulations) involving it, see Fig. 6.17.

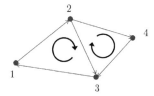

Figure 6.17. Residue pairs generation. We assume that the phase variation estimated from the wrapped data corresponding to the edge connecting the node pair $(2, 3)$ differs from the true phase variation by a 2π multiple. The two clockwise circulation involving this edge sum the wrong phase variation with opposite signs. Considering that residues arising from phase gradients estimated from wrapped data have $k_c \in \{-1; 0; +1\}$ it results that the residues in the two curls are -2π and 2π.

To demonstrate this concept let refer to Fig. 6.17 and assume that all the variations satisfy the Itoh condition, i.e. $|\varphi_p - \varphi_l| < \pi$, but for that relevant to the common edge ($l = 2, p = 3$). We consider $\varphi_3 - \varphi_2 = k\pi + \epsilon$, with $k \in \mathbb{N}^+$ and $\epsilon > 0$: the case $\varphi_3 - \varphi_2 = -k\pi - \epsilon$, with $k \in \mathbb{N}^+$ and $\epsilon > 0$, follows by symmetry. Remembering that $\langle \phi_3 - \phi_2 \rangle = \langle \varphi_3 - \varphi_2 \rangle$, it results $\langle \phi_3 - \phi_2 \rangle = \varphi_3 - \varphi_2 - 2k_{2,3}\pi$, with $k_{2,3} \in \mathbb{N}^+$. Accordingly we obtain the two following left and right circulations residues:

$$\begin{cases} Q_l = \langle \phi_2 - \phi_1 \rangle + \langle \phi_3 - \phi_2 \rangle + \langle \phi_1 - \phi_3 \rangle \\ \quad = \varphi_2 - \varphi_1 + \varphi_3 - \varphi_2 - 2k_{2,3}\pi + \varphi_1 - \varphi_3 = -2k_{2,3}\pi \\ Q_r = \langle \phi_4 - \phi_2 \rangle + \langle \phi_3 - \phi_4 \rangle + \langle \phi_2 - \phi_3 \rangle \\ \quad = \varphi_4 - \varphi_2 + \varphi_3 - \varphi_4 + \varphi_2 - \varphi_3 + 2k_{2,3}\pi = 2k_{2,3}\pi \end{cases}$$

(6.100)

respectively. By observing once more that Q_l and Q_r are by definition limited to the $(-3\pi, 3\pi]$, see (6.95), (6.100) provides $Q_l = -2\pi$, $Q_r = 2\pi$ or $Q_l = 2\pi$, $Q_r = -2\pi$, q.e.d.

Inconsistencies, and corresponding residues, can accumulate over adjacent edges leading also to residues pairs cutting more edges. The name "branch cuts" derives from the fact that the network is "cut" by lines that connect opposite residues: the edges crossed by the cuts must be not included in the integration path in order to avoid the propagation of the error without any attenuation up to the crossing of an edge with an opposite error.

An alternative solution to the phase unwrapping problem is to find the solution of (6.99) in the least square (LS) sense:

$$\boldsymbol{\varphi} = \underset{\varphi_1,\ldots,\varphi_{N-1}}{\operatorname{argmin}} \; \| \mathbf{D}\boldsymbol{\varphi} - \Delta\hat{\boldsymbol{\varphi}} \|_2 \qquad (6.101)$$

where $\| \cdot \|_2$ is the square norm operator. The advantage associated with the LS approach is that the solution can be in principle written in a closed form: the large dimensionality of the problem in the presence of a high number of pixels can however lead to practical difficulties in the implementation of procedures based on direct matrix inversion. For a full rectangular grid of points, the LS solution can be achieved in the spectral domain [51]: in this case, by using a Green formulation to the solution of the problem, it has been demonstrated that the error propagation from inconsistent measurements reduces with the inverse of the distance [51,77] thus achieving an improvement with respect to the error propagation that might be generated from a direct integration of the measurements. The LS solution has the advantage of a closed form for the retrieval of the unwrapped phase, but has the disadvantage to neglect the 2π multiple nature of the error in the phase variation measurements. It also does not relax, differently from the brunch cut method, the assumption on the "continuity", i.e., the presence of jumps larger than π in modulus.

An approach inspired by the brunch cut algorithm, which paved the way for the modern procedure is known as Minimum Discontinuity phase-unwrapping (MD-PhU) [78]. As suggested by the name MD-PhU looks for the solution that shows the minimum discontinuities, i.e.:

$$\boldsymbol{k} = \underset{k_1,\ldots,k_{P-1}}{\operatorname{argmin}} \sum_{q=1}^{Q} w_q |\phi_p + 2\pi k_p - \phi_l - 2\pi k_l| \qquad k_P = 0 \quad (6.102)$$

where (l, p) is the node pair corresponding to the q-th branch, ϕ_l and ϕ_p the associated wrapped (known) phase values, w_q are positive weights corresponding to the q-th arc and $k_P = 0$ as a condition for the reference node. Once $\boldsymbol{k}$ (wrap count vector) is retrieved, the vector $\boldsymbol{\varphi}$ collecting the unwrapped phase values can be simply retrieved by summing the 2π jumps counted in $\boldsymbol{k}$ to the wrapped phase values collected in the vector $\boldsymbol{\phi}$, i.e. $\boldsymbol{\varphi} = \boldsymbol{\phi} + 2\pi\boldsymbol{k}$.

A popular method, somehow similar to the MD-PhU, is referred to as Minimum Cost Flow (MCF)-PhU [50]. It takes inspiration by the problem to find the minimum flux generated from residues, considering positive residue as "sources" and negative residue as "sinks". More precisely the PhU problem is formulated as the problem to find proper corrections to the Q estimated phase gradients such that the corrected field is irrotational over closed circuits:

$$
\begin{cases}
\boldsymbol{k} = \underset{\boldsymbol{k}}{\arg\min} \sum_{q=1}^{Q} w_q |k_q| & \text{subject to} \\
\mathbf{L}(\Delta\hat{\boldsymbol{\varphi}} + 2\pi\boldsymbol{k}) = \mathbf{0} \\
k_1, \ldots.k_Q & \text{integers}
\end{cases}
\tag{6.103}
$$

or equivalently:

$$
\begin{cases}
\boldsymbol{k} = \underset{\boldsymbol{k}}{\arg\min} \sum_{q=1}^{Q} w_q |k_q| & \text{subject to} \\
-\mathbf{L}\Delta\hat{\boldsymbol{\varphi}} = 2\pi\mathbf{L}\boldsymbol{k} \\
k_1, \ldots.k_Q & \text{integers}
\end{cases}
\tag{6.104}
$$

where $w_1, \ldots w_Q$ are a set of proper positive and $\mathbf{L}$ is the matrix with elements in $\{-1, 0, 1\}$ that defines the oriented circulations over the elementary circuits (triangles), thus having only three non-null elements on the generic row. The second formulation clarifies that the vector $\boldsymbol{k}$ correcting the estimated phase variations ($\Delta\hat{\boldsymbol{\varphi}}$) must compensate for the measured residues. Very efficient solvers framed in the context of Linear Programming, specifically the MCF tools, can be used to solve the problem in (6.103) or (6.104). Once $\boldsymbol{k}$ is found, the corrected phase variation $\Delta\boldsymbol{\varphi} = \Delta\hat{\boldsymbol{\varphi}} + 2\pi\boldsymbol{k}$ can be integrated over any integration path reaching all the nodes, thank to his irrotational property.

Thanks to the use of a (weighted) L^1 norm [79], the solution provided by (6.103) or (6.104) archives less dispersed, i.e. more compressed corrections, i.e. a vector $\boldsymbol{k}$ with "less" nonzero values. As in the previous Minimum Discontinuity PhU solution, a key role in the PhU process is played by the weights. A possibility is to set them according to the length of the edge: the choice to set weights inversely proportional to the edge length, allows longer edges to be more likely accommodating 2π multiple corrections. In the case of availability of multiple interferograms, that is the subject of the next chapter, selecting a common sparse grid for the PhU of all the interferograms can allow the determination of the weights of each arc based on multiple measurements.

In Fig. 6.18 the unwrapped phase subsequent to the phase unwrapping operation, achieved by implementing the MCF procedure in (6.104) on the Sentinel-1 interferogram provided in Fig. 6.3 is reported. A second colorbar representing the approximated estimated height, evaluated by inverting the relationship between the unwrapped phase and the corresponding height through the coefficient $4\pi b/(\lambda r \sin\theta)$ in (6.21) evaluated at scene center, is also provided. Obviously it should be remarked that, being the acquisitions characterized by time differences of the order of several days, the estimated heights are biased by the atmospheric contribution.

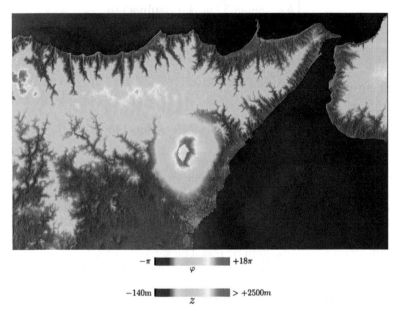

Figure 6.18. Unwrapped interferometric phase for the Sentinel-1 pair acquired on 4th and 10th January 2020 over the area of the Etna volcano, South Italy. The z colorbar approximately expresses the estimated height (but for the atmospheric propagation delay phase bias) associated with the unwrapped phase.

A last remark is now dedicated to generalization of the PhU formulation in (6.104). To this end we refer again to the problems in (6.98) and (6.99). It is evident that, by construction, the true phase variation vector $\Delta\boldsymbol{\varphi}$ in (6.98) belongs to the *range* of the matrix $\mathbf{D}$, $\mathcal{R}(\mathbf{D})$, i.e. to the P-dimensional subspace of $\mathbb{R}^Q$ spanned by the $P < Q$ columns of $\mathbf{D}$. The effect of errors, i.e. inconsistencies in the estimation of the phase variations is to bring $\Delta\hat{\boldsymbol{\varphi}}$ outside $\mathcal{R}(\mathbf{D})$, or equivalently to have a component in the *left nullspace* of $\mathbf{D}$, $\mathcal{N}(\mathbf{D}^T)$. If we indicate with $\mathbf{Z}$ the $Q \times (Q - P)$ matrix whose columns form a basis of the $(Q - P)$-dimensional subspace $\mathcal{N}(\mathbf{D}^T)$, the phase unwrapping problem in (6.98) and (6.99) can be formulated as the problem to find the $2\pi k$ correction vector that brings $\varphi = \hat{\varphi} + 2\pi k$ back into $\mathcal{R}(\mathbf{D})$, that is to show a null component in $\mathcal{N}(\mathbf{D}^T)$. In formulas [80]:

$$\begin{cases} k = \underset{k}{\mathrm{argmin}} \sum_{q=1}^{Q} w_q |k_q| \quad \text{subject to} \\ \mathbf{Z}^T (\Delta\hat{\varphi} + 2\pi k) = \mathbf{0} \\ k_1, ...k_Q \quad \text{integers} \end{cases} \tag{6.105}$$

The basis of the left null space of $\mathbf{D}$ can be easily found by applying the Singular Value Decomposition (SVD) tool [80]. The advantage of the formulation in (6.105) is related to the fact that the matrix $\mathbf{D}$ can describe any network, not necessarily defined by triangulation, and even by increasing the number of edges ($Q \gg P$). The disadvantage is the computational time increase to find the solution due to the dimensionality increase, but also to the fact that, although based on Linear Programming, the solution cannot be achieved by very efficient MCF tools that are strictly associated with the choice to define the network with triangulation tools.

7

Multitemporal SAR interferometry

The previous section highlighted the information content of a SAR interferogram. Assuming parallel or almost parallel tracks, the presence of an orbit offset, i.e. of a nonnegligible orthogonal baseline (spatial diversity), makes the interference signal to be characterized by a phase component associated with the difference of the propagation path (range parallax), which is dependent on the offset of the target in the elevation direction, i.e. the direction orthogonal to the azimuth and range. This feature gives the 3D sensitivity to interferometric SAR systems and allows the determination of the Digital Elevation Model (DEM) of the observed scene. Moreover, if the acquisition is carried out at different times (temporal diversity) and the target on the ground is moving, a further signal contribution is added to the phase component, i.e. a factor proportional to the component of the displacement along the l.o.s. In a single interferogram, i.e. generated by a single acquisition pair, the separation between the two components cannot be easily carried out unless one factor is negligible with respect to the other. In Differential SAR Interferometry (DInSAR) the so-called "zero baseline steering" (ZBS) is carried out starting from an available DEM to isolate the deformation component in the interferometric phase. Nevertheless the DEM might be not accurate enough: available DEM are generally unable to describe the building topography in urban areas. Moreover, repeat pass interferograms are affected by another important error source, whose compensation is nontrivial, that is the atmospheric propagation delay (APD) component, commonly also referred to as Atmospheric Phase Screen (APS). APS is characterized by spatial correlation properties on a scale of the order of a hundred of meters: it shows patterns that can be typically confused with the useful deformation terms, thus limiting the application of two-pass DInSAR only to measure large magnitude displacements, such as those by major earthquakes. Differently from the decorrelation noise, which is spatially uncorrelated, APD cannot be mitigated simply by operating local spatial

Multi-Dimensional Imaging with Synthetic Aperture Radar
https://doi.org/10.1016/B978-0-12-821655-2.00011-5

averaging, i.e. multilook. However, APD generally changes significantly over the time and therefore the only possibility to distinguish it from the useful signal component is to analyze the phase signal over different epochs corresponding to a sequence of images acquired at different times. This task is the subject of the so-called Advanced Differential Interferometry SAR (A-DInSAR) methods, also known as Multi-Temporal Differential Interferometry SAR (MT-DInSAR), which are of interest of this chapter.

7.1 Signal phase model over multiple interferograms

Satellites are able to regularly repeat orbits over the time for most of the areas of the Earth surface, thus providing stacks of multipass interferometric acquisitions characterized by angular and temporal diversity worldwide. Such a diversity can be exploited to improve, with respect to the standard single pair Interferometry, the separation of the signal components of interest, i.e. topographic and deformation ones, as well as for reducing at the same time the inaccuracies due to the APD, thus accessing also weak deformation signals as those generated, for instance, by slow landslides.

The Advanced DInSAR (A-DInSAR) techniques, also referred to as Multitemporal DInSAR (MT-DInSAR) techniques, process data acquired over repeated passes in order to generate very accurate deformation measurements and obtain a regular monitoring of the observed scene surface displacements over the time. A-DInSAR in fact allows mitigating the atmospheric phase delay (APD), which limits the applicability of the classical single pair DInSAR. Single pair Interferometry especially fails in the applications for the identification of small, albeit persistent, deformation phenomena, such as those associated with slow landslides [55].

7.1.1 Generalization of the two-antenna interferometry model

We start by reformulating the geometry described in the previous chapter to perform a first step toward the generalization of the interferometric phase signal modeling. We make reference to the case of absence of deformation, so that the interferometric system is sensitive only to the scatterers height. The discussion can be easily extended to include possible deformations occurring over the time. To this end, let us refer to Fig. 7.1 where it is shown the geometry in the plane orthogonal to the azimuth for an interferometric system with two illuminating antennas located in $\mathbf{S}_1$ and $\mathbf{S}_2$.

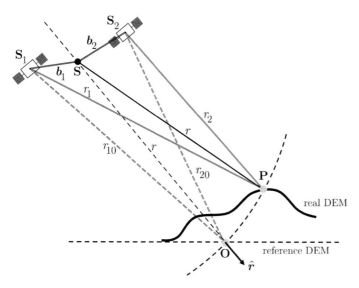

Figure 7.1. Reference geometry for the two antenna case. The geometry is drawn in the plane orthogonal to the azimuth. The reference system, which includes the origin for the range, is located in the point **S** around the mid-point between the two antennas. The point **P** represents the position of the illuminated scatterer on the surface, whereas the point **O** is the reference point, at the same range of **P**, located on the reference surface, which is generally associated to the known external DEM. In the example the reference surface is the flat Earth.

According to what stated in the previous chapters, the reference system should be cylindrical and with the origin located in the position of one of the two antennas acquiring the interferometric images (referred to as master image). In order to generalize to the case of multiple antennas, we move the origin in a point **S** located around the midpoint between the two antennas. We also assume **P** to be the point scatterer under investigation located on the real surface at range r measured from the reference system origin (i.e. **S**).

The (unrestricted) phase φ_n corresponding to the n-th acquisition is, according to (6.14), given by:

$$\varphi_n = \underline{/\gamma_n} - \frac{4\pi}{\lambda} r_n \qquad n = 1, 2 \tag{7.1}$$

where r_n is the distance between the antenna $\mathbf{S}_n$ and the point **P**, γ_n is the backscattering of **P**, and $\underline{/\cdot}$ is the operator that extract the phase of a complex number. Note that, as usually done also in the previous chapter, to lighten the notation here we do not distinguish primed from nonprimed range coordinates.

In multibaseline SAR processing, including the tomographic SAR processing addressed in the next chapter, the phase values are "referred" to the phase resulting from the knowledge of a reference DEM. The latter, typically termed "external DEM", can be even the flat Earth (ellipsoid) in the case of the absence of any information. Without loss of generality, we refer precisely to the flat Earth. Accordingly, we subtract from the absolute phase signal the phase corresponding to the point **O** located on the flat Earth (reference DEM) at the same range of **P**, see Fig. 7.1. This operation in the multiacquisition or tomographic processing is commonly referred to as deramping.

Letting r_{n0} be the distance of the generic antenna $\mathbf{S}_n$ to the point **O**, we have:

$$\varphi_n = \underline{/\gamma_n} + \frac{4\pi}{\lambda}(r_n - r_{n0}) \tag{7.2}$$

where, with a slight abuse of notation, the same symbol φ_n has been adopted in (7.1) and (7.2) to denote the absolute phase before and after the deramping, respectively. Eq. (7.2) can be rewritten as:

$$\varphi_n = \underline{/\gamma_n} + \frac{4\pi}{\lambda}[(r_n - r) - (r_{n0} - r)] \tag{7.3}$$

Referring to Fig. 7.2, which highlights the geometry relevant to the first antenna, it is clear that the term in the square brackets of (7.3)

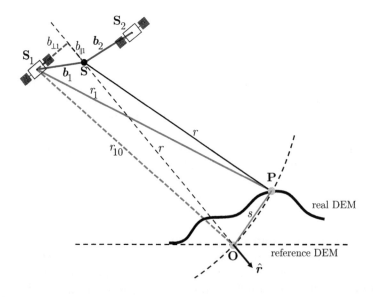

Figure 7.2. Highlight on the first antenna in Fig. 7.1.

represents the variation from the point **O** to **P** of the path difference measured by an ideal (i.e., noise-free) interferometric system with a slave antenna $\mathbf{S}_1$ and a virtual master antenna located in **S**. According to what discussed in the previous chapter, and specifically to (6.31), the latter variation is approximately proportional (in the far field region where the cylindrical reference system can be locally rectified) to the separation (denoted by s) between **O** and **S** in the slant height (or elevation) direction orthogonal to the slant range ($\hat{r}$ in Fig. 7.2). Such a phase variation is thus the topographic phase variation induced by the presence of a difference between the real DEM and the reference DEM on the ground. Similar considerations can be carried out for the second antenna. We therefore have:

$$\varphi_n = \underline{/\gamma_n} + \frac{4\pi}{\lambda}\delta r_{T_n} + \varphi_{w_n} \quad i = 1, 2 \tag{7.4}$$

where the term φ_{w_n} has been introduced to account for the unavoidable additive noise w at the receiver and:

$$\delta r_{T_n} = -\frac{b_n}{r}s \quad n = 1, 2 \tag{7.5}$$

where b_n is the orthogonal baseline length at the acquisition n, see (6.31). Finally, the elevation s can be easily related to the vertical topographic variation between **O** and **P**. To this end it is sufficient to refer to Fig. 7.3 to explain the following slant to vertical height conversion:

$$s = \frac{\Delta z}{\sin\alpha} \tag{7.6}$$

where, to generalize, we relaxed the reference to the flat Earth so that the height is now intended as a height error (Δz) with respect to the available external DEM. According to this relaxation, we have also considered the incidence angle with respect to the vertical direction in place of the look angle to account for the Earth curvature, see Section 4.1 thus considering the vertical direction in 7.3 not necessarily parallel the nadir direction, see Fig. 4.2.

A comment is now in order: The above formulation could be directly related to what presented in the previous chapter by assuming the origin of the reference system **S** on one of the two antennas $\mathbf{S}_1$ or $\mathbf{S}_2$. With this assumption the baseline, and hence the topographic phase contribution, corresponding to the reference antenna would be zeroed. Conversely, the topographic contribution to the phase of the other antenna would be associated with the total baseline vector between the two antennas, i.e. the vector $\mathbf{S}_1\mathbf{S}_2$.

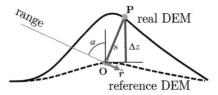

Figure 7.3. Slant to vertical height conversion. The slant height, measured for the two points **O** and **P** with the same range and belonging to the reference (external) DEM and true DEM, is converted to the vertical separation. α is the incidence angle, i.e. the (local) angle between the vertical direction and the direction of the incoming illumination (-$\hat{r}$).

7.1.2 Multiacquisition SAR interferometry model

The formulation in (7.4) for a single acquisition pair can be easily extended to the more general case of N antennas, shown in Fig. 7.4, acquiring at times t_n, $i \in \{1, \ldots, N\}$ with spatial baselines b_n, $i \in \{1, \ldots, N\}$ defined with respect to a master acquisition. Following the notation simplification adopted in Chapter 6, the orthogonal baselines, which provide the angular diversity and hence the topography sensitivity as explained in Chapter 6, have been indicated by suppressing the $\perp$ subscript.

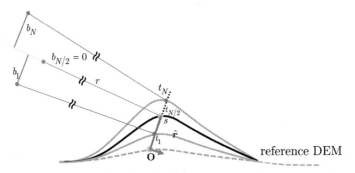

Figure 7.4. Reference geometry for the multiacquisition phase model in the presence of multiple baselines and times. The dashed line represents the reference surface, s is the residual topography. An inflation, exaggerated for drawing reasons, of the terrain is schematically represented over the N acquisition.

The master image is generally selected as the geometric center among the acquisitions once they are represented in the acquisition domain, that is a domain where all the images are represented in the plane associated with the temporal and spatial (orthogonal) baselines, see Fig. 7.5.[1] The spatial baselines are in turn defined as

[1] In this chapter we assume for simplicity that all images are acquired with the same Doppler Centroid. The presence of significant variations of the Doppler

the offsets of the sensor positions in the direction orthogonal to the LOS corresponding to the master acquisition, generally at the scene center, see Fig. 7.4. Operationally, a generic image is chosen as master; the spatial (orthogonal to the LOS) and temporal baselines are then calculated. The final master image is selected as the one closest to the geometric center: this choice allows minimizing, on average, both the temporal and spatial separation with respect to the master image. Without loss of generality we assume the acquisition $n = N/2$ to be the master image, also referred to as *supermaster* for a reason that will be clear afterword, see Fig. 7.4.

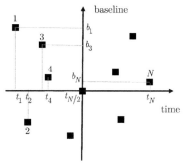

Figure 7.5. Representation in the baseline domain of N acquisitions. The reference, zero baseline, acquisition is the one with index $N/2$.

The reference DEM in Fig. 7.4 is represented by the dashed line. An inflation from time t_1 to time t_N is schematically represented. While the interferometric phase is sensitive to the variation of the residual topography of the order of a few meters (depending on the spatial baseline), deformations occurring on the scale of a fraction of the wavelength (i.e., cm) can be detected by multipass DInSAR, see also Chapter 6 for the two-pass DInSAR. The inflation occurring on such scale in Fig. 7.4 has been magnified and made comparable to the topography variations only for visualization purposes.

Extending the model in (7.4), the phase signal at the generic n-th acquisition for the case of N antennas can be therefore written as:

$$\varphi_n = \underline{/\gamma_n} + \frac{4\pi}{\lambda}\delta r_n + \varphi_{w_n} \quad n = 1, \ldots, N \tag{7.7}$$

where δr_n is the total path difference associated to the slant height (elevation) and the possible deformation of the analyzed scatterer; φ_{w_n} provide the contribution to the phase of the receiver additive

Centroid associated to different azimuth beam pointings, would require the extension of the baseline domain from 2D to 3D, to account also for the azimuth baseline offset corresponding to the Centroid, or equivalently to the beam rotation.

noise, which generally impacts more areas with low backscattering.

Some remarks are in order. Similar to the two antenna case we have assumed that an external DEM has been used to implement the deramping operation, see (7.3). This operation, which eliminates from the phase the topographic contribution associated with the external DEM, has been referred to as Zero Baseline Steering (ZBS) in the previous chapter. In the case of perfect knowledge of the external DEM, it allows producing a phase stack that would be equivalent to the one collected over repeated passes with zero orthogonal baselines. In general, depending on the accuracy of the external DEM, ZBS emphasizes the contribution of possible deformation, particularly on small baseline interferograms. Even in the case where the "simple" flat Earth is used as external DEM, the deramping contributes to eliminate the fringe contribution associated with the variation of the look angle from the near to the far range that, especially for large baselines, induces generally high fringe rates. The success of topographic cancellation depends on the orthogonal baseline span magnitude associated with the acquisitions, as well as on the accuracy of the external DEM. Generally, especially for the application of high-frequency (short wavelength) SAR Interferometry (e.g. at X-Band) to urban areas, topographic variations corresponding f.i. to buildings are typically present and not mapped in the external DEM. Accordingly the contribution associated with the residual topography s (see for instance Fig. 7.1 for the two antenna case) must be always considered in the interferogram signal phase model. An example of the presence of residual topography contribution in interferograms corresponding to buildings is provided in Fig. 6.7. As shown in the following, the presence of the residual topography in the phase model will allow, when the baseline span is large enough, gathering the information for the accurate localization of the monitored scatterers on the ground especially in urban areas. Additionally, as discussed in the previous chapter, when the images are acquired at different times the phase model should include also the presence of a pattern related to the propagation delay through the atmosphere. According to what has been discussed so far, the multibaseline interferometric phase model in (7.7) can be better specified as:

$$\varphi_n = \varphi_{\gamma_n} + \frac{4\pi}{\lambda}\delta r_n + \frac{4\pi}{\lambda}\frac{b_n}{r}\frac{\Delta z}{\sin\alpha} + \varphi_{a_n} + \varphi_{w_n} \qquad (7.8)$$

where the total path difference δr_n in (7.7) has been divided into the contribution associated with the possible deformation, which is still denoted by δr_n, and the contribution associated with the topography Δz. Furthermore, in (7.8), φ_{a_n} is atmospheric phase screen contribution, $\varphi_{\gamma_n} = \angle\gamma_n$ and $b_{N/2} = 0$.

It is also worth at this point a comment on the first term at the right hand side of (7.8). This term is a random variable that models the phase of the backscattering coefficient at the n-th antenna, which accounts for the so-called radar echo decorrelation discussed in Section 6.4. Although physically different from the thermal noise, both terms show generally a high spatial variability and, moreover, can be collected in a single factor as explained in the following.

7.1.3 Interferometric data covariance model

We consider the model for the complex acquisitions y_n corresponding to (7.8), that is:

$$y_n = a_n\gamma_n + w_n = a_n(\gamma_n + a_n^* w_n), \quad i = 1, \dots, N \qquad (7.9)$$

with w_n and γ_n being the complex noise and the complex backscattering coefficient at the n-th acquisition,[2] respectively, and a_n being a pure phase complex quantity accounting for the remaining phase factors in (7.8):

$$a_n = e^{j\frac{4\pi}{\lambda}\delta r_n + j\frac{4\pi}{\lambda}\frac{b_n}{r}\frac{\Delta z}{\sin\alpha} + j\varphi_{a_n}} \qquad (7.10)$$

A remark on the notation adopted in (7.9) is now in order. With the symbol y_n we intend here the complex value representative of the received signal at the n-th antenna in a given pixel after the SAR focusing, coregistration and possibly ZBS with respect to a reference DEM. It therefore represents the estimation (at the n-th antenna) of the reflectivity function integrated over the fixed azimuth-range resolution cell, which was indicated with the symbol $\hat{\gamma}$ in the previous chapters. Here we preferred to use the symbol y for two reasons. The first refers to the simplification of the notation. The second is related to the multibaseline tomographic context, discussed in the next chapter, in which this signal describes again a "raw signal" to be further focused along the elevation direction.

Let now $\boldsymbol{a} = [a_1, \dots, a_N]^T$, $\boldsymbol{\gamma} = [\gamma_1, \dots, \gamma_N]^T$ and $\boldsymbol{n} = [w_1, \dots, w_N]^T$ be the vectors collecting the a_n, γ_n, and w_n factors, respectively. Eq. (7.9) can be then written in the vector form:

$$\boldsymbol{y} = \boldsymbol{a} \odot \boldsymbol{\gamma} + \boldsymbol{w} = \boldsymbol{a} \odot (\boldsymbol{\gamma} + \boldsymbol{a}^* \odot \boldsymbol{w}) \qquad (7.11)$$

where $\odot$ is, as usual, the Hadamard (element by element) product and $*$ denotes the element by element conjugate operator.

---

[2] Note that $\varphi_{w_n} \neq \angle w_n$.

Assuming w to be formed by all independent, zero mean, circularly Gaussian complex random variables, with covariance matrix $\mathbf{C}_w = \sigma_w^2 \mathbf{I}_N$, $\mathbf{I}_N$ being the N element identity matrix, it follows that the statistic of w is the same of that of the vector $a^* \odot w$.

SAR Interferometry implies the presence of a sufficient degree of correlation between the signal backscattered at the different antennas. Vector γ can be therefore modeled as a complex zero mean and circularly Gaussian random vector, with a covariance matrix $\mathbf{C}_\gamma$. The latter is generally expressed as:

$$\mathbf{C}_\gamma = \sigma_\gamma^2 \mathbf{M}_\gamma \tag{7.12}$$

where σ_γ^2 is the variance of each element of γ and $\mathbf{M}_\gamma$ backscattering coherence matrix.[3] The case $\mathbf{M}_\gamma = \mathbf{I}_N$ corresponds to a completely decorrelated scatterer, whereas the case $\mathbf{M}_\gamma = \mathbf{1}_N$, with $\mathbf{1}_N$ being the $N \times N$ all-ones matrix, represents the situation of an ideal scatterer fully correlated along all the acquisitions. More in general, the element $\chi_\gamma(n, i)$ of the coherence matrix $\mathbf{M}_\gamma$ is:

$$\chi_\gamma(n, i) = \begin{cases} 1 & \text{for } i = n \\ 0 \leqslant \dfrac{E[\gamma_n \gamma_i^*]}{\sqrt{E[|\gamma_n|^2]}\sqrt{E[|\gamma_i|^2]}} \leqslant 1 & \text{for } i \neq n \end{cases} \tag{7.13}$$

which generalizes the coherence addressed in the previous chapter for a single interferometric pair, i.e. (6.48), to the multiple pairs case. With the above definition, it follows that:

$$\mathbf{C}_y = (aa^H) \odot \mathbf{C}_\gamma + \mathbf{C}_w = (aa^H) \odot (\mathbf{C}_\gamma + \mathbf{C}_w) \tag{7.14}$$

In analogy to (7.12) the data covariance matrix can be therefore expressed as:

$$\mathbf{C}_y = \sigma_y^2 (aa^H) \odot \mathbf{M}_y \tag{7.15}$$

with $\sigma_y^2 = \sigma_\gamma^2 + \sigma_w^2$ and $\mathbf{M}_y$ being the data coherency matrix expressed as

$$\chi_y(n, i) = \begin{cases} 1 & \text{for } i = n \\ \dfrac{r_\gamma(n, i)}{1 + SNR^{-1}} & \text{for } i \neq n \end{cases} \tag{7.16}$$

being $SNR = \sigma_\gamma^2/\sigma_w^2$ the Signal to Noise Ratio (SNR). Eq. (7.16) highlights the effect of the noise in reducing the off-diagonal elements of the data coherence matrix and therefore as an additional

---

[3]The assumption that the backscattering vector does not alter the steering translates in the hypothesis that $\mathbf{M}_\gamma$ is real and nonnegative valued.

contribution to the coherence loss. Above derivation suggests that equivalently one could absorb the noise term in the backscattering signal as follows:

$$\gamma \leftarrow \gamma + a^* \odot w = \gamma + w \qquad (7.17)$$

where the last equality should be interpreted in the statistical sense as an equality in distribution. The following substitutions should be also carried out as a consequence of (7.17)

$$\sigma_\gamma^2 \leftarrow \sigma_\gamma^2 + \sigma_w^2$$
$$\mathbf{M}_\gamma \leftarrow \mathbf{M}_y \qquad (7.18)$$

thus leading, under the Gaussian assumption for both data and noise, to the following simplified data model with respect to (7.11):

$$y = a \odot \gamma \qquad (7.19)$$

where the vector of the additive noise is absorbed in the vector of the backscattering coefficients. Bearing this in mind we can rewrite (7.8) as:

$$\varphi_n = \frac{4\pi}{\lambda}\delta r_n + \frac{4\pi}{\lambda}\frac{b_n}{r}\frac{\Delta z}{\sin\alpha} + \varphi_{a_n} + \varphi_{w_n} \qquad (7.20)$$

so that φ_{w_n} accounts for all decorrelation terms, including that associated to the additive noise affecting areas of low backscattering, i.e. low signal to noise ratio.

7.2 Phase component separation

Before proceeding with the description of the rationale of MT-InSAR, it is at this stage instructive to interpret what hereto derived in the previous sections with reference to the classical two pass ($N = 2$), i.e. single pair Interferometry described in Chapter 6. Letting without loss of generality $i = 1$ to be the master image, i.e. $b_1 = 0$, from what stated about the effect of the backscattering phase, we consider the phase difference $\varphi = \varphi_2 - \varphi_1$:

$$\varphi = \frac{4\pi}{\lambda}\delta r_2 + \frac{4\pi}{\lambda}\frac{b_2}{r}\frac{\Delta z}{\sin\alpha} + \varphi_{a2} - \varphi_{a1} + \varphi_{\gamma 2} - \varphi_{\gamma 1} \qquad (7.21)$$

where, according to the above assumptions ($b_1 = 0$), it results $\delta r_1 = 0$.

Eq. (7.21) highlights that it is not possible achieving just from φ any separation between the different components, unless the variation over the image of one contribution is far larger than the others. This is for instance the case of application to large earthquakes where the first term in φ contains the variation of the range between the slave, say after the earthquake, and the master, say before the earthquake, related to the coseismic displacement: in this case, it can be the predominant term, provided that the backscattering (and the noise) variation is small enough, that is $|\varphi_{\gamma 2} - \varphi_{\gamma 1}| \ll 1$, or, equivalently, that the decorrelation is limited. However, especially for deformations related to subsidence phenomena, landslides, or for deformations of volcanoes as well as of seismic areas in the post and interseismic phases, the first term turns out to provide frequently a contribution which is on the same basis as both the second term, i.e. the phase corresponding to the DEM error (which depends on the magnitude of b_2), and the variation of the atmosphere between the master and slave image.

Multitemporal SAR Interferometry achieves the separation of the components in (7.21) by exploiting the availability of N measurements, generally with $N \gg 2$. In this case the measurements (unrestricted phase values) φ_n, $n = 1, \ldots, N$, can be stacked into vectors thus allowing rewriting (7.20) as:

$$\boldsymbol{\varphi} = \boldsymbol{\varphi}_d + \boldsymbol{k}_z \Delta z + \boldsymbol{\varphi}_a + \boldsymbol{\varphi}_w \tag{7.22}$$

where:

$$
\begin{aligned}
\boldsymbol{\varphi} &= [\varphi_1, \ldots, \varphi_N]^T \\
\boldsymbol{\varphi}_d &= \frac{4\pi}{\lambda}[\delta r_1, \ldots, \delta r_N]^T \\
\boldsymbol{k}_z &= \frac{4\pi}{\lambda}\left[\frac{b_1}{r \sin \alpha}, \ldots, \frac{b_N}{r \sin \alpha}\right]^T \\
\boldsymbol{\varphi}_a &= [\varphi_{a_1}, \ldots, \varphi_{a_N}]^T \\
\boldsymbol{\varphi}_w &= [\varphi_{w_1}, \ldots, \varphi_{w_N}]^T
\end{aligned}
\tag{7.23}
$$

with $\boldsymbol{\varphi}_d$ stacking the phase values associated with the deformation terms δr_n, $n = 1, \ldots, N$, or equivalently

$$\boldsymbol{\varphi} = \boldsymbol{\varphi}_d + \boldsymbol{k}_s s + \boldsymbol{\varphi}_a + \boldsymbol{\varphi}_w \tag{7.24}$$

with

$$\boldsymbol{k}_s = \frac{4\pi}{\lambda}\left[\frac{b_1}{r}, \ldots, \frac{b_N}{r}\right]^T. \tag{7.25}$$

The separation between the phase vectors in (7.22) can be carried out thanks to the different behavior, in a general statistical meaning, of what are the useful contributions, deformation and topographic terms, and the atmospheric contribution which is generally classified as an error source. The algorithms that perform such operation are generally divided in two classes that will be referred to as Stacking of Coherent Interferograms (SCI) and Persistent Scatterers Interferometry (PSI). In addition to the possibility of estimating and removing the atmosphere, methods of both the classes are also capable of extracting the information about the displacements at the different times (epochs): the latter are usually referred to as Time Series (TS).

7.2.1 Stacking of coherent interferograms

SCI analyses stacks of interferograms showing sufficiently high coherence values which are typically associated with limited (small) spatial baseline and temporal separation (temporal baseline). The popularity of such approaches is essentially due to the continuity with the classical, two-pass DInSAR. These approaches take benefit of the use of the (coherent) multilook carried out for the measurement of the spatial coherence and the mitigation of the noise, as discussed in Chapter 6. The noise mitigation is achieved at the expense of a spatial resolution loss. The loss of spatial resolution also allows the possibility of operating a spatial data subsampling along the azimuth and range thus reducing the computational demand, which is generally a problem for the analysis of large SAR data stacks.

SCI algorithms are generally suitable for the analysis of wide areas at lower resolution, i.e. small scale analysis. Although the frequent adoption in the DInSAR literature of an opposite terminology, "small scale" is here associated to "wide areas" in accordance with the definition in cartography and Geographical Information System (GIS) for which a small scale map is the one in which a given part of the Earth is represented by a small area on the map.

Hereafter we focus on the Small BAseline or Small Baseline Subset (SBAS) [81] method introduced as a generalization of the Least Square interferometric database analysis approach in [82].

SBAS adopts a strategy for which interferograms are obtained by pairing acquisitions with a (hard) upper bound on the spatial and the temporal separations (baselines). This allows limiting the decorrelation associated with distributed scattering and scattering changes over the time, usually occurring in rural areas with the presence of a sparse vegetation.

Let M be the number of interferograms selected among all the possible $N(N-1)/2$ interferograms. The M phase values extracted in a given pixel can be stacked, similarly to the N acquisition phase values in (7.22), in M-dimensional phase vectors. In particular, in analogy to (7.22) and (7.23), we have:

$$\Delta\boldsymbol{\varphi} = \Delta\boldsymbol{\varphi}_d + \Delta\boldsymbol{k}_z\Delta z + \Delta\boldsymbol{\varphi}_a + \Delta\boldsymbol{\varphi}_w \tag{7.26}$$

where, letting (i, n) be the indexes of the master and slave acquisitions involved in the generic m-th interferometric pair, the m-th (out of M) element of $\Delta\boldsymbol{\varphi}$ is given by:

$$
\begin{aligned}
\Delta\varphi_m &= \varphi_n - \varphi_i = \\
&= \varphi_{d_n} - \varphi_{di} + (\Delta k_{z_n} - \Delta k_{z_i})\Delta z + \varphi_{a_n} - \varphi_{ai} + \varphi_{w_n} - \varphi_{w_i}
\end{aligned}
\tag{7.27}
$$

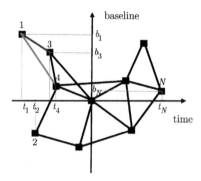

Figure 7.6. An example of representation of $N = 10$ acquisitions (squares) and $M = 16$ interferograms (lines) in the baseline domain. The representation corresponds to graph, which should be oriented according to the choice of master and slave in the generation of each interferogram. Without lack of generality we adopt the choice that the master is temporally preceding the slave so that the *head* of the edges should be located in the right of each edge.

As shown in Fig. 7.6, the acquisition pairing can be represented in the baseline domain. It is evident that the pairing operation can be represented in a graph. Each pairing (interferogram) involves a couple of images and therefore a connection (edge) between vertexes. Being each the interferograms generated by choosing a master and a slave image, the representation is carried out more precisely via a oriented graph with edges characterized by a *tail* and a *head*. For each fixed pair, there is no requirement on the choice of the master and of the slave image: for convenience we will choose the master (head of the edge) temporally following the slave (tail of the edge). Consequently the edges in Fig. 7.6, which have been drawn without any orientation, should be characterized by a head in the temporally preceding vertex. To distinguish the master involved in the generic interferogram from the master of the interferometric stack previously mentioned, i.e., the acquisition corresponding to zero baseline, the latter is generally referred to as "supermaster".

The oriented graph representing the pairing of the acquisitions of the stack in the generation of the interferograms can be described by an incidence matrix, similarly to what discussed in Section 6.7 in Chapter 6. In the following we describe the graph with a matrix $\mathbf{D}$ having the number of rows equal to the number of interferograms, i.e. M, and the number of columns equal to the number of acquisitions, i.e. N, and having on the k-th row all zero elements but for $+1$ and -1 on the n-th and i-th columns, where (n, i) indicates the master and slave pair involved in the m-th interferogram.[4] With reference to Fig. 7.6 we have:

$$\mathbf{D} = \begin{bmatrix} -1 & 0 & 0 & 1 & 0 & \ldots & 0 \\ -1 & 0 & 1 & 0 & 0 & \ldots & 0 \\ . & . & . & . & . & \ldots & . \\ 0 & 0 & 0 & 0 & 0 & \ldots & 1 \end{bmatrix} \quad (7.28)$$

where the first and second rows are relevant to the interferograms associated with the green and red (gray and mid gray in print version) arcs, respectively; furthermore, according to the paring order assumption, the first image has been assumed as slave for both interferograms.

The matrix $\boldsymbol{D}$ allows easily writing the relation between the vector (7.26) collecting the interferometric phase measurements on the edges of the graph and the vector (7.22) collecting the acquisition phase values in vertexes of the graph, i.e.:

$$\mathbf{D}\boldsymbol{\varphi} = \Delta\boldsymbol{\varphi} \quad (7.29)$$

In principle there is no requirement on the ordering of the elements of $\boldsymbol{\varphi}$. Without any loss of generality we hereafter assume that $\boldsymbol{\varphi}$ is chronologically ordered. Alternatively, a chronological ordering of the elements of $\boldsymbol{\varphi}$ is required after the inversion of (7.29) to achieve the sequence of phase values (phase series) corresponding to the different ordered times (epochs).

Some considerations about the incidence matrix in (7.29) are now in order. First of all, interferometric paring is typically carried out in a redundant way, see Fig. 7.6 where, f.i., the interferometric pair (1,4) can be achieved also as combination of the interferometric pairs (1,3) and (3,4). Here the term redundant indicates that the exploited edges are more than those strictly necessary to connect all the vertexes, that is those forming a tree.[5] More precisely we

[4]As mentioned in the previous chapter, the incidence matrix considered in this book to describe a network is actually the transpose of that generally considered in network theory bibliography.

[5]A tree is a graph with $N - 1$ edges (*bridges*) connecting all the nodes: removing even a single edge of a tree generates a disconnected graph.

assume, in a first instance, that the graph is completely and redundantly connected, that is with all vertexes connected and with a set of edges additional to the $N-1$ necessary to build a tree. This choice to deal with a redundant graph is dictated first of all by the need to perform proper checks of the unwrapped phase values: the phase unwrapping operation is in fact a prerequisite for establishing the linear relation in (7.29). The redundancy translates in the feature that the linear equation system in (7.29) results to be in the practice overdetermined,[6] i.e. characterized by a number of equation (i.e. the number of interferograms, M) larger than the number of unknowns (i.e. the number of acquisition phase values, N): generally M is at least equal to $3N$. As a consequence, the solution of the system (7.29) can be only evaluated in the Least Squares (LS) sense.

Secondly, it is well known that the incidence matrix is rank deficient by construction, even in the case of pairing of all the acquisitions, i.e. $M = N(N-1)/2$. The rank deficiency is a consequence of the fact that the vector $\Delta\varphi$, essentially built by performing differences between the elements of φ, is invariant with respect to additive constant values. Accordingly, the Least Squares solution of the system (7.29) is not unique. Such an intrinsic rank deficiency can be overcome by considering an additional equation setting the reference for the phase values. For instance, assuming the first SAR image to be the first in the chronological order, the condition $\varphi_1 = 0$ can be imposed by extending the matrix $\mathbf{D}$ in (7.29) with the row given by the vector $d_{M+1} = [1, 0, 0, 0, \ldots, 0]^T$, thus leading to the full rank matrix:

$$\mathbf{D}_e = \begin{bmatrix} \mathbf{D} \\ d_{M+1}^T \end{bmatrix} \tag{7.30}$$

and extending the vector $\Delta\varphi$ with a zero element:

$$\Delta\varphi_e = [\Delta\varphi^T, 0]^T \tag{7.31}$$

The linear system in (7.29) can be therefore substituted by:

$$\mathbf{D}_e\varphi = \Delta\varphi_e \tag{7.32}$$

The LS solution of (7.32), which represents an estimate of the vector φ collecting the phase values at the different passes, is this time unique and given by

$$\hat{\varphi} = \mathbf{D}_e^{\dagger}\Delta\varphi_e \tag{7.33}$$

---

[6]More precisely the redundancy of the graph translates in the property that the dimension of the left null space of the system in (7.29), i.e. the nullity of $\mathbf{D}^T$, is larger than 0.

where $\mathbf{D}_e^{\dagger}$ is the generalized inverse of the extended matrix $\mathbf{D}_e$: in this case, being $\mathbf{D}_e$ a full rank matrix, it results $\mathbf{D}_e^{\dagger} = (\mathbf{D}_e^T \mathbf{D}_e)^{-1} \mathbf{D}_e^T$.

Alternatively, the minimum L^2 (Euclidean) norm LS solution of the linear system (7.29) can be achieved by mean of the generalized inverse $\mathbf{D}^{\dagger}$ of the rank deficient incidence matrix $\mathbf{D}$, evaluated by resorting to the Singular Value Decomposition (SVD). The condition $\varphi_1 = 0$ can be subsequently imposed by subtracting a constant phase values from the achieved minimum norm solution.

It should be noted that, due to the constraints on the maximum spatial and temporal baseline, it can happen that the graph in Fig. 7.6 could be not connected, in the sense that it is not possible to reach any node starting form a different one. Generally the baseline constraints generate a disconnected graph formed by subgraphs (corresponding to subsets of acquisitions), each one being redundant. In this case, even the extended matrix $\mathbf{D}_e$ in (7.30) turns out to be rank deficient and the LS solution of the extended system (7.32) is again not unique. Once more, the minimum L^2 norm LS solution can be achieved by exploiting in (7.33) the generalized inverse of the (now rank deficient) system matrix $\mathbf{D}_e$, evaluated through the use of the SVD. However, this solution may be unsatisfactory. To clarify this point, in Fig. 7.7 it is reported an example of acquisitions characterized by the presence of two subsets in the baseline domain. We assume a linear deformation, for instance an increase of the range with a rate of 1.6 cm/yr. The minimum norm LS solution is depicted in Fig. 7.8. It can be noticed that the part of solution corresponding to the second subset has been centered around zero to provide the lowest L^2 norm of the whole solution: the result achieved in this way is evidently unsatisfactory.

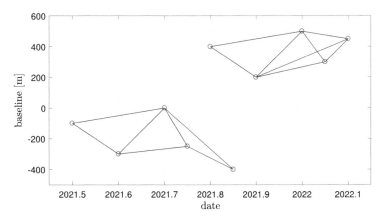

Figure 7.7. Baseline domain corresponding to acquisitions clustered in two different subsets.

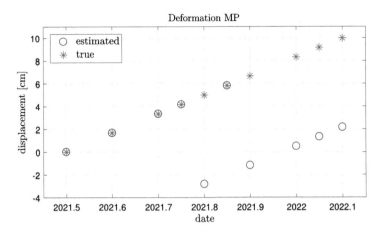

Figure 7.8. Comparison between the simulated phase (red (mid gray in print version) stars) and the minimum norm solution (blue (dark gray in print version) circles) achieved in the presence of the two subsets shown in Fig. 7.7.

In order to improve the estimate of the phase values, it could be observed that the presence of the discontinuities in the solution depicted in Fig. 7.8 leads to an increase of the magnitude of the phase differences between successive times. Accordingly, instead of searching the minimum norm solution of the linear system (7.32), i.e. with respect the N-length phase vector $\boldsymbol{\varphi}$, we can search a minimum norm solution with respect to the $(N-1)$-length vector $\boldsymbol{\varphi}_\Delta = [\varphi_2 - \varphi_1, \ldots, \varphi_N - \varphi_{N-1}]^T$ collecting the variations between consecutive times. To this aim we rewrite $\boldsymbol{\varphi}$ as:

$$\boldsymbol{\varphi} = \begin{bmatrix} 0 & 0 & 0 & \ldots & 0 \\ 1 & 0 & 0 & \ldots & 0 \\ 1 & 1 & 0 & \ldots & 0 \\ . & . & . & \ldots & 0 \\ 1 & 1 & 1 & \ldots & 1 \end{bmatrix} \boldsymbol{\varphi}_\Delta = \mathbf{B}\boldsymbol{\varphi}_\Delta \qquad (7.34)$$

where $\mathbf{B}$ is the $N \times (N-1)$ full rank matrix converting the unknowns from $\boldsymbol{\varphi}$ to $\boldsymbol{\varphi}_\Delta$. Note that the first null row forces to zero the value of the first (reference) phase φ_1.

The substitution of (7.34) in (7.32) leads to the following linear system:

$$\mathbf{D}_e \mathbf{B}\boldsymbol{\varphi}_\Delta = \Delta\boldsymbol{\varphi}_e \qquad (7.35)$$

which can be solved in the LS sense, thus providing an estimate

$$\hat{\boldsymbol{\varphi}}_\Delta = (\mathbf{D}_e \mathbf{B})^\dagger \Delta\boldsymbol{\varphi}_e \qquad (7.36)$$

of the unknown vector $\boldsymbol{\varphi}_\Delta$. Subsequently, a new estimate $\hat{\boldsymbol{\varphi}}$ of the former unknown vector $\boldsymbol{\varphi}$ can be achieved by substituting (7.36) in (7.34)

$$\hat{\boldsymbol{\varphi}} = \mathbf{B}(\mathbf{D}_e\mathbf{B})^\dagger \Delta\boldsymbol{\varphi}_e \tag{7.37}$$

or, equivalently, by sequentially integrating the elements $\hat{\varphi}_{\Delta_l}$ of $\hat{\boldsymbol{\varphi}}_\Delta$ over adjacent times:

$$\begin{aligned} \hat{\varphi}_1 &= 0 \\ \hat{\varphi}_i &= \sum_{l=1}^{i-1} \hat{\varphi}_{\Delta_l} \quad \text{for} \quad i = 2, \dots, N \end{aligned} \tag{7.38}$$

In the absence of any ill-conditioning for the matrix $\mathbf{D}_e$ (i.e., for a full rank matrix $\mathbf{D}_e$), the system matrix $\mathbf{D}_e\mathbf{B}$ is still full rank and, therefore, the estimate $\hat{\boldsymbol{\varphi}}_\Delta$ in (7.36) corresponds to the unique LS solution of the system (7.35). In this case, the estimates of $\boldsymbol{\varphi}$ given by (7.33) and (7.37) turn out to be the same.[7]

In the presence of a rank deficiency, f.i. for the case of multiple subsets as in Fig. 7.7, the LS inversion of the system (7.35) provides the LS estimate of the phase differences with the minimum L^2 norm. In this case, by means of (7.37), a more continuous phase signal can be reconstructed, as depicted in Fig. 7.9, which refers to the case of the two subsets shown in Fig. 7.7.

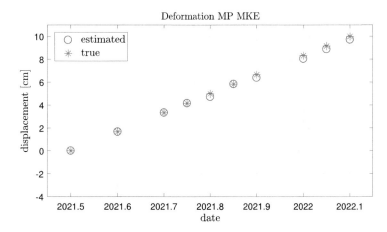

Figure 7.9. Comparison between the simulated phase (red (mid gray in print version) stars) and the velocity minimum norm solution (blue (dark gray in print version) circles) achieved in the presence of the two subsets shown in Fig. 7.7 by solving (7.35) in place of (7.32).

---

[7]The matrices relating the estimate $\hat{\boldsymbol{\varphi}}$ and the data vector $\Delta\boldsymbol{\varphi}_e$ in (7.33) and (7.37) are actually different only for the last column. Such a difference however does not play any role in the evaluation of $\hat{\boldsymbol{\varphi}}$ since $\Delta\boldsymbol{\varphi}_e$ has, by construction, the last component equal to zero.

To understand the physical meaning of the minimum norm LS estimate of the vector $\boldsymbol{\varphi}_\Delta$, let us focus only on the deformation contribution of the phase signal. In this case, the phase variations φ_{Δ_l} and the (approximated) samples v_l of the (los component) of the discrete instantaneous deformation velocity of the observed scatterer are related as follows:

$$\varphi_{\Delta_l} = \frac{4\pi}{\lambda}(t_{l+1} - t_l)v_l \quad l = 1, \ldots, N-1 \tag{7.39}$$

Eqs. (7.39) can be rewritten in the more compact vector form:

$$\boldsymbol{\varphi}_\Delta = \frac{4\pi}{\lambda}\mathbf{F}\boldsymbol{v} \tag{7.40}$$

where $\boldsymbol{v} = [v_1, \ldots, v_{N-1}]^T$ and $\mathbf{F}$ is the $(N-1) \times (N-1)$ diagonal matrix with elements $t_{k+1} - t_k$ along the diagonal. The linear system achieved by substituting (7.40) in (7.35) can be solved in the LS sense with respect to the vector $\boldsymbol{v}$. In case of multiple subsets, that is for a rank deficient system matrix, it is possible evaluating the minimum norm LS solution, given by:

$$\hat{\boldsymbol{v}} = \frac{\lambda}{4\pi}(\mathbf{D}_e\mathbf{BF})^\dagger \Delta\boldsymbol{\varphi}_e \tag{7.41}$$

On the other hand, an estimate of $\boldsymbol{v}$ can be also achieved by substituting in (7.40) the vector $\boldsymbol{\varphi}_\Delta$ with the minimum norm LS solution $\hat{\boldsymbol{\varphi}}_\Delta$ given by (7.36), and then inverting with respect $\boldsymbol{v}$:

$$\hat{\boldsymbol{v}} = \frac{\lambda}{4\pi}\mathbf{F}^{-1}\hat{\boldsymbol{\varphi}}_\Delta = \frac{\lambda}{4\pi}\mathbf{F}^{-1}(\mathbf{D}_e\mathbf{B})^\dagger \Delta\boldsymbol{\varphi}_e \tag{7.42}$$

It is worth noting that the two estimates (7.41) and (7.42) are equivalent, since $(\mathbf{D}_e\mathbf{BF})^\dagger = \mathbf{F}^{-1}(\mathbf{D}_e\mathbf{B})^\dagger$ according to the Theorem 1 in [83].[8] As a consequence, the (indirect) estimate (7.42), achieved by starting from the minimum norm estimate of $\hat{\boldsymbol{\varphi}}_\Delta$, turns out to be itself the minimum norm estimate of $\boldsymbol{v}$. In other words, the minimum norm estimate $\hat{\boldsymbol{\varphi}}_\Delta$ is the one minimizing the kinetic energy of the scatterer deformation.

An alternative estimate of the unknown phase vector $\boldsymbol{\varphi}$ in presence of separate acquisition subsets can be conceived by observing Fig. 7.9. Indeed, the main reason for the incorrect combination

[8]The Theorem 1 in [83] asserts that, given two arbitrary matrices $\mathbf{A}$ and $\mathbf{B}$ such that the product $\mathbf{AB}$ is defined, $(\mathbf{AB})^\dagger = \mathbf{B}^\dagger\mathbf{A}^\dagger$ if and only if both the conditions:

$$\mathbf{A}^\dagger\mathbf{ABB}^H\mathbf{A}^H = \mathbf{BB}^H\mathbf{A}^H$$

$$\mathbf{BB}^\dagger\mathbf{A}^H\mathbf{AB} = \mathbf{A}^H\mathbf{AB}$$

are verified.

of the solution between the subsets is associated with the presence of a trend in the true phase (positive in the figure), which causes a deformation accumulation during the epochs. As explained in the following it is possible to estimate and subtract the linear phase component directly from the measured interferograms. To this aim we model the unknown phase vector as:

$$\boldsymbol{\varphi} \approx \frac{4\pi}{\lambda} v \boldsymbol{t} \tag{7.43}$$

where $\boldsymbol{t} = [t_1, \cdots, t_N]^T$ and v is the (unknown) mean deformation velocity. Substitution of (7.43) in (7.29) leads to the following linear system:

$$\frac{4\pi}{\lambda} \mathbf{D} \boldsymbol{t} v = \Delta \boldsymbol{\varphi} \tag{7.44}$$

whose LS solution with respect the scalar unknown v provides an estimate of the mean deformation velocity:

$$\hat{v} = \frac{\lambda}{4\pi} (\boldsymbol{t}^T \mathbf{D}^T \mathbf{D} \boldsymbol{t})^{-1} \boldsymbol{t}^T \mathbf{D}^T \Delta \boldsymbol{\varphi} \tag{7.45}$$

This estimate can be used to "flatten", i.e., *"detrend"* the (known) measured data $\Delta \boldsymbol{\varphi}$, thus achieving the new "flattened" data vector $\Delta \boldsymbol{\varphi}_F$, whose elements are given by

$$\Delta \varphi_{F_m} = \Delta \varphi_m - \frac{4\pi}{\lambda} \hat{v} \Delta t_m \quad m = 1, \dots, M \tag{7.46}$$

being Δt_m the time separation involved in the generic m-th interferogram. The new data vector $\Delta \boldsymbol{\varphi}_F$ can be used to evaluate an estimate of the flattened (unknown) phase vector $\boldsymbol{\varphi}_F$ by solving, again in a LS sense, the following system of equations:

$$\mathbf{D}_e \boldsymbol{\varphi}_F = \Delta \boldsymbol{\varphi}_F \tag{7.47}$$

which, exactly as the system (7.32), leads to the estimate $\hat{\boldsymbol{\varphi}}_F = \mathbf{D}_e^\dagger \Delta \boldsymbol{\varphi}_F$. Obviously, after solving (7.47), the linear component must be added back for achieving the original (i.e., not detrended) estimate:

$$\hat{\boldsymbol{\varphi}} = \hat{\boldsymbol{\varphi}}_F + \frac{4\pi}{\lambda} \hat{v} \boldsymbol{t} \tag{7.48}$$

Fig. 7.10 shows the reconstructed phase when the solution is achieved via the detrending. It is clear that the detrending operation allows avoiding the discontinuity present in Fig. 7.8 due to the minimization of the norm of the original phase vector.

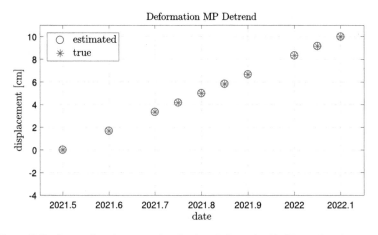

Figure 7.10. Comparison between the simulated phase (red (mid gray in print version) stars) and the minimum norm solution (blue (dark gray in print version) circles) achieved in the presence of the two subsets shown in Fig. 7.7 after detrending.

A last remark is dedicated to the presence of the residual topography. It is clear that the mean deformation velocity estimation is influenced by disturbances, mainly noise and atmosphere, which can be filtered out only after the inversion. In the presence of a large baseline excursion, also the residual topography may affect the mean deformation velocity estimation. A further improvement for the deformation estimation can be achieved by the joint estimation of the residual topography Δz and v. This can be performed by substituting to (7.44) the following linear system:

$$\mathbf{D}_e[\frac{4\pi}{\lambda}\boldsymbol{t}\ \boldsymbol{c}_z][v, \Delta z]^T = \Delta\boldsymbol{\varphi} \tag{7.49}$$

whose unknown is the two elements vector $[v, \Delta z]^T$, and where the topography coefficient vector $\boldsymbol{c}_z$ is the one described in (7.23).

7.2.2 Persistent scatterers interferometry

Differently from the approach described in the previous section, the second class of A-DInSAR algorithms performs the analysis of interferograms at maximum spatial resolution. The aim is to determine the deformation of possible dominant scatterers, typically associated with man-made structures (dihedral and trihedral corresponding to walls edges, corners, masts, gratings, etc.) which show persistency of the scattering properties even with large time separation (temporal baseline). Moreover, unlike SCI, no limitation on the spatial baseline is introduced in order to achieve higher sensitivity to the topography. The latter feature allows performing an accurate evaluation of the residual (i.e., with respect to the external DEM) topography and achieving a precise localization of the scatterer even for tall buildings.

The Persistent Scatterers (PS) technique, originally Permanent Scatterers technique in [84,85] and now known as PSInSAR, belongs to this class of algorithms and was the first technique to demonstrate the extraordinary capability of multitemporal interferometric SAR analysis for monitoring ground deformations on a spatially dense grid, opening a new era for the development and the application of SAR to environmental monitoring.

To describe the PS SAR data processing approach, we refer again to Fig. 7.4 where N images are supposed to be acquired at times $t_1, \cdots, t_N$ and with (spatial) orthogonal baselines $b_1, \cdots, b_N$ (with respect to the reference supermaster acquisition). As for CSI, all images are assumed again to be coregistered with respect to the supermaster image, so that the response of a ground target is located in the same pixel along the different images.

Generally, the coherence is exploited in all A-DInSAR methods to assess the degree of reliability of the measurement. CSI exploits the spatial multilook to measure the coherence of the response of a pixel from the adjacent pixels; because of the constraint of retaining the highest possible resolution, (spatial) multilook is, however, undesired in PSI. To provide a discrimination between the signal and the noise and, at the same time to assess the "reliability" of the phase information over the multitemporal acquisition, PSI exploits the multitemporal coherence, a index normalized in the [0, 1] interval that will be defined in the following. The temporal coherence is measured by directly exploiting the phase model in (7.20) that, after a proper phase calibration, can be assumed to be dependent on a set of parameters, whose cardinality is significantly lower that the number of independent phase measurements, i.e. $N - 1$.

We start by analyzing the unknowns in the $N - 1$ phase measurements in (7.20): $N - 1$ unknowns are related to the noise (φ_{w_n}), $N - 1$ unknowns are related to the atmospheric disturbance, $N - 1$ unknowns to the deformation and 1 unknown to the residual topography, the last two items being the wanted contributions. We have evidently a heavily underdetermined problem and therefore the phase information cannot be used at this stage to operate a separation of the different contributions. PS technique adopts an inherent two-step strategy. In the original PSInSAR approach, the so-called PS Candidates (PSCs) are initially selected by analyzing pixel by pixel the stability of the amplitude of the complex acquisitions y_i, referred to as Amplitude Dispersion Index (ADI). The ADI, defined as:

$$D_A = \frac{\sqrt{\dfrac{1}{N}\displaystyle\sum_{n=1}^{N}\left(|y_n| - \dfrac{1}{N}\sum_{n=1}^{N}|y_n|\right)^2}}{\dfrac{1}{N}\displaystyle\sum_{n=1}^{N}|y_n|} \qquad (7.50)$$

represents the sample (temporal) standard deviation of the amplitude normalized to its sample mean. It can be shown that for a scatterer for which the pdf of the amplitude follows the Rice distribution, which is the simplest model of the statistic of the response pertinent to a dominant scatterer immersed in a weak noise clutter, the standard deviation of the phase is approximately equal to the ADI [84,85]. Pixels with D_A below a fixed threshold (typically set to 0.25) are identified as candidate PS (PSC) and, therefore, as proxy indicators of PS, that is of scatterers showing a persistent phase information across the multitemporal acquisition.

The set of PSC defines a sparse grid of pixels characterized by a phase information which is less affected by noise. After the phase unwrapping on the sparse PSC grid of all the interferograms (in the original version [84,85] generated by beating all the acquisition with the reference, i.e. supermaster image), the phase (at all available acquisitions) has to be decomposed into the three contributions of topography, deformation and atmosphere. To this end the phase signal is spatially low pass filtered to cancel the residual topography contribution, which is generally highly variable in space especially in urban areas. Relying on their different temporal behaviors (differently from the deformation, the atmosphere is highly variable in time), a separation between the deformation and atmosphere is then carried out. The estimated deformation and atmosphere are then cancelled from the phase signal stack thus achieving a phase calibrated stack that, for the generic pixel, can be written as:

$$\varphi_c = \varphi - \hat{\varphi}_{dSLP} - \hat{\varphi}_a = \varphi_{dSHP} + k_z \Delta z + \varphi_w \qquad (7.51)$$

where the deformation term φ_d has been split in a Spatial Low-Pass (SLP) φ_{dSLP} and spatial High-Pass (SHP) φ_{dSHP} component. Evidently, due to the presence of the above mentioned spatial low pass filtering necessary for the estimation of the atmospheric component, the spatially high-pass component of the deformation cannot be estimated at this stage.

The calibrated, or preferably the residual phase stack in (7.51) can be exploited to estimate the SHP components by operating at full resolution. It is worth noting that this problem is still under-determined because in (7.51), apart for the noise, we have $N - 1$ useful measurements, with N unknowns, i.e. the elements of vector φ_{dSHP} from the second to the last one and Δz. PSI at this stage adopts a model for the residual SHP deformation φ_{dSHP}. This model is generally based on a linear deformation component, that is:

$$\varphi_{dSHP} = \varphi_{dSHP0} + \frac{4\pi}{\lambda} v t \tag{7.52}$$

with t being the vector collecting the acquisition times and v the (unknown) mean deformation velocity. Substitution of (7.52) in (7.51) leads to the following model for the residual phase:

$$\boldsymbol{\varphi}_c = \varphi_{dSHP0} + \frac{4\pi}{\lambda} v t + \boldsymbol{k}_z \Delta z + \boldsymbol{\varphi}_w \tag{7.53}$$

and the problem therefore reduces to the estimation of the two unknowns v and Δz: the phase constant φ_{dSHP0} is useless because of the already discussed significance of the phase measurements up to a constant phase.

All the acquisitions and all possible interferometric measurements should be used to estimate the two unknowns. In the PS technique this estimation is achieved by minimizing, in the complex domain, the difference between the observed (calibrated) phase and the modeled phase, i.e.:

$$(\hat{v}, \hat{\Delta z}) = \arg\max_{v, \Delta_z} |\rho_t(v, \Delta z)| \tag{7.54}$$

with:

$$\rho_t(v, \Delta z) = \frac{1}{N} \sum_{n=1}^{N} e^{j\left(\varphi_{cn} - \frac{4\pi}{\lambda} v t_n - k_{z_n} \Delta z\right)} \tag{7.55}$$

where φ_{c_n}, t_n, and k_{z_n} are the n-th elements of the vectors $\boldsymbol{\varphi}_c$, t, and $\boldsymbol{k}_z$, respectively.

The quantity in (7.55) defines what is called Multitemporal Coherence (MTC), an extremely powerful statistic for the discrimination of PS from noise. The maximization in (7.54) allows the accurate estimation of the residual topography of the PS and of its possible linear deformation component. Interestingly, the quantity in (7.55) can be compactly written as a scalar product:

$$\rho_t(v, \Delta z) = \frac{1}{N} \boldsymbol{a}^H(z, v) \boldsymbol{y}_\varphi \tag{7.56}$$

with:

$$
\begin{aligned}
\boldsymbol{a}(z, v) &= \left[e^{j\left(vt_1 + k_{z_1}\Delta z\right)}, \ldots, e^{j\left(vt_N + k_{z_N}\Delta z\right)} \right]^T \\
&= \left[e^{j\left(vt_1 + k_{s_1}s\right)}, \ldots, e^{j\left(vt_N + k_{s_N}s\right)} \right]^T
\end{aligned}
\tag{7.57}
$$

and:

$$
\boldsymbol{y}_\varphi = \left[e^{j\varphi_{c_1}}, \ldots, e^{j\varphi_{c_N}} \right]^T
\tag{7.58}
$$

The vector $\boldsymbol{a}(z, v)$ in (7.57) is referred to as steering vector, borrowing the terminology from the tomographic context addressed in the next chapter. By definition in (7.57), it results $\boldsymbol{a}^H \cdot \boldsymbol{a} = N$; sometimes a scaling factor $1/\sqrt{N}$ is considered to normalize $\boldsymbol{a}$, that is to guarantee $\boldsymbol{a}^H \cdot \boldsymbol{a} = 1$. Moreover in the following notation $\boldsymbol{a}(s, v)$ can be adopted in the following to change the dependence from the height z to the slant height s, according to the relation (7.6). As a last remark we note that the scalar product with the steering vector is carried out by considering only the phase of the received signal, see (7.58). This is obviously a consequence of the interferometric approach, which essentially considers only the phase, or better the phase difference between the SAR images. As shown in the next chapter, the choice to neglect the amplitude information leads to a decrease of the detection performances.

An example of results of PS estimation derived from the processing of Sentinel-1A and Sentinel-1B in the period 2014-2021 is provided in Fig. 7.11. The processing involved all the step described above, i.e. the accurate registration of all the acquisitions, the estimation and compensation of the atmosphere and of the low-pass deformation components and the final PS identification with the use of the MIC. The monitored area is in the Calabria region in South Italy, heavily affected by the presence of landslides. The image on the top shows the entire monitored area; a zoom is provided along the main highway in the proximity of the Lagonegro town. It is evident the efficacy of the methodology for the wide area monitoring.

7.3 Multilook in multitemporal SAR interferometry: SqueeSAR and CAESAR methods

The interferometric multilook operation exploited in the SBAS approach can be further revisited in the context of Persistent Scatterers Interferometry to improve the coverage of monitored points.

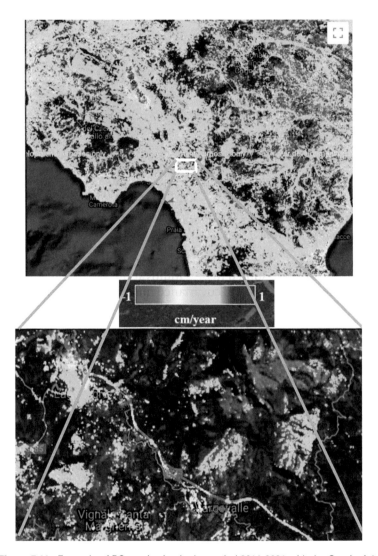

Figure 7.11. Example of PS monitoring in the period 2014-2021 with the Sentinel-1 data: the identified PS are overlaid to a Google Earth image and colored according to the deformation velocity in cm/yr.

We refer to Fig. 7.12 where it is schematically represented the N-length data vectors collecting the available acquisitions, exploited (on the left) in the Single Look (SL) and (on the right) in the Multi-Look (ML) cases: in latter case we assume to have L looks. The data vectors are modeled as in (7.9)-(7.11) and assumed to be statistically independent and homogeneous, i.e., sampled from the same

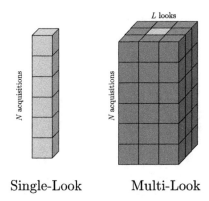

Figure 7.12. Schematic representation of the multivariate observations for the Single-Look and Multi-Look case.

statistic, which is assumed to be a multivariate zero mean circular Gaussian distribution, with covariance matrix defined as in (7.14).

7.3.1 The SqueeSAR approach

The SqueeSAR approach [86,87] founds on the observation that scattering can be roughly divided in two classes: Persistent Scatterers (PS) and Distributed Scatterers (DS).

PS are scatterers for which the backscattering coefficient can be considered as a deterministic value or, at most, a random variable that does not change, i.e. assumes the same realization over the different acquisitions. This model stems from the observation that, in the presence of a dominant scatterer, the backscattering received at the different antennas changes only according to the variation of the distance between the scatterer and the antenna induced by the spatial baselines (range parallax) or the possible deformation. In other words, the signal model in (7.9) simplifies as:

$$y_n = a_n \gamma + w_n, \quad n = 1, \ldots, N \tag{7.59}$$

where, as usual, a_n is a pure phase factor composed by the phase term related to the wanted distance (dependent on the scatterer topography and deformation) and the unwanted phase disturbance (mis-calibrating) terms as the APD. According to what described also in Section 7.1.3 the covariance matrix simplifies in this case in $\mathbf{C}_\gamma = \sigma_\gamma^2 \mathbf{1}\mathbf{1}^T$.

On the other hand, when there is the presence of multiple and almost equally powerful scatterers, the most appropriate physical model of a DS must be considered, compliant with the presence of a variation of the backscattering contribution from antenna to antenna. The model in (7.9) should be therefore considered in place of (7.59). To consider the most general case, we assume the presence of say K independent scatterers:

$$y_n = a_n \sum_{k=1}^{K} v_{nk} e^{-j\frac{4\pi}{\lambda}\delta r_{nk}} + w_n = a_n \gamma_n + w_n \quad n = 1, \ldots, N \quad (7.60)$$

where v_{nk} is the backscattering coefficient of the k-th scatterer to the n-th antenna (pass), and δr_{nk} accounts for the variation of the distance of the n-th antenna from the k-th scatterer and the pixel center. The presence of geometrical and/or temporal decorrelation is accounted for in (7.60) by assigning the correlation properties of v_{nk} along the antennas (n index).

Assuming without loss of generality a symmetry of the scatterers distribution around the pixel center, it follows that a_n contains the phase associated to the distance of the n-th antenna from pixel center.

The backscattering coefficient at the N different antennas must be seen as N different realizations γ_n, $n = 1, \ldots, N$, of the same random variable or, equivalently, as a realization $\boldsymbol{\gamma}$ of a N-length random vector with identically distributed components. Evidently the backscattering covariance matrix is $\mathbf{C}_\gamma = \sigma_\gamma^2 \mathbf{M}_\gamma$ with $\mathbf{M}_\gamma \neq \mathbf{1}\mathbf{1}^T$.

In the case of a multivariate zero mean circular Gaussian distribution, that is:

$$f_{\boldsymbol{\gamma}}(\boldsymbol{\gamma}) = \frac{1}{(2\pi)^N |\mathbf{C}_\gamma|} \exp\left(-\boldsymbol{\gamma}^H \mathbf{C}_\gamma^{-1} \boldsymbol{\gamma}\right) \quad (7.61)$$

the statistical characterization of $\boldsymbol{\gamma}$ is fully described by the scattering covariance matrix $\mathbf{C}_\gamma$, see also (7.12)-(7.13).

In the SL case, under the assumed hypothesis and according to the model in (7.11), the data vector $\boldsymbol{y}$ can be as well as fully characterized by the multivariate Gaussian distribution:

$$f_{\boldsymbol{y}}(\boldsymbol{y}) = \frac{1}{(2\pi)^N |\mathbf{C}_y|} \exp\left(-\boldsymbol{y}^H \mathbf{C}_y^{-1} \boldsymbol{y}\right) \quad (7.62)$$

where the $N \times N$ covariance matrix $\mathbf{C}_y$, expressed by (7.14)-(7.16), depends on the phase vector $\boldsymbol{a}$, see (7.14), and collects all the $N(N-1)/2$ (expected) interferometric values. In the ML case, the L independent and homogeneous looks are jointly distributed as:

$$f_{\boldsymbol{y}_1, \cdots, \boldsymbol{y}_L}(\boldsymbol{y}_1, \cdots, \boldsymbol{y}_L) = \frac{1}{(2\pi)^N |\mathbf{C}_y|} \exp\left(-\sum_{l=1}^{L} \boldsymbol{y}_l^H \mathbf{C}_y^{-1} \boldsymbol{y}_l\right) \quad (7.63)$$

where $\mathbf{C}_y$ is the same covariance matrix as in (7.62), which is common to all the exploited looks.[9]

The probability density function in (7.63) allows to evaluate the maximum likelihood estimate of the vector $\boldsymbol{a}$ containing the useful interferometric phase components:

$$\hat{\boldsymbol{a}} = \arg\max_{\boldsymbol{a}} f_{\boldsymbol{y}_1,\cdots,\boldsymbol{y}_L}(\boldsymbol{y}_1,\cdots,\boldsymbol{y}_L) \tag{7.64}$$

By resorting to the eigenvalues decomposition of $\mathbf{C}_y$ and according to the expression (7.14), it can be easily shown that the determinant $|\mathbf{C}_y|$ is independent from $\boldsymbol{a}$.

This allows rewriting (7.64) as:

$$\hat{\boldsymbol{a}} = \arg\min_{\boldsymbol{a}} \sum_{l=1}^{L} \boldsymbol{y}_l^H \mathbf{C}_y^{-1} \boldsymbol{y}_l \tag{7.65}$$

that is

$$\hat{\boldsymbol{a}} = \arg\min_{\boldsymbol{a}} \operatorname{tr}\big(\widehat{\mathbf{C}}_y \mathbf{C}_y^{-1}(\boldsymbol{a})\big) \tag{7.66}$$

where the dependence of the covariance matrix $\mathbf{C}_y$ on $\boldsymbol{a}$ has been, this time, explicitly expressed and

$$\widehat{\mathbf{R}}_y = \frac{1}{L} \sum_{l=1}^{L} \boldsymbol{y}_l \boldsymbol{y}_l^H = \widehat{\mathbf{C}}_y \tag{7.67}$$

is the sample data covariance matrix. According to the adopted simple statistical model, we have used the zero mean hypothesis in (7.67) regardless of the fact that in the reality the data may have a nonzero mean.

Taking benefit of the expression (7.15) of $\mathbf{C}_y$, the inverse of the data covariance matrix can be written as:

$$\mathbf{C}_y^{-1}(\boldsymbol{a}) = \frac{1}{\sigma_y^2}\big(\boldsymbol{a}\boldsymbol{a}^H \odot \mathbf{M}_y^{-1}\big) \tag{7.68}$$

which finally leads to:

$$\hat{\boldsymbol{a}} = \arg\min_{\boldsymbol{a}} \operatorname{tr}\Big[\widehat{\mathbf{C}}_y(\boldsymbol{a}\boldsymbol{a}^H \odot \mathbf{M}_y^{-1})\Big] \tag{7.69}$$

A final consideration is now in order. The estimate (7.69) depends on the backscattering coherence matrix $\mathbf{M}_y$, which usually is not known: therefore an estimation of such a matrix is required. The

[9]For notation convenience, in this chapter we have used the subscript to indicate the l-th look.

assumption of $\mathbf{M}_y$ to be a real and nonnegative valued matrix and the expression (7.12) of the covariance matrix $\mathbf{C}_y$ lead to the relation:

$$\mathbf{M}_y = \sigma_y^{-2} \text{ABS}(\mathbf{C}_y) \tag{7.70}$$

which can be exploited to achieve the following estimate of the coherence matrix based on the sample covariance matrix (built from the data):

$$\widehat{\mathbf{M}}_y \leftarrow \mathbf{D}_y^{-1/2} \text{ABS}(\widehat{\mathbf{C}}_y) \mathbf{D}_y^{-1/2} \tag{7.71}$$

where $\mathbf{D}_y$ is the diagonal matrix obtained from the diagonal elements of $\widehat{\mathbf{C}}_y$, that is $\mathbf{D}_y = \widehat{\mathbf{C}}_y \odot \mathbf{I}_N$. The substitution of (7.71) in (7.69) makes it possible to evaluate $\hat{a}$, which however deviates from the maximum likelihood estimate. This substitution, which is almost always necessary in practice, reduces the estimation (filtering) performances.

7.3.2 The CAESAR approach

The Component Analysis and Selection SAR (CAESAR) [88,89] approach is based on the Karhunen-Loeve expansion, which allows expanding the covariance matrix in a sum of dyadic (rank 1) orthogonal matrices. We start from the Eigenvalue Decomposition (EVD) of the expected data covariance matrix $\mathbf{C}_y$, which is given by:

$$\mathbf{C}_y = \sum_{k=1}^{N} \lambda_k \boldsymbol{u}_k \boldsymbol{u}_k^H \tag{7.72}$$

where λ_k and $\boldsymbol{u}_k$ are the k-th eigenvalue and the corresponding eigenvector, respectively; without loss of generality, the eigenvalues (which are nonnegative, being $\mathbf{C}_y$ a positive semidefinite matrix) are supposed to be sorted in a decreasing order.

This decomposition is optimal in the sense that, among the say K rank estimates of the covariance matrix ($K < N$), the one achieved by (7.72) as:

$$\mathbf{C}_K = \sum_{i=1}^{K} \lambda_k \boldsymbol{u}_k \boldsymbol{u}_k^H \tag{7.73}$$

minimize the distance (induced by the Frobenius norm [90]) from $\mathbf{C}_y$. Accordingly, by analyzing the eigenvalues dynamic and identifying the few, say again K, most powerful components (assumed to be correspondent to the useful, i.e. noiseless, part of the signal), the samples of the original multivariate zero mean circular Gaussian distribution corresponding to $\mathbf{C}_y$ are well approximated by the samples of the reduced multivariate zero mean circular Gaussian distribution corresponding to $\mathbf{C}_K$. Accordingly, the data vector y, which is a sample of the original distribution, can be structured along orthogonal directions where the variance-covariance expected characteristics can be compacted in a decreasing order starting from the most variable (i.e. powerful) direction, up to the last (i.e. the K-th) significant component. The two extreme cases are the white case and the case in which only one component is significant. To better explain this aspect, we refer to the form of $\mathbf{C}_y$ in (7.15).

The case of a fully coherent scatterer in absence of noise implies that the data coherence matrix is given by:

$$\mathbf{M}_y = \mathbf{1}\mathbf{1}^T \tag{7.74}$$

and, therefore, $\mathbf{C}_y$ can be written as:

$$\mathbf{C}_y = \sigma_y^2 \boldsymbol{a}\boldsymbol{a}^H \tag{7.75}$$

As a consequence, supposing $\boldsymbol{a}$ to be a unit norm vector ($\|\boldsymbol{a}\| = 1$), the expansion in (7.72) is characterized by:

$$\begin{aligned} \lambda_1 &= \sigma_y^2 \\ \lambda_2 &= \cdots = \lambda_N = 0 \end{aligned} \tag{7.76}$$

thus reducing to a single dyadic component. Interestingly, the first eigenvector associated with the only nonnull eigenvalue provides the interferometric phase factor $\boldsymbol{a}$, i.e. the phase only complex vector embedding the wanted (deformation and topography component) and the unwanted (atmospheric) interferometric components. It can be easily shown that, in the presence of noise with power σ_w^2 (on the single image), we have:

$$\begin{aligned} \lambda_1 &= \sigma_y^2 = \sigma_\gamma^2 + \sigma_w^2 \\ \lambda_2 &= \cdots = \lambda_N = \sigma_w^2 \end{aligned} \tag{7.77}$$

with:

$$\begin{aligned} &\boldsymbol{u}_1 = \boldsymbol{a} \\ &\boldsymbol{u}_2, \dots, \boldsymbol{u}_N \text{ any orthonormal basis of } \mathcal{A}^\perp \end{aligned} \tag{7.78}$$

being $\mathcal{A}^\perp$ the $(N-1)$-dimensional subspace of $\mathbb{C}^N$ orthogonal to $\boldsymbol{a}$. It is clear that, whatever the SNR, $\boldsymbol{a}$ can be always exactly evaluated as the eigenvector associated with the first (dominant) eigenvalue.

Note that the exact estimate can be achieved thanks to the availability of the expected data covariance matrix. This specific case is tantamount to stating that the data model in (7.11), that is

$$y = a \odot \gamma + w \tag{7.79}$$

can be substituted with

$$y = a\gamma + w \tag{7.80}$$

emphasizing that the scatterer provides a contribution at the different antennas structured as a deterministic vector, i.e. a, scaled by the realization of a random variable, i.e. γ.

Finally, for a completely decorrelating scattering, i.e. a scattering vector γ having:

$$\mathbf{M}_\gamma = \mathbf{I} \tag{7.81}$$

that is:

$$\mathbf{C}_y = \sigma_y^2 \mathbf{I} \tag{7.82}$$

we have:

$$\lambda_1 = \cdots = \lambda_N = \sigma_y^2 \tag{7.83}$$

that is all constant eigenvalues indicating the absence of concentration of power along any direction. The eigenvectors are in this case provided by any orthonormal basis of $\mathbb{C}^N$. It is evident that, due to its uncorrelated nature at the different antennas, the response of the same scatterer over the array acts on with the same statistic of the noise and therefore the deterministic structure associated with the phase vector a cannot anymore be inferred from the data.

The above reasoning concerning the expected data covariance matrix can be applied to the sample covariance matrix estimated by averaging say L data looks, that is $\widehat{\mathbf{C}}_y$ provided by (7.67). Note that such a matrix collects all the possible multilook interferograms that can be generated by the N acquisitions. In the presence of a fully coherent scattering, for a finite L, the eigenvector associated with the dominant eigenvalue in the EVD of $\widehat{\mathbf{C}}_y$ provides an estimate of a that is the more and more accurate, the greater L and SNR. What discussed above can be interpreted in the context of the Principal Component Analysis. We start by stacking the L looks composing the sample covariance matrix in (7.67) within the $N \times L$ matrix $\mathbf{Y}$, hereafter referred to as data look matrix:

$$\mathbf{Y} = [y_1, \ldots, y_L] \tag{7.84}$$

Accordingly, it results:

$$\widehat{\mathbf{C}}_y = \frac{1}{L}\mathbf{Y}\mathbf{Y}^H \tag{7.85}$$

Generally it is $L \ll N$: this condition, although not necessary, will be assumed in the following to specify the matrix form. Moreover, again without loss of generality, we assume that $\mathbf{Y}$ has the maximum rank, i.e. L. Let us consider the Singular Value Decomposition of the data look matrix $\mathbf{Y}$:

$$\mathbf{Y} = \mathbf{U}\mathbf{\Sigma}\mathbf{V}^H = \sum_{k=1}^{L} \sigma_k \boldsymbol{u}_k \boldsymbol{v}_k^H \tag{7.86}$$

where $\mathbf{U}$ is the $N \times L$ matrix collecting the left singular vectors $\boldsymbol{u}_k$, $\mathbf{V}$ is the $L \times L$ matrix collecting the right singular vectors $\boldsymbol{v}_k$, and $\mathbf{\Sigma} = \mathrm{diag}(\sigma_1, \ldots, \sigma_L)$ is the $L \times L$ diagonal matrix collecting the singular values σ_k. It is worth noting that:

$$\mathbf{Y}\mathbf{Y}^H = \mathbf{U}\mathbf{\Sigma}^2\mathbf{U}^H \tag{7.87}$$

so, according to the expression (7.85) of the sample covariance matrix, we conclude that the eigenvectors of $\widehat{\mathbf{C}}_y$ are the left singular vectors of $\mathbf{Y}$, whereas the eigenvalues of $\widehat{\mathbf{C}}_y$ are the squared singular values of $\mathbf{Y}$, scaled by L. Based on the expansion (7.86), it can be easily derived the following synthesis expression for $\mathbf{Y}$:

$$\mathbf{Y} = \left[\sum_{k=1}^{L} \sigma_k \boldsymbol{u}_k v_{k,1}^*, \sum_{k=1}^{L} \sigma_k \boldsymbol{u}_k v_{k,2}^*, \ldots, \sum_{k=1}^{L} \sigma_k \boldsymbol{u}_i v_{k,N}^* \right] \tag{7.88}$$

where $v_{k,n}$, $n = 1, \ldots, N$, is the n-th entry of the vector $\boldsymbol{v}_k$ and $(\cdot)^*$ is the conjugate operator. The expression (7.88) highlights the fact that the different data vectors can be all synthesized as a combination of the L eigenvectors $\boldsymbol{u}_k$ of $\widehat{\mathbf{C}}_y$ through the coefficients given by the rows elements of $\mathbf{V}$. The case for which only the first eigenvalue $\lambda_1 = \sigma_1^2/L$ of $\widehat{\mathbf{C}}_y$ is significant, i.e.:

$$\mathbf{Y} = \left[\sigma_1 \boldsymbol{u}_1 v_{1,1}^*, \sigma_1 \boldsymbol{u}_1 v_{1,2}^*, \ldots, \sigma_1 \boldsymbol{u}_1 v_{1,N}^* \right] \tag{7.89}$$

corresponds to a situation where the look data vectors are one scaled with respect to all the others, that is in the presence of a fully coherent scatterer. The above reasoning allows, to identify the dominant eigenvectors $\boldsymbol{u}_k$ as the principal data components that can be extracted to filter, similarly to SqueeSAR, the data. Differently from SqueeSAR CAESAR can also separate multiple interfering components due to layover.

spatial multilook filtering   CAESAR filtering

building topography separation with CAESAR

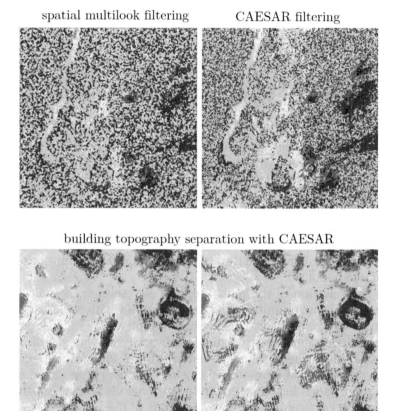

Figure 7.13. Example of filtering achieved with the CAESAR approach. The images in the upper row show a comparison between the result of the filtering of an interferogram achieved via the classical multilook and that obtained by filtering the whole stack with the CAESAR approach. The images in the lower row show the capability of CAESAR to separate interfering scatterers directly at the level of interferograms generation, prior to any atmospheric calibration. After the separation of two dominant scattering mechanisms, the height difference is evaluated to order according to the height.

Fig. 7.13 presents an example of results of CAESAR filtering of the dominant component and separation of layover components over built structures. The images in the upper row show how the CAESAR approach, by filtering the whole stack, allows recovering reliable phase information in areas which result to be decorrelated to a degree that cannot be handled by the standard (complex) multilook discussed in Section 6.5. The lower row provides an example of separation of layover components, which have been ordered according to the height. Interferograms having the topography emphasized or de-emphasized in the layover areas can be generated: topography mitigation can be helpful to facilitate the unwrapping step [91].

7.4 Phase unwrapping with multiple acquisitions

Section 6.7 has described the problem to retrieve the unrestricted interferometric phase values on a generally sparse grid of P pixels of a single interferogram (stacked in the vector φ of P elements), by adding to the corresponding wrapped values ϕ "proper" 2π multiples.

It has been also pointed out that this problem is intrinsically ill-posed (infinitely many solutions are compatible with the measured wrapped data) and thus requires additional assumptions in order to be solved.

The selection of the most "appropriate" solution is typically carried out by minimizing the discontinuities between adjacent pixels. More in details, considering a graph with Q edges generated, commonly via triangulation, from the P vertexes of the sparse grid of measurements, almost all the PhU implement a first step aimed at estimating the Q unrestricted phase variations on the edges (stacked in the Q-dimensional vector $\Delta\hat{\varphi}$) starting from the corresponding phase variations of the wrapped phase, see (6.92) and (6.93).[10] This can be easily carried out, under the Itoh assumption (i.e. by assuming that true phase has variations in modulus limited to π) by restricting the wrapped phase variations. Violations of the Itoh condition are almost always present and as a consequence the estimated unrestricted phase variation field does not "curls to zero" over closed loops (circuits) formed by edges, see (6.103).

Thanks to the fact that the unwrapping is carried out on a 2D grid, it is possible to operate (2π multiple) corrections of the estimated phase variation to eliminate circulation residues. It should be evident that, in unwrapping a single interferogram, the formation of a redundant spatial network connecting the pixels, which allows identifying circuits, is a prerequisite to the possibility to perform a correction of the estimated gradients making the unwrapped solution independent from the chosen integration path.

---

[10]Note that differently from this chapter, in which the symbol Δ represents the variation of the signal between two acquisitions in the baseline domain, i.e. an interferogram, in Chapter 6 (where only a single interferogram was considered) the same symbol indicated the spatial variation over the edges of the spatial graph.

It should be also rather intuitive at this point thinking that the availability of multiple ($M > 1$) interferograms gives an opportunity to further improve the unwrapping process, which remains a critical step in interferometric processing. As described in the following, a check on the unwrapped phase or even implementation of algorithms that performs "better"[11] than the classical 2D unwrapping are in fact possible when working with a stack of interferogram.

We start from the possibility to perform a check of the goodness of the unwrapping, that is to detect pixels where the 2D (independent) phase unwrapping of the M interferograms generates inconsistent results.

Let us assume the m-th interferogram to be formed by the n-th and i-th images has been unwrapped on the spatial grid formed by the $p = 1, \ldots, P$ pixels. With reference to the p-th pixel, the unwrapped interferometric phase $\Delta\varphi_{m_p}$[12] is related to the unrestricted phase φ_{n_p} and φ_{i_p} corresponding to the acquisitions n and i (involved in the m-th interferogram) by:

$$\Delta\varphi_{m_p} = \varphi_{n_p} - \varphi_{i_p} \tag{7.90}$$

Collecting as usual all the M unwrapped interferometric phase values in the vector $\Delta\boldsymbol{\varphi}_p$ and the unrestricted phases of the N acquisitions in the vector $\boldsymbol{\varphi}_p$, if the unwrapping process has produced consistent 2π corrections, according to (7.29), $\Delta\boldsymbol{\varphi}_p$ should be in the range of the operator $\mathbf{D}$, that is, in the space spanned by the columns of $\mathbf{D}$. Consequently, the component of $\Delta\boldsymbol{\varphi}_p$ belonging to the orthogonal complement of the range of $\mathbf{D}$, i.e. to the left null space of $\mathbf{D}$, provides an indication of the consistency of the phase unwrapping. By resorting to the definition of orthogonal projector on the range of a matrix, such a component is given by:

$$\boldsymbol{d}_p = (\mathbf{I}_M - \mathbf{D}\mathbf{D}^\dagger)\Delta\boldsymbol{\varphi}_p = \Delta\boldsymbol{\varphi}_p - \mathbf{D}\hat{\boldsymbol{\varphi}}_p \tag{7.91}$$

where $\hat{\boldsymbol{\varphi}}_p = \mathbf{D}^\dagger \Delta\boldsymbol{\varphi}$ is the least squares estimate of the vector $\boldsymbol{\varphi}_p$, given by the inversion discussed in Section 7.2.1 and collecting the phase series in the pixel p. Frequently, in order to have a normalized index, the following quantity is considered as an index of

---

[11]It should be underlined that the adverb "better" relates to the fact that the solution is closer, according to a weighted norm criteria, to the wanted "least discontinuous" (see (6.103)) solution, which is however generally not the true one in all the pixels.

[12]We explicitly indicate the dependence of the quantities on the pixel p to highlight the presence of two different domains, the spatial domain (with pixels) and the baseline domain with acquisitions.

consistency of the unwrapping process in the pixel p:

$$C_p = \frac{1}{M} \left| \sum_{m=1}^{M} e^{j d_m} \right| = \frac{1}{M} \left| \sum_{m=1}^{M} e^{j\left[\Delta\phi_m - \{\mathbf{D}\hat{\varphi}\}_m\right]} \right| \qquad (7.92)$$

$$= \frac{1}{M} \left| \sum_{m=1}^{M} e^{j\left[\Delta\phi_m - j(\hat{\varphi}_n - \hat{\varphi}_i)\right]} \right|$$

where as usual $\{\cdot\}_m$ is the operator that extract the m-th element and the subscripts n and i depend obviously on m. It is worth noting that the last equality in the index C_p emphasizes that it can be interpreted as the average of the coherent sum of the residuals between the starting (wrapped) interferograms and the interferograms built from the estimated phase series, that is the phase at all acquisitions achieved after the inversion of (7.29).

Obviously the index in (7.92) is significant only in the presence of redundant network of interferograms.[13] If the independent 2D unwrapping of all interferograms provides consistent results, all the elements of $\Delta\boldsymbol{\varphi}$ and $\mathbf{D}\hat{\boldsymbol{\varphi}}$ are equal[14]: the index C_p in this case is 1 or close to 1 if a multilook is implemented. In real cases, especially moving away from the starting point of the integration during the unwrapping process, the vector $\boldsymbol{d}$ in (7.91) is characterized by more and more elements different from zero, thus generating residual differences between the started interferogram and the interferogram generated from the phase series after the inversion: as a result the consistency index C_p decreases. Accordingly, by selecting pixels with the consistency index larger than a suitable threshold, one can identify areas where unwrapping results to be high consistent, according to (7.92).

It is possible to exploit the baseline graph to operate a correction of the unwrapped phase values generated with independent 2D phase unwrapping. More specifically one could look for the vector $\boldsymbol{k}_p$ of integers values, with some minimum weighted norm, such that the vector of the corrected unwrapped interferometric values:

$$\Delta\boldsymbol{\varphi}_p + 2\pi \boldsymbol{k}_p = \mathbf{D}\boldsymbol{\varphi}_p \qquad (7.93)$$

[13] If the graph representing the interferogram network is a tree (that is if removing a single edge disconnects the graph) C_p would achieve values compressed around its maximum (1).

[14] Actually, if a multilook has been adopted as discussed in Section 6.5 the measured wrapped interferometric phase value does not triangulate and therefore the difference between $\Delta\boldsymbol{\varphi}$ and $\mathbf{D}\hat{\boldsymbol{\varphi}}$ will be close but not exactly zero. Only the use of a SqueeSAR like algorithm allows extracting wrapped phases characterized by the triangulation property.

is in the range of the operator $\mathbf{D}$, that is having no component in the left null space of the operator $\mathbf{D}$. It should be clear that the above correction is a problem dual to the spatial domain unwrapping formulated in (6.103), (6.104), or (6.105). It should be also clear that this correction makes the solution to depart from the one which guarantees the space domain optimization (6.103), or equivalently (6.104), or (6.105). Actually, as discussed in [80] one can solve jointly the unwrapping problem considering the graphs built in the two domains, i.e. the spatial and the baseline domain. It should be however noticed that, such a solution will make the consistency index saturated to 1 thus preventing the possibility to use (7.92) to detect areas where the unwrapping in one domain (generally spatial) with the given weighted norm provides inconsistencies in the other domain. Some intermediate solutions based on a sequence of a phase unwrapping in the baseline and space domain are discussed in [92].

8

SAR tomography

InSAR and DInSAR techniques, described in Chapters 6 and 7, assume that a single dominant scatterer is backscattering the incidence radiation in each range-azimuth resolution cell. These techniques can not therefore provide reliable results in the presence of volume scattering; moreover, they may not be fully appropriate in areas with the presence of layover phenomena, e.g. urban areas, where several scatterers may be located in the same range-azimuth resolution cell at different heights.

More recently, Tomographic SAR (TomoSAR) techniques have extended interferometric methods to multi-dimensional imaging. These techniques exploit multibaseline acquisitions in interferometric configurations to perform the fully 3D focusing of the ground reflectivity profile along range, azimuth, and elevation coordinates. When the data are acquired at different dates (multitemporal data) it is possible to obtain the image focusing in the 4D (3D space + deformation velocity) or 5D (3D space + deformation velocity + thermal dilation) spaces.

Tomographic 3D imaging provides a reconstruction of the scene scattering structure along the vertical direction, which is not possible with InSAR techniques. This is a very important feature that has been widely exploited in applications involving volume-distributed scattering along the vertical direction, like in the case of forested areas or ice regions, where the incident wave can penetrate the surface especially when lower frequency sensors, such as P-band or L-band SAR systems, are used. For the forest case, the backscattered signal is mainly the superposition of contributions from the vegetation layer and from the ground, that have to be separated for reconstructing the vertical structure of the forest. TomoSAR imaging has also proved to be very useful also when surface scattering is dominant, such as in urban areas reconstruction. These complex scenes are typically characterized by the presence of buildings with different heights, and thus layover between the facades of the higher structures, the rooftop of the smaller buildings and the ground is possible. Also in this case, the backscattered signal is the superposition of different contributions, from the different man-made structures present in the same range-azimuth

Multi-Dimensional Imaging with Synthetic Aperture Radar
https://doi.org/10.1016/B978-0-12-821655-2.00012-7

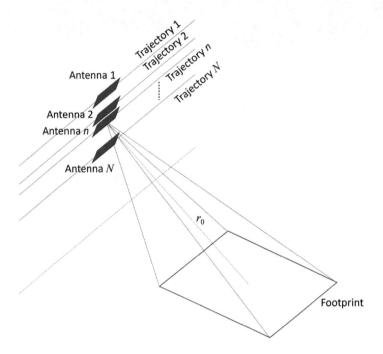

Figure 8.1. TomoSAR geometry.

resolution cell and from the ground surface, that have to be separated for the correct reconstruction of the vertical profile of the observed scene.

8.1 TomoSAR data model

TomoSAR can be essentially viewed as an extension in terms of modelling and processing of the multibaseline-multipass SAR Interferometry described in Chapter 7.

Consider the geometry of multiview SAR acquisitions of Fig. 8.1, where a given scenario on the ground is observed from N different trajectories of the SAR system. After the SAR image focusing process, a stack of N 2D range-azimuth images is produced (one for each trajectory). Consider now the 2D geometry of Fig. 8.2, obtained fixing the azimuth coordinate x, where the usual Cartesian reference system determined by the range r and elevation s is represented. Moreover, for notation symmetry with the azimuth case, we have introduced an axis s', parallel to s to describe the antenna positions in terms of orthogonal baselines. We remind that the latter are necessary to provide the sensitivity to the elevation (slant height) s in interferometric systems, see (7.7) and (7.6).

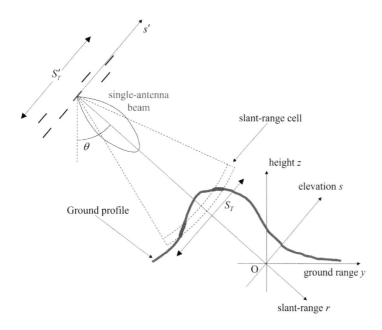

Figure 8.2. Range cell at fixed azimuth in TomoSAR geometry.

If we consider a possible ground profile, a part of it, which is depicted in blue (dark gray in print version) in Fig. 8.2, is imaged in the same range-azimuth cell. Consequently, the value of the n-th SAR image in the generic (x', r') pixel is given by the coherent superposition of the echoes back-scattered by all scatterers contained in the considered range-azimuth resolution cell (in the blue (dark gray in print version) part), located at different elevations s. By understanding the dependence on the coordinates (x', r') for notation simplicity, the complex image pixel value can be expressed as:

$$y\left(s_n'\right) = \int_{-S_T/2}^{S_T/2} \gamma\left(s\right) \exp\left(j\frac{4\pi}{\lambda} R\left(s_n', s\right)\right) ds,$$

$$\text{with} - S_T'/2 \leqslant s_n' \leqslant S_T'/2 \tag{8.1}$$

where λ is the working wavelength, $\gamma\left(\cdot\right)$ is the reflectivity function, s_n' for $n = 1, ..., N$, are the elevation coordinates of the different SAR trajectories along the axis s', i.e. the orthogonal baselines indicated with b_n in the previous chapter, $R(s_n', s)$ is the distance between the scatterer at elevation s and the n-th antenna positions, and S_T is the extension of the range-azimuth cell along the elevation. The notation $y(s_n')$ in place of y_n, which was adopted in the previous chapter to indicate the signal in a given azimuth-range pixel at the generic acquisition n, highlights the substitution of b_n with s'_n, that is the n-th sample along the axis s'. A comment is now in order on the sign of the exponential factor in (8.1). It can be noted that, differently from the previous chapters (see f.i. (5.16)), this term has been indicated with a positive sign. The choice has been done to obtain in the following a direct FT relation between the data and the reflectivity function. This change of sign does not impact the analysis because the final aim of SAR tomography is the reconstruction of the reflectivity amplitude or intensity profile along the elevation. Even adopting the convention of the previous chapters one can consider the conjugate of the data to achieve the direct FT relation.

The different trajectories, as it is shown in Fig. 8.2, are characterized by different coordinates s'_n along the s' axis, corresponding to different orthogonal baselines respect to a reference (or master) antenna, allows viewing a ground point from slightly different view angles. The synthetic antenna, composed by N TX/RX (monostatic) elements at different s'_n, can be also viewed as an array of size S'_T along the elevation direction. The principle behind SAR Tomography is the exploitation of the information acquired along the elevation direction to synthesize a large array in elevation, similarly to the azimuth direction, which focus the data and allows achieving meter-order resolution for the reconstruction of the three-dimensional reflectivity of the ground scene.

If the overall baseline span S'_T is small compared to the distance from the ground r_0 (condition always verified in space-borne systems), the distance function can be expanded in Taylor series so that $R(s'_n, s) \cong r_0 + (s'_n - s)^2/(2r_0) + b_{\|n}$, where $b_{\|n}$ is the parallel baseline, and (8.1) becomes:

$$y\left(s'_n\right) = e^{j\frac{4\pi}{\lambda}\left(r_0 + b_{\|n}\right)} \int_{-S_T/2}^{S_T/2} \gamma(s) \exp\left(j\frac{2\pi}{\lambda r_0}(s'_n - s)^2\right) ds \tag{8.2}$$

$$\text{with} - S'_T/2 \leqslant s'_n \leqslant S'_T/2$$

With a slight abuse of notation, we operate the following substitutions:

$$y\left(s'_n\right) \rightarrow y\left(s'_n\right) \exp\left(-j\frac{4\pi}{\lambda}\left(r_0 + b_{\|n}\right)\right) \exp\left(-j\frac{2\pi}{\lambda r_0}s'^2_n\right)$$

$$\gamma(s) \rightarrow \gamma(s) \exp\left(j\frac{2\pi}{\lambda r_0}s^2\right) \tag{8.3}$$

thus rewriting (8.2) as:

$$y\left(s_n'\right) = \int_{-S_T/2}^{S_T/2} \gamma\left(s\right) \exp\left(-j\frac{4\pi}{\lambda r_0}s_n's\right) \mathrm{d}s \tag{8.4}$$
$$\text{with} - S_T'/2 \leqslant s_n' \leqslant S_T'/2$$

Note that the first substitution in (8.3) implies a multiplication by a quadratic phase factor in s'. This operation is equivalent to the deramping procedure described in Chapter 7, specifically (7.2) which is typically implemented with respect to a reference height profile, generally provided by an external DEM. From (8.4) it can be seen that the signal $y\left(s_n'\right)$, obtained by phase compensating each range-azimuth pixel of the image $y(s_n')$ acquired in the position s_n', is the Fourier transform of the phase corrected reflectivity $\gamma\left(s\right)$ computed in the spatial frequency:

$$\zeta_n = 2s_n'/(\lambda r_0) \tag{8.5}$$

Eq. (8.4) is the key equation for the mathematical formulation of TomoSAR. It states that the SAR multibaseline data and the elevation reflectivity profile form a Fourier pair. Each acquisition, in a given azimuth range pixel, provides (after deramping) a sample of the scene reflectivity spectrum at a (spatial) frequency related to the baseline, see (8.5). Hence, $\gamma(s)$ can be retrieved by taking the inverse Fourier Transform of the data with respect to the orthogonal baseline.

From a qualitative point of view, by viewing the imaging system along elevation as a nonuniform array, we can claim that the larger the array size S_T', the better the resolution of the recovered reflectivity profile along s on the ground. This point will be discussed more in detail in Section 8.1.2.

In several practical situations, as in urban applications, the acquisition geometry is such that the part of scenario imaged in a single range-azimuth resolution cell can be very well represented by a set of discrete scatterers, as shown in Fig. 8.3. In this case, the multiple-antenna system (dashed box), once detected the presence of scatterers inside the cell, can recover their positions along s, by coherently processing the signals received by the different antennas. As a result, a larger antenna along s is synthesized along the elevation, and narrower antenna beams are obtained. This antenna synthesis along the elevation allows to image the targets T_1 and T_2 as separate from each other.

In the following part of this chapter, we will consider first the discrete TomoSAR model obtained by discretizing (8.4), then we will describe the approaches that can be used to recover the re-

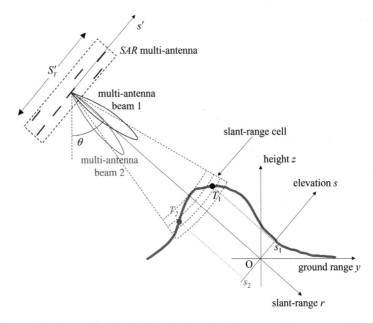

Figure 8.3. Range cell at fixed azimuth in TomoSAR geometry for a topographic profile generating a layover.

flectivity profile along the elevation and we will analyze their performance.

8.1.1 The discrete TomoSAR data model

From an operational point of view, the inversion of (8.4) can be implemented by uniformly sampling the elevation coordinate s with a constant step Δs. Without lack of generality we assume that the origin of the s axis is such that the values of s are not negative. The discrete data model can be therefore expressed as follows:

$$y\left(s_n'\right) \cong \sum_{m=0}^{M-1} \gamma\left(m\Delta s\right)\exp\left(-j\frac{4\pi}{\lambda r_0}s_n'm\Delta s\right)\Delta s \qquad (8.6)$$

where M is the number of cells (tomographic bins) in which the ground scene is subdivided along the elevation direction, i.e. $S_T = (M-1)\Delta s$.

In a vector form, (8.6) can be written as:

$$y = \mathbf{A}\boldsymbol{\gamma} \qquad (8.7)$$

where

$$\boldsymbol{y} = [y(s_1'), y(s_2'), \cdots, y(s_N')]^T$$
$$\boldsymbol{\gamma} = [\gamma(0), \gamma(\Delta s), \cdots, \gamma((M-1)\Delta s)]^T$$
$$\mathbf{A} = [\boldsymbol{a}(0), \boldsymbol{a}(\Delta s), \cdots, \boldsymbol{a}((M-1)\Delta s)]$$
$$\boldsymbol{a}(s) = \left[\exp\left(-j\frac{4\pi}{\lambda r_0}s_1's\right), \exp\left(-j\frac{4\pi}{\lambda r_0}s_2's\right), \cdots, \exp\left(-j\frac{4\pi}{\lambda r_0}s_N's\right) \right]^T$$

$$\tag{8.8}$$

with $s_m = m\Delta s$ being the elevation bin and $\mathbf{A}$ a matrix of size $N \times M$ usually referred to as the steering matrix; its columns are the steering vectors to the generic elevation s introduced in the previous chapter, see (7.57), with reference to the 2D domain including the height and the mean deformation velocity. Here we do not consider the velocity: $\boldsymbol{a}(s)$ is equivalent to $\boldsymbol{a}(z = s\sin\theta, v = 0)$, see also Fig. 8.3.

The recovery of the sampled reflectivity profile can be performed by inverting the linear system (8.7), which relates the unknown vector $\boldsymbol{\gamma}$ (belonging to the unknown space $\mathbb{C}^M$) and the data vector $\boldsymbol{y}$ (belonging to the data space $\mathbb{C}^N$) through the linear operator $\mathbf{A}$.

To address the inversion of the system (8.7), following the subspace notation introduced also in Chapter 7, let us denote by $\mathcal{R}(\mathbf{A})$ the range (also referred to as image) and $\mathcal{N}(\mathbf{A})$ the null space (also referred to as kernel) of $\mathbf{A}$. Furthermore, let $K \leqslant \min\{N, M\}$ be the rank of $\mathbf{A}$: the dimensions of the range and null space are therefore given by $\dim(\mathcal{R}(\mathbf{A})) = K$ and $\dim(\mathcal{N}(\mathbf{A})) = M - K$, respectively [93].

The system (8.7) admits a unique solution when it is well-posed, i.e., when $\boldsymbol{y} \in \mathcal{R}(\mathbf{A})$ and $\dim(\mathcal{N}(\mathbf{A})) = 0$, the latter condition being satisfied when $K = M$. If one of these two conditions is not verified, the system is said to be ill-posed. In this case it can admit either infinite solutions (when $\boldsymbol{y} \in \mathcal{R}(\mathbf{A})$ but $\dim(\mathcal{N}(\mathbf{A})) > 0$) or no solutions (when $\boldsymbol{y} \notin \mathcal{R}(\mathbf{A})$). No solutions are due to the presence of noise, which brings the data vector $\boldsymbol{y}$ outside $\mathcal{R}(\mathbf{A})$. In this case, although no exact solutions exist, the system (8.7) can be solved in the least squares sense, by looking for the solutions whose images minimize the Euclidean distance from the data vector $\boldsymbol{y}$. The ill-posed system admits a unique least squares solution when $\dim(\mathcal{N}(\mathbf{A})) = 0$; otherwise, if $K < M$, infinite least squares solutions exist. In both cases, a least squares solution can be found by inverting the system (8.7) through the generalized inverse (already introduced in Chapter 7) of the matrix $\mathbf{A}$. The generalized inverse matrix has a simple closed form expression when the least squares solution is unique; in case of infinite solutions, on the other hand, it can be evaluated by resorting to the SVD and leads to the minimum norm solution.

Accordingly, the following different cases may occur:

1. The dimension of the unknown space is smaller that the dimension of the data space ($M < N$). In this case the system in (8.6) is said to be overdetermined and the condition $K = M$ (full rank system matrix) can be verified: when $K = M$, a unique least squares solution exists. However the system matrix $\mathbf{A}$ can be ill-conditioned[1] even when it is full rank: in this case the unique least squares solution results to be unstable. This condition can occur when a large number N of multibaseline acquisitions, with very irregular baseline distribution, is available. Accordingly, in both cases of infinite ($K < M$) or unique ($K = M$) but unstable least squares solutions, a regularization is required.

2. The dimension of the data space is smaller that the dimension of the unknown space ($N < M$). In this case the system in (8.6) is said to be underdetermined and the condition $K < M$ is verified even in the case of full rank system matrix ($K = N$). Therefore, in this case, the system admits always an infinite number of exact ($K = N$) or least squares ($K < N$) solutions. This condition occurs, for instance, when few acquisitions are available and the ground scene is rather extended in the elevation direction. It is the most frequent case because of the need to avoid loss of information in the sampling of the unknown, and particularly if one aims to achieve a degree of superresolution in the tomographic imaging process. This point will be analyzed in deeper details in the following sections.

3. The data space and the unknown space have the same dimension ($N = M$), and the system matrix is full rank ($K = N$). In this case $\mathbf{A}$ is a square and invertible matrix, and the system in (8.6) has a unique exact solution for every data vector $\mathbf{y}$. This situation, however, rarely occurs in practical cases. Moreover, as for an overdetermined system, the matrix $\mathbf{A}$ can be ill conditioned and, in this case, the solution is unstable.

Due to the presence of measurement noise and reflectivity changes related to the variations of the view angle (spatial decorrelation) and to the temporal evolution of the scene between each pass (temporal decorrelation), there are some fluctuations that can be modeled by describing the N observations as a complex-valued random vector.

---

[1]A matrix is said to be ill-conditioned when has a large condition number, the latter being the ratio between the maximum and the minimum singular value [93].

If the decorrelation phenomena are negligible, this perturbation is simply modeled by considering a complex valued additive component, which mainly takes into account thermal noise and incoherent clutter components, so that Eq. (8.7) becomes:

$$y = A\gamma + w \tag{8.9}$$

with

$$w = [w_1, \dots, w_N]^T \tag{8.10}$$

being the noise vector.

The decorrelation phenomena can also be accounted for, although this requires a slight modification of the data model in (8.9). To this aim, let us rewrite (8.9) as:

$$y = \sum_{m=0}^{M-1} a_m \gamma_m + w \tag{8.11}$$

where, for sake of notation simplicity, we put $\gamma_m = \gamma(m\Delta s)$ and $a_m = a(m\Delta s)$, according to the definitions in (8.8).

A consideration is now in order. Eq. (8.11) highlights that each element of the vector a_m (m fixed), forming the kernel of the FT pair described above, is weighted by the same backscattering coefficient γ_m, which has been assumed so far a deterministic unknown parameter. This assumption is tantamount assuming that the scatterer does not decorrelate from one acquisition to the other. Even if we model γ_m as a random variable, the fact that the same realization is assumed on each acquisition still corresponds to the assumption of absence of decorrelation. The presence of decorrelation phenomena translates in the fact that the backscattering coefficient varies along the acquisitions. The most general data model accounting for decorrelation is achieved by modifying (8.11) as:

$$y = \sum_{m=0}^{M-1} a_m \odot \gamma_m + w \tag{8.12}$$

where γ_m is a complex random vector, generally assumed zero mean with a given correlation matrix. The latter matrix describes the degree of correlation held by the scatterer at the elevation s_m along the different acquisition pairs. It is important to stress here the difference between γ_m in (8.12) and γ in (8.9): the former is a random vector (or its realization) collecting the values of the backscattering coefficient characterizing the scatterer imaged at mth sample of the elevation along the different N acquisitions; the latter is the deterministic vector collecting values of the backscattering coefficient of the scatterers imaged by each antenna along the M samples of elevation axis. Decorrelation is important for acquisitions collected over different times, with large temporal separation. For airborne systems used for 3D mapping of forest, typically the acquisitions are collected at closed times. Accordingly, in the following the discussion will be carried out by assuming absence of decorrelation phenomena: the problem of reconstructing the (deterministic) vector γ from the noiseless or noisy data model in (8.7) or (8.9), respectively, will be thus addressed.

8.1.2 Fourier tomography and Rayleigh elevation resolution

When the involved signals $y(s')$ and $\gamma(s)$ are both uniformly sampled at a sampling rate satisfying the Nyquist condition, the inversion of (8.9) can be efficiently implemented through Fast Fourier Transform (FFT) algorithms.

In fact, assuming a uniform baseline distribution with step $\Delta s'$, so that $s'_n = n\Delta s'$ and $s_m = m\Delta s$ with:

$$\frac{2}{\lambda r_0}\Delta s'\Delta s = \frac{1}{N} \tag{8.13}$$

and $N = M$, substituting the discretized values of s' and s in (8.8), we obtain:

$$\mathbf{A} = \begin{bmatrix} 1 & 1 & \dots & 1 \\ 1 & e^{-j\frac{2\pi}{N}} & \dots & e^{-j\frac{2\pi}{N}(N-1)} \\ 1 & e^{-j\frac{2\pi}{N}2} & \dots & e^{-j\frac{2\pi}{N}2(N-1)} \\ . & . & \dots & . \\ 1 & e^{-j\frac{2\pi}{N}(N-1)} & \dots & e^{-j\frac{2\pi}{N}(N-1)(N-1)} \end{bmatrix} \tag{8.14}$$

that is the $N \times N$ Discrete Fourier Transform (DFT) matrix.

In other words, under the assumption of a uniform baseline distribution and uniform reflectivity sampling, the steering matrix becomes a DFT transformation matrix, provided that the baseline and reflectivity sampling steps are chosen according to (8.13). In this hypothesis, $\mathbf{A}$ is a scaled unitary matrix and the inversion of (8.7) can be efficiently performed by means of an inverse FFT carried out via the matrix $\mathbf{A}^H$, where as usual $(\cdot)^H$ is the hermitian (transpose conjugate) operator.

The elevation resolution of the reflectivity profile is given by the 3 dB DFT spectral resolution and then is expressed as:

$$\rho_s = \frac{\lambda r_0}{2N\Delta s'} = \frac{\lambda r_0}{2S_T'}, \tag{8.15}$$

where $S_T' = N\Delta s'$ is the total baseline span (see Fig. 8.2). The quantity ρ_s in (8.15) is commonly denoted as Rayleigh resolution along the elevation direction. The corresponding height resolution ρ_z is given by:

$$\rho_z = \rho_s \sin\theta = \frac{\lambda r_0}{2S_T'}\sin\theta \tag{8.16}$$

where θ is the look angle (see Fig. 8.2). We again underline that the Earth surface curvature has been neglected for simplicity: in the presence of Earth curvature the look angle should be substituted by the incidence angle with respect to the vertical direction.

When using DFT techniques, the operator in (8.4) has to be discretized by uniformly sampling the ground elevation coordinate s with a sampling interval lower or equal to the value given by (8.15), while the corresponding in orbit baseline spacing is given by:

$$\Delta s' = \frac{\lambda r_0}{2N\Delta s} = \frac{\lambda r_0}{2S_T}, \tag{8.17}$$

which represents the maximum baseline spacing, corresponding to the Nyquist rate, required for reconstructing an alias free reflectivity profile over an elevation extension equal to S_T. When the extension of the ground scene along the elevation direction increases, a smaller baseline spacing is required.

Nonuniform sampling can also be applied in accordance to the extended Shannon theorem, provided that the average sampling rate satisfies the Nyquist condition [94]. The reconstruction will be still good, but more difficult to be implemented. It consists in the evaluations of the data that would be acquired with uniformly spaced baselines obtained by interpolating the nonuniformly sampled data by means of nonuniform sampling interpolation functions [95]. After interpolation, DFT techniques can be applied for the elevation reflectivity profile reconstruction. We remind again that this interpolation based processing scheme would require a large number of acquisitions, with an average baseline spacing lower that the Nyquist one (8.17).

Other kind of processing techniques can be applied when the acquisition spatial density is lower than that required by the nonuniform sampling theorem. Before introducing the most used techniques, some specifications on the adopted data model are given.

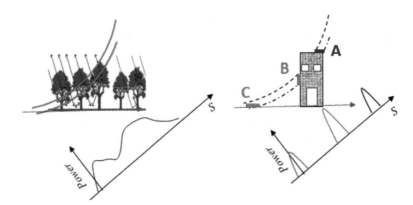

Figure 8.4. Typical reflectivity elevation profiles for (left) forested and (right) urban areas.

8.1.3 Reflectivity models

In (8.9), each element of the vector γ defines the complex reflectivity of the elementary scattering volume at elevation s_i, with $s_i = i \Delta s$. Many different models can be adopted for the reflectivity, depending on the particular nature of the ground scene and on the system configuration.

When the observed scene is a forested area, the incident electromagnetic wave penetrates in the vegetation layer, interacting with the multiple scatterers at different elevation inside the layer, all contributing to the backscattered signal. Then, the reflectivity profile along the elevation s is typically a smooth continuous profile extending over a wide interval, corresponding to the elevation extension of the forested scene (see Fig. 8.4 left image). In urban areas, instead, the incident waveform does not penetrate in the structures on the ground and the scattering mechanism is a surface scattering one, so that only few scatterers at different elevations contribute to the backscattering toward the SAR antenna, so that the resulting reflectivity profile is of a discrete type (see Fig. 8.4 right image). As it will be discussed in the following section, different methods can be applied for the tomographic reconstruction of different reflectivity profile types.

8.2 3D TomoSAR imaging techniques

In the previous section, it was shown that for uniform baseline distribution the steering matrix **A** reduces to a DFT matrix and the reconstruction of the unknown reflectivity vector γ amounts essentially to solve a spectral estimation problem.

Anyway, in practical cases, baselines are usually unevenly distributed. In some cases the number of acquisition is lower than the number required by the sampling theorem. Therefore, generally, the TomoSAR reconstruction requires the inversion of an underdetermined linear system.

In this section we recall the most important and well-known spectral estimators used for TomoSAR [96].

8.2.1 Classical beamforming

One of the simplest estimators that can be used to retrieve the reflectivity profile is given by the single-look beamforming estimator, which performs the matched filtering that maximizes the signal to noise ratio under the assumption of additive white noise. The estimated reflectivity profile can then be expressed as:

$$\hat{\boldsymbol{\gamma}}_{BF} = \mathbf{A}^H \boldsymbol{y} = \mathbf{A}^H \mathbf{A} \boldsymbol{\gamma} + \mathbf{A}^H \boldsymbol{w} \qquad (8.18)$$

where the $M \times M$ matrix $\mathbf{A}^H \mathbf{A}$ describes the Point Spread Function of the overall (acquisition and imaging) system along the elevation.

If the baselines are uniformly spaced, this estimator simply performs the inverse DFT of the data vector $\boldsymbol{y}$.

Although fast to compute, this approach generally provides a resolution of the order of (8.15) and high side-lobes and spurious peaks, depending on the specific baseline distribution and system configuration.

For improving robustness to noise, a multilook beamforming estimation of the backscattering intensity density (which is of interest, see (3.19)), can be considered:

$$\hat{p}_{BF}(s_m) = E\left[|\hat{\gamma}_{BF}(s_m)|^2\right] \cong \frac{1}{L} \sum_{l=1}^{L} \boldsymbol{a}^H(s_m) \boldsymbol{y}_l \boldsymbol{y}_l^H \boldsymbol{a}(s_m)$$
$$= \boldsymbol{a}^H(s_m) \hat{\mathbf{R}}_y \boldsymbol{a}(s_m) = \text{tr}\left[\boldsymbol{a}\boldsymbol{a}^H \hat{\mathbf{R}}_y\right] \qquad (8.19)$$

where

$$\hat{\mathbf{R}}_y = \frac{1}{L} \sum_{l=1}^{L} \boldsymbol{y}_l \boldsymbol{y}_l^H, \qquad (8.20)$$

is the (data) sample correlation matrix, L is the cardinality of the set of statistically homogeneous pixels (generally surrounding) to the pixel under investigation, see also (7.67), and tr[·] the trace operator. Note that the latter, when applied to a product between two matrices with the same size as in (8.19), can be interpreted as a scalar product under the Frobenius norm.[2]

Accordingly, (8.19) simply states that the reconstruction in a generic elevation bin s_m is given by the scalar product between the corresponding steering matrix $\boldsymbol{aa}^H$ and the sample correlation matrix.

With this estimator, L samples are combined to compute an estimate of the $N \times N$ correlation matrix $\mathbf{R}_y$. This operation introduces a spatial averaging for reducing speckle fluctuations at expenses of a loss of resolution along azimuth and range directions.

Eq. (8.19) can be also written in a vector form as:

$$\hat{\boldsymbol{p}}_{BF} = \text{diag}\left(\mathbf{A}^H \hat{\mathbf{R}}_y \mathbf{A}\right), \tag{8.21}$$

where $\hat{\boldsymbol{p}}_{BF}$ is the $M \times 1$ vector containing the intensity of the estimated reflectivity in the considered discrete elevation values, and $\text{diag}(\cdot)$ is the operator extracting the diagonal entries of a square matrix.

Single-look and multilook beamforming exhibit an elevation resolution which is limited by the overall baseline extent according to (8.15).

As an example, we consider 28 TerraSAR-X acquisitions over Barcelona (Spain), with an overall baseline extension of about 500 m and an elevation resolution, according to (8.15), of about 19 m, corresponding to a height resolution of about 11 m. Azimuth and range resolutions are about 3 m and 1.5 m, respectively. The SAR intensity image averaged over the 28 acquisitions is shown in Fig. 8.5. The presence of two tall buildings can be clearly noted. We consider the tomographic reconstruction of the facade of the Marpfre Tower (on the right), illuminated by the SAR sensor as shown in Fig. 8.6(a).

The single look beamforming reconstruction of the range line on that facade is shown in Fig. 8.6(b). The layover effect, induced by the mechanism sketched in Fig. 8.3 is clearly visible in Fig. 8.6(a): the top of the building is located at a range lower than the range of its base, producing in the tomogram the typical slanted image of the facades that are vertical respect to the $x - y$ (azimuth - ground range) plane. The layover effect can be compensated by means of the geocoding operation, that is the conversion from the radar coordinates (x, r, s) to a ground reference system, typically WGS84 [97]. In our simplified framework we

---

[2] Given two $N \times M$ matrix $\mathbf{A}$ and $\mathbf{B}$, the trace of the hermitian product between them, i.e. $\text{tr}(\mathbf{A}^H \mathbf{B})$ defines the scalar product between the matrixes under the Frobenius norm [90].

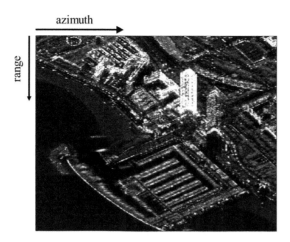

Figure 8.5. TerraSAR-X intensity image of an area of Barcelona (Spain), averaged over 28 acquisitions.

simply convert the range to the ground range so to that the tomogram 8.6(c) is obtained. Looking at the tomograms (b) and (c) it can be clearly noted that the ground range resolution (b) is worse than the (slant) range resolution (a), see (4.9) and the discussion therein. Moreover, in the layover area occurring approximately in the range interval (−50 m, 30 m), the response of the ground beneath the facade is clearly overwhelmed by the facade backscattering. The tomograms obtained using multilook beamforming with 3×3 looks and 15×15 look are own in Figs. 8.6 (d)-(e) and (f)-(g), respectively. The effect of multilook averaging at expenses of a reduction of range resolution is clearly visible.

8.2.2 Truncated singular value decomposition (T-SVD)

The Singular Value Decomposition (SVD) is widely used in signal processing for implementing an effective regularized and "controlled" inversion of linear problems. It allows decomposing a matrix into several (rank-1 or dyadic) component matrices, exposing many of the properties of the original matrix. Using SVD, it is in fact possible to determine the rank of a matrix, to quantify the sensitivity of a linear system to numerical errors, or to obtain an optimal lower-rank approximation of the decomposed matrix.

The singular value decomposition of the $N \times M$ steering matrix **A** can be expressed as:

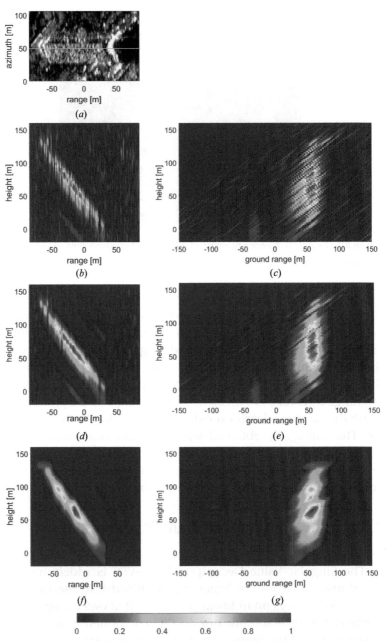

Figure 8.6. (a) TerraSAR-X intensity image of Marpfre Tower, Barcelona (Spain), with the considered range line highlighted; (b)-(c) single look beamforming reconstruction before and after geocoding; (d)-(e) 3 × 3 multilook beamforming reconstruction before and after geocoding; (f)-(g) 15 × 15 multilook beamforming reconstruction before and after geocoding. All tomographic reconstructions refer to the normalized intensity. Range is referred to the center of the patch.

$$\mathbf{A} = \mathbf{U}\mathbf{\Sigma}\mathbf{V}^H = \sum_{n=1}^{K} \sigma_n \boldsymbol{u}_n \boldsymbol{v}_n^H \qquad (8.22)$$

where K is the rank of $\mathbf{A}$, with $K \leqslant \min(N, M)$. In (8.22) $\mathbf{U}$ is a $N \times N$ matrix with orthonormal columns $\boldsymbol{u}_n$, that are called left singular vectors; $\mathbf{\Sigma}$ is a $N \times M$ with only the first K elements of the main diagonal different from zero containing the singular values σ_n of $\mathbf{A}$ (generally in a decreasing order), $\mathbf{V}$ is a $M \times M$ matrix with orthonormal columns $\boldsymbol{v}_n$, that are called right singular vectors. Due to the cited orthonormality property, it follows that $\mathbf{U}$ is a unitary matrix and its columns form a basis for the space spanned by the columns of $\mathbf{A}$, while $\mathbf{V}$ is a unitary matrix and its columns form a basis for the space spanned by the rows of $\mathbf{A}$ (i.e. the columns of $\mathbf{A}^T$), and:

$$\mathbf{U}^H \mathbf{U} = \mathbf{I}_N \qquad\qquad \mathbf{V}^H \mathbf{V} = \mathbf{I}_M \qquad (8.23)$$

where $\mathbf{I}_K$ denotes the identity matrix of size $K \times K$.

Substituting (8.22) in (8.7) we have:

$$\boldsymbol{y} = \sum_{n=1}^{K} \sigma_n \boldsymbol{u}_n \boldsymbol{v}_n^H \boldsymbol{\gamma}, \qquad (8.24)$$

thus emphasizing the powerfulness of SVD in the representation of a matrix, and more in general of a linear operator. SVD allows in fact identifying triplets, $(\sigma_n, \boldsymbol{u}_n, \boldsymbol{v}_n)$ $n = 1, \ldots, K$, which explain how the unknown ($\boldsymbol{\gamma}$), usually termed as "object", is imaged through the matrix ($\mathbf{A}$) in the formation of the measurement ($\boldsymbol{y}$). More specifically [90] [93]:

1. the right singular vectors $\boldsymbol{v}_1, \ldots, \boldsymbol{v}_K$ form a (orthogonal) basis of the so-called "visible object space", that is the (K-dimensional) subspace of all $\boldsymbol{\gamma} \in \mathbb{C}^M$ imaged in a nonnull data vector $\boldsymbol{y} \in \mathbb{C}^N$, i.e. $\mathbf{A}\boldsymbol{\gamma} \neq \mathbf{0}$;

2. the right singular vectors $\boldsymbol{v}_{K+1}, \ldots, \boldsymbol{v}_M$ form a (orthogonal) basis of the so-called "invisible object space", orthogonal to the visible object space and known as "null space" of $\mathbf{A}$, indicated with $\mathcal{N}(\mathbf{A})$ in Section 8.1.1. Mathematically, it is the ($(M - K)$-dimensional) subspace of all $\boldsymbol{\gamma} \in \mathbb{C}^M$, with $\boldsymbol{\gamma} \neq \mathbf{0}$, imaged in the null vector of $\mathbb{C}^N$, i.e. $\mathbf{A}\boldsymbol{\gamma} = \mathbf{0}$;

3. the left singular vectors $\boldsymbol{u}_1, \ldots, \boldsymbol{u}_K$ form a (orthogonal) basis of the measurable space known as "range space" (not to be confused with the radar range direction), that is the (K-dimensional) subspace of $\mathbb{C}^N$ spanned by the columns of $\mathbf{A}$, indicated with $\mathcal{R}(\mathbf{A})$ in Section 8.1.1. Mathematically, it is the subspace of all $\boldsymbol{y} \in \mathbb{C}^N$ for which $\exists \boldsymbol{\gamma} \in \mathbb{C}^M : \mathbf{A}\boldsymbol{\gamma} = \boldsymbol{y}$;

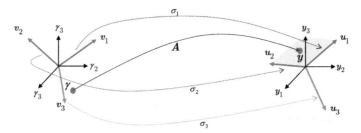

Figure 8.7. Schematization of the forward mapping through the matrix/operator **A** as represented by the SVD decomposition. In the figure it is considered a mapping from a 3D space to a 3D space. The lighter colors of the mapping through the third direction σ_3, u_3, v_3 indicate a lower value of σ_3 compared to the other singular values. According the measured data have a lower component along u_3 so that the mapping point is close to the (u_2, u_3) highlighted in the right with the light gray. As a consequence inversion along this direction may be critical and emphasize the noise.

4. the left singular vectors $u_{K+1}, \ldots, u_N$ form a (orthogonal) basis of the unreachable object space, known as "left null space", that is the $((N - K)$-dimensional) subspace of $\mathbb{C}^N$ that cannot be imaged by the operator, and hence orthogonal to the range space. Mathematically, it is the subspace of all $y \in \mathbb{C}^N$ for which $\nexists \gamma \in \mathbb{C}^M : A\gamma = y$.

In light of this basis identifications, we can interpret (8.24) (forward model) in the process of measurement formation. The (noise free) data, achieved by imaging γ through **A**, is provided by a combination of the range space (left) singular vectors u_n, $n = 1, \ldots, K$, through coefficients which are given by the projection of the (unknown) object along the associated (right) singular vector in the object space (i.e. the scalar product $v_n^H \gamma$) scaled by the corresponding singular values σ_n. The process is depicted in Fig. 8.7.

The system (8.9) can be therefore solved in the LS sense, i.e. by minimizing the norm $||y - A\gamma||^2$. The use of the SVD of the matrix **A** leads to the solution:

$$\hat{\gamma} = \sum_{n=1}^{K} \frac{1}{\sigma_n} v_n u_n^H y. \tag{8.25}$$

Considering that $K \leqslant \min(N, M)$ and, generally, $N \leqslant M$ as previously stated, for $N < M$ this is the LS solution among the ∞^{M-K} ones with minimum L^2 norm. As already stated in the previous chapter, specifically in Section 7.2.1, the matrix that multiply y in

(8.25) is the Moore-Penrose pseudo-inverse $\mathbf{A}^\dagger$ of $\mathbf{A}$:

$$\mathbf{A}^\dagger = \sum_{n=1}^{K} \frac{1}{\sigma_n} \boldsymbol{v}_n \boldsymbol{u}_n^H. \tag{8.26}$$

Let us now consider the effect of noise. In this case $\boldsymbol{y} = \mathbf{A}\boldsymbol{\gamma} + \boldsymbol{w}$, and (8.25) becomes:

$$
\begin{aligned}
\hat{\boldsymbol{\gamma}} &= \sum_{n=1}^{K} \frac{1}{\sigma_n} \boldsymbol{v}_n \boldsymbol{u}_n^H \mathbf{A}\boldsymbol{\gamma} + \sum_{n=1}^{K} \frac{1}{\sigma_n} \boldsymbol{v}_n \boldsymbol{u}_n^H \boldsymbol{w} = \\
&= \mathbf{A}^\dagger \mathbf{A}\boldsymbol{\gamma} + \sum_{n=1}^{K} \frac{1}{\sigma_n} \boldsymbol{v}_n \boldsymbol{u}_n^H \boldsymbol{w}.
\end{aligned}
\tag{8.27}
$$

We see that the reconstruction $\hat{\boldsymbol{\gamma}}$ is the sum of the estimate of the true vector $\boldsymbol{\gamma}$ associated with the data projected in the range space of $\mathbf{A}$ (i.e. $\mathcal{R}(\mathbf{A})$) and the term due to the noise. If some of the singular values σ_n are small (f.i., σ_3 in Fig. 8.7), the associated (random) noise component becomes very large, often completely overwhelming the component of $\boldsymbol{\gamma}$ in the direction associated to the singular value. In other words, we can say that the small singular values correspond to directions for which the data contain very little information about the unknown vector. Low singular values are generally present in the case of nonuniform baseline distribution. For reconstructing the components of $\boldsymbol{\gamma}$ which lie along these directions, we have to amplify the small signal buried in the data by dividing by the small singular value, but in this way we strongly amplify also the noise that is always present on the data. Thus, when there are small singular values, Eq. (8.25) leads to a bad solution. To avoid this problem, a simple choice is to restrict the solution space to that obtained by considering only the singular vectors corresponding to the large singular values, i.e. by eliminating the vector basis corresponding to the small singular values. The approach is referred to as Truncated Singular Value Decomposition (T-SVD) thus obtaining:

$$\hat{\boldsymbol{\gamma}} = \sum_{n=1}^{K_T} \mu_n \boldsymbol{v}_n \boldsymbol{u}_n^H \boldsymbol{y}, \tag{8.28}$$

where

$$\mu_n = 1/\sigma_n \quad n = 1, \ldots, K_T \tag{8.29}$$

and $K_T \leqslant K$ is the number of the first singular values (ordered in a decreasing sequence) that are greater than a fixed threshold.

As an alternative the coefficient μ_n can be defined as:

$$\mu_n = \frac{\sigma_n}{\sigma_n^2 + \alpha^2} \qquad n = 1, \ldots, K \qquad (8.30)$$

which regularizes (in the sense of Tikhonov) the inversion by imposing a minimum square norm constraint on the solution, i.e.:

$$\hat{\gamma} = \underset{\gamma}{\mathrm{argmin}} \ \|\mathbf{A}\boldsymbol{\gamma} - \boldsymbol{y}\|_2^2 + \alpha^2 \|\boldsymbol{\gamma}\|_2^2. \qquad (8.31)$$

This method, referred to as Tikhonov regularization, has a parallelism in the statistical estimation theory framework to what is known as minimum mean square error solution.

8.2.3 Capon beamforming

Capon Beamforming is derived by imposing the maximum attenuation of the power from all elevation values different from the considered one. It does not require the a priori knowledge on the noise distribution, but requires an estimate of the data correlation matrix. The strength of Capon's method lies in the fact that the method performs a robust data fitting even when the spatial sampling of data is below the Nyquist one.

Letting $\mathbf{R}_y = E\left[\boldsymbol{y}\boldsymbol{y}^H\right]$ be the data correlation matrix, for each spatial frequency s_n we want to find the filter $\boldsymbol{h}_n$ that minimize the output power, subject to unitary gain along the direction $\boldsymbol{a}(s_m)$:

$$\boldsymbol{h}_m = \underset{\boldsymbol{h}}{\mathrm{argmin}} \ E\left[|\boldsymbol{h}^H \boldsymbol{y}|^2\right] = \underset{\boldsymbol{h}}{\mathrm{argmin}} \ \boldsymbol{h}^H \mathbf{R}_y \boldsymbol{h},$$
$$\text{subject to } |\boldsymbol{h}^H \boldsymbol{a}(s_n)| = 1 \qquad (8.32)$$

The solution of the constrained problem in (8.32) can be achieved by using the Lagrangian multiplier and provides the Capon filter [98] [99]:

$$\boldsymbol{h}_m = \frac{\mathbf{R}_y^{-1} \boldsymbol{a}(s_m)}{\boldsymbol{a}^H(s_m) \mathbf{R}_y^{-1} \boldsymbol{a}(s_m)} \qquad (8.33)$$

The corresponding Capon beamforming estimate of the power spectrum is:

$$\hat{p}_C(s_m) = \frac{1}{\boldsymbol{a}^H(s_m) \hat{\mathbf{R}}_y^{-1} \boldsymbol{a}(s_m)} = \frac{1}{\mathrm{tr}\left[\boldsymbol{a}(s_m) \boldsymbol{a}^H(s_m) \hat{\mathbf{R}}_y^{-1}\right]}, \qquad (8.34)$$

where the unknown correlation matrix $\mathbf{R}_y$ has been by the sample correlation matrix in (8.20) and tr[·] is again the trace of a square matrix.

It is interesting to compare the Capon beamforming with the classical beamforming given by (8.19) in the presence of a concentrated scatterer along s in the presence of sufficient signal to noise ratio and absence of decorrelation. In this case the classical beamforming should produce a peak corresponding to a maximum of the scalar product between the steering matrix $a(s_m) a^H(s_m)$ and the sample correlation matrix. Accordingly, the classical beamforming works by measuring the matching in the subspace of $\mathbb{C}^N$ corresponding to the signal space. Eq. (8.34) highlights, on the contrary, that Capon beamforming produces a peak when the quadratic form at the denominator is minimum. By resorting the interpretation of SVD, or better EVD, one can note that the operation of inverting $\hat{\mathbf{R}}_y$ amplifies the directions corresponding to low eigenvalues, that is the noise subspace and attenuate the directions corresponding to the signal subspace. Accordingly the minimum of the denominator is achieved when the scalar products between the steering matrix and the noise subspace matrix generate the highest mismatch. As the dimension of the noise subspace is typically larger than the signal subspace, Capon beamforming can outperform the classical beamforming. Obviously, as (8.34) involves the inverse of a matrix, attention should be paid when the sample correlation matrix is critical, e.g. in the case of $L < N$, i.e. few looks.

Capon beamforming has empirically shown improved performances with respect to other tomographic estimators when distributed scatterers are encountered due to its capability to better suppress the sidelobes [100]. It is regarded as a high-resolution estimator since it exhibits better resolution and better interference rejection capability than the standard beamformer, provided that the steering vector structure corresponding to the signal of interest is accurately known. However, when the steering vector is not accurate, due for instance to inaccurate baseline knowledge, its performance may become worse than that of the standard beamformer.

Different methods have been developed to manage this problem, as for instance diagonal loading, which is a popular approach to improve the robustness of the Capon beamformer [101] and achieve a regularization of the inversion of the sample correlation matrix.

An example of a tomographic profile reconstruction of a forested area is shown in Fig. 8.8.

In Fig. 8.8(a) the intensity of the P-band SAR image of a forested area in the French Guiana is shown. The tomogram corresponding to the range line highlighted in Fig. 8.8(a) and obtained using the 15×15 multilook beamforming is shown in Fig. 8.8(b),

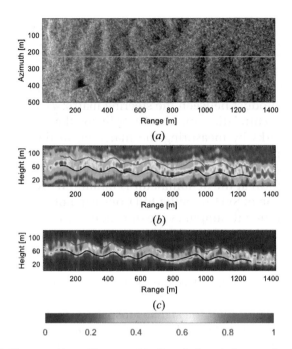

Figure 8.8. Tomographic profile reconstruction of a forested area using 6 interferometric Tropisar images: (a) SAR intensity image; (b) Classical multilook beamforming reconstruction; (c) Capon beamforming reconstruction. All tomographic reconstructions refer to the normalized intensity.

while the tomogram obtained by using Capon beamforming is shown in Fig. 8.8(c). The better height resolution and the lower side-lobe levels of Capon reconstruction are evident. Anyway, it has to be reminded that Capon filter involves the inversion of the correlation matrix, and then requires its reliable estimate that is obtained averaging on extended spatial neighborhoods of each range-azimuth pixel. This operation, to achieve reliable results, requires the observed scene to be slowly spatially varying, in such a way to consider it to be locally stationary. This assumption is commonly satisfied for forested areas, while it is not in the case of urban scenes, where the man-made structure on the ground has the extension of only few meters. For instance, we consider the tomographic reconstruction of the Marpfre Tower from TerraSAR-X data, already considered in Fig. 8.6. The Capon reconstruction obtained by computing the correlation matrix on a 15×15 moving window is shown in Fig. 8.9 (b) and (c), before and after geocoding. It can be noted that the reconstruction is worse than the one obtained using classical beamforming, shown in Fig. 8.6. This effect is probably related to the fact that, in this case, the scene cannot

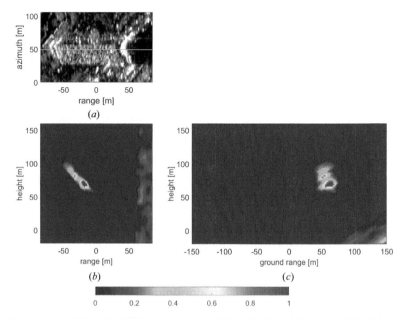

Figure 8.9. (a) TerraSAR-X intensity image of Marpfre Tower, Barcelona (Spain), with the considered range line highlighted; (b)-(c) Capon (normalized intensity) reconstruction obtained computing correlation matrix on 15×15 looks.

be considered stationary in the 15×15 window adopted for the estimation of the correlation matrix.

8.2.4 MUltiple SIgnal classification (MUSIC)

MUltiple SIgnal Classification (MUSIC) is a parametric spectral estimation technique with superresolution capabilities, i.e. the capability to achieve a higher (numerically lower) height resolution than that of the DFT given by (8.15), specifically designed to retrieve and localize a finite number K of discrete scatterers from a mixture corrupted by additive white noise.

MUSIC estimates the frequency content of the signal samples using a method separating the signal and noise subspaces by means of the eigendecomposition of the sample correlation matrix, which can be expressed as:

$$\mathbf{R}_y = \begin{bmatrix} \mathbf{U}_s & \mathbf{U}_w \end{bmatrix} \begin{bmatrix} \mathbf{\Lambda}_s & \underline{0} \\ \underline{0} & \mathbf{\Lambda}_w \end{bmatrix} \begin{bmatrix} \mathbf{U}_s^H \\ \mathbf{U}_w^H \end{bmatrix} \qquad (8.35)$$

where subscripts s and w refer to the signal and noise components. Moreover,

$$\mathbf{\Lambda} = \begin{bmatrix} \mathbf{\Lambda}_s & \underline{0} \\ \underline{0} & \mathbf{\Lambda}_w \end{bmatrix} \tag{8.36}$$

is the matrix containing the eigenvalues of $\mathbf{R}_y$ in descending order and $\mathbf{U}_s$ and $\mathbf{U}_w$ are the matrices containing the corresponding eigenvectors:

$$\mathbf{U}_s = \begin{bmatrix} \boldsymbol{u}_1 & \boldsymbol{u}_2 & \cdots & \boldsymbol{u}_K \end{bmatrix} \qquad \mathbf{U}_w = \begin{bmatrix} \boldsymbol{u}_{K+1} & \boldsymbol{u}_{K+2} & \cdots & \boldsymbol{u}_N \end{bmatrix} \tag{8.37}$$

Since $\mathbf{R}_y$ is a Hermitian matrix, all of its N eigenvectors $\{\boldsymbol{u}_1, \boldsymbol{u}_2, \ldots, \boldsymbol{u}_N\}$ are orthogonal to each other. If the eigenvalues of $\mathbf{R}_y$ are sorted in decreasing order, the eigenvectors $\{\boldsymbol{u}_1, \ldots, \boldsymbol{u}_K\}$ corresponding to the first K eigenvalues span the signal subspace, while the remaining $N - K$ eigenvectors span the noise subspace, which is orthogonal to the signal one.

Typically, only few scatterers dominate the backscattered signal in a given resolution cell, then usually K is much lower than the number of acquisitions N and the corresponding eigenvectors span a subspace of small dimension.

Since the vector $\boldsymbol{\gamma}(s)$ belongs to the signal space, it has to be orthogonal to each eigenvector $\boldsymbol{u}_i$, with $i = K + 1, \ldots, N$, spanning the noise subspace.

The estimation of the height s of each scatterer can be done by measuring how much orthogonal the steering vector $\boldsymbol{a}(s)$ is to the subspace $\mathbf{U}_n$ associated with the $N - K$ smallest eigenvalues of $\mathbf{R}_y$:

$$\hat{p}(s_m) = \frac{1}{\boldsymbol{a}^H(s_m)\mathbf{U}_w\mathbf{U}_w^H\boldsymbol{a}(s_m)} = \frac{1}{\sum_{m=K+1}^{N} |\boldsymbol{a}^H(s_m)\boldsymbol{u}_i|^2}, \tag{8.38}$$

where the "^" highlights that, similarly to the Beamforming and Capon beamforming the estimate, $\hat{\mathbf{R}}_y$ in (8.20) has to be used in place of $\mathbf{R}_y$ in a real operative scenario. In (8.38) The scatterers positions are then obtained by identifying the K largest peaks in the MUSIC pseudo-spectrum $\hat{p}(s)$.

Although only the noise-related eigenvectors are used in (8.38), it has been shown that MUSIC exhibits significant superresolution capabilities and generally provides better results than classic beamforming or Capon beamforming.

8.2.5 Compressive sensing

More recently, the Compressive Sensing (CS) approach has been successfully employed for separating multiple scatterers in urban areas and the contributions of ground and canopy in forested areas [102] [103].

This technique does not rely on statistical models and does not exploit the data correlation matrix, therefore allowing full resolution analysis, as it simply searches for a sparse reflectivity vector, i.e. a vector with few elements different from zero and most elements equal to zero, that best matches to the data.

As we will see, one of the main advantages of CS technique is that it allows a noticeable reduction of the number of measurements required for the reflectivity reconstruction and achieves a better elevation resolution respect to beamforming and T-SVD. A drawback of this technique, instead, is that it relies on the unknown vector sparsity assumption, which requires the individuation of a proper basis in which the unknown vector can be represented by a sparse vector.

CS theory states that a signal having a sparse representation in one basis, can be reconstructed from a small number of measurements collected in a second basis, that is incoherent with the first one [104] [105]. Incoherence property expresses the idea that objects having a sparse representation in a given basis must be spread out in the domain in which they are acquired.

In SAR Tomography, this approach can be particularly effective, and it can be used very favorably either in urban or in vegetated environments. In both cases, CS technique allows the recover of the elevation reflectivity profile at an increased resolution and starting from a reduced number of acquisitions. Anyway, a fundamental difference exists between the two cases as regards the sparsity representation basis of the unknown reflectivity vector. As it can be seen from Fig. 8.4, while in the case of urban environment the sparse nature of the reflectivity elevation profile appears evident (concentrated scatterers A, B, and C in Fig. 8.4(b)), in the case of forests and trees this characteristic of sparsity may not be verified, as shown in Fig. 8.4(a), and must be "forced" in some way.

In order to model the above mentioned sparsity property, we can begin to introduce it mathematically. In particular, let us consider the complex vector γ of dimension $M \times 1$, and a $M \times M$ unitary transformation matrix Ψ, so that $\Psi^{-1} = \Psi^H$. The vector γ is K-sparse in the orthonormal basis represented by Ψ, if the $M \times 1$ vector α:

$$\alpha = \Psi\gamma \qquad (8.39)$$

has, at most, K nonzero elements. All the remaining elements of α are negligible or null. In the framework of Compressive Sensing, the $M \times M$ transformation matrix Ψ is denoted as the sparsity matrix.

Inverting (8.39), and being $\mathbf{\Psi}$ a unitary matrix, we get $\gamma = \mathbf{\Psi}^H \alpha$. According to (8.39), model (8.7) can be written as:

$$y = \mathbf{A}\mathbf{\Psi}^H \alpha \tag{8.40}$$

In the framework of Compressive Sensing theory the steering matrix $\mathbf{A}$ is also referred as measurement or sensing matrix. The above model links the N-dimensional data vector y (the SAR image data at a fixed range-azimuth cell on the N trajectories) to the M-dimensional K-sparse unknown vector α. In other words, the model matrix $\mathbf{A}\mathbf{\Psi}^H$ maps the K nonzero elements of the unknown representation to the available data.

The CS reconstruction of the sparse vector α can be found by solving the following constrained problem [106]:

$$\hat{\alpha} = \underset{\alpha}{\operatorname{argmin}} \ ||\mathbf{A}\mathbf{\Psi}^H \alpha - y||_2^2 \ \text{subject to} \quad ||\alpha||_0 \leqslant K \tag{8.41}$$

where $|| \cdot ||_0$ denotes the L_0 pseudo-norm, which counts the nonzero elements in the vector, while the generalized expression of the L_p norm, from which definition of other relevant norms can be obtained, is given by:

$$||\alpha||_p = \left(\sum_{m=1}^{M} |\alpha_m|^p \right)^{1/p}. \tag{8.42}$$

The solution of the problem (8.41) exhibits a combinatorial complexity and then is impractical in most of cases. Anyway, under certain conditions, an equivalent solution of (8.41) can be determined by solving the following convex problem:

$$\hat{\alpha} = \underset{\alpha}{\operatorname{argmin}} \ ||\mathbf{A}\mathbf{\Psi}^H \alpha - y||_2^2 + \mu ||\alpha||_1 \tag{8.43}$$

where μ is a parameter that is used to control the trade-off between the data fitting and the sparsity constraint, and then has to be adjusted according to the noise level.

Eq. (8.43) can be solved by using solvers from linear programming. Different solver formulations have been proposed in literature, such as Basis Pursuit (BP) proposed by Chen et al. [107], which provides faithful reconstructions for noise-free measurements. Other solvers for robust signal recovery from noisy measurements are Dantzig Selector [108], Basis Pursuit Denoising (BPDN) [107], which is the same as Least Absolute Shrinkage Selection operator (LASSO) [109], and Total Variation (TV) minimization [110].

Once the CS estimation of the sparse vector $\hat{\boldsymbol{\alpha}}$ has been obtained, the corresponding tomographic profile $\hat{\boldsymbol{\gamma}}$ is obtained by:

$$\hat{\boldsymbol{\gamma}} = \boldsymbol{\Psi}^H \hat{\boldsymbol{\alpha}} \tag{8.44}$$

CS theory gives some sufficient conditions on the sensing matrix that guarantee a stable sparse recovery of the unknown vector in presence of additive noise [104] [105]. Two sufficient conditions are the Restricted Isometry Property (RIP) and the incoherence property between the sensing and the sparsity matrices. The first condition is equivalent to state that the transformation performed by the matrix $\mathbf{A}\boldsymbol{\Psi}^H$ preserves the distances. The latter condition states that for faithful reconstruction, the measurement basis $\boldsymbol{a}_i$ and sparse basis $\boldsymbol{\psi}_j$ must be incoherent from each other. The coherence between two matrices measures the maximum correlation among any pair of columns taken from the two matrices, and is given in by:

$$\rho(\mathbf{A}, \boldsymbol{\Psi}) = \operatorname*{argmax}_{1 \leqslant i,j \leqslant M} |\boldsymbol{a}_i^H \boldsymbol{\psi}_j|. \tag{8.45}$$

The range of coherence is $\rho(\mathbf{A}, \boldsymbol{\Psi}) \in \left[1, \sqrt{M}\right]$.

Depending on orthogonal baseline distribution and on sampling rate along the elevation direction, these conditions may be or not be satisfied. This may occur when the sampling interval Δs, adopted for the elevation discretization, is much smaller of the Rayleigh resolution (8.15). Therefore, when trying to recover a superresolution reflectivity profile, a particular attention to the sensing matrix properties has to be given.

In case of urban scenarios, where the sparsity of the elevation profile is likely and as such it can be assumed, the sparsity matrix reduces to an identity one, so that $\boldsymbol{\Psi} = \mathbf{I}$. It follows that $\boldsymbol{\gamma} = \boldsymbol{\alpha}$. In this case, the sparsity property is equivalent to the requirement that only a small number of stable targets is present in the same range-azimuth resolution cell.

In case of forested scenarios, the assumption of sparsity of the elevation reflectivity profile is not likely. Then, a proper sparsity basis has to be found. The choice of the orthonormal basis $\boldsymbol{\Psi}$ requires some special considerations. First, the signal $\boldsymbol{y}$ must have a sparse expansion $\boldsymbol{\Psi}\boldsymbol{y}$. Second, the coherence between the measurement basis $\mathbf{A}$ and the sparsity basis $\boldsymbol{\Psi}$ has to be as small as possible. A sparsifying basis that satisfies these two requirements is the Discrete Wavelet Transform (DWT) basis, that has been successfully applied for the tomographic reconstruction of the reflectivity profiles of forested areas [103].

8.3 Application of SAR tomography to surface scattering scenarios for PS detection

As stated in previous sections, TomoSAR techniques process multibaseline interferometric signals in each range-azimuth pixel to recover an estimate of the complex reflectivity profile in the corresponding 3D (range-azimuth-elevation) cell associated with the observed volume on the ground. In the presence of surface scattering, the image pixel is generally characterized along the vertical direction by the presence a single scattering mechanism located at an unknown height. SAR Tomography can be exploited, extending multibaseline Interferometry, to estimate the height position of the scatterer by using both the amplitude and phase information.

In urban scenarios generally a high density of single scatterer is observed, which also behave as PS in the multitemporal case. As specified in the following, TomoSAR can be still used in the multitemporal case to improve the performances of multitemporal Interferometry. However, due to the imaging geometry, it can happen that a superposition of different scattering mechanisms may be present in a given azimuth-range pixel, see Fig. 8.3. As a consequence, urban areas are characterized by a backscattering profile along the height, or better slant height direction, showing the presence of single or multiple peaks. Compressive sensing, previously described, is a nonparametric approach particularly suitable for the reconstruction of such typology of scattering profiles. MUSIC, also previously addressed, is an approach capable to achieve excellent reconstructions of lines spectra and therefore turns out to be suitable to the urban scenario. Compared to the CS, MUSIC requires the evaluation of the data correlation matrix and therefore, in its natural form, implies the use of a multilook, thus leading to a reduction of the spatial resolution. Such a spatial (azimuth-range) resolution loss can represent a main disadvantage for the application to the built environment.

Having said this, it should be pointed out that, regardless of the application scenario, we have to face a problem in which we are typically interested in a limited and fixed number of, say K, scattering mechanisms. The data model in (8.11) can be therefore simplified in:

$$y = \sum_{k=1}^{K} a(s_k)\gamma_k + w \qquad (8.46)$$

where $a(s_k)$ is the steering vector corresponding to the k-th scatterer, i.e. the k-th column of $\mathbf{A}$; the latter being dependent on the (slant) height of the k-th scatter s_k. So far, equations have been referred to the 3D case, assuming that the N acquisitions are carried at the same, or close times or, conversely by assuming the scene to be "frozen" in the acquisition interval. If the first situation is rather rare and limited to the airborne case, the second case hardly corresponds to a realistic situation because the acquisitions are commonly collected from current satellites over months or years.

Similarly to the Persistent Scatterers Interferometry treated in the previous chapter, we therefore let each scatterer also to be characterized by possible linear deformation over the time, see (7.57).

$$ y = \sum_{k=1}^{K} a(s_k, v_k) \gamma_k + w \tag{8.47} $$

In the multitemporal case the problem to be faced is generally the discrimination in each pixel of the K scatterers from the noise: the problem is pertinent, with different degrees of complexity, to urban as well as rural areas cases. As for urban areas, although beamforming and Compressive Sensing may provide peaked profiles, the issue of separation between PS and noise can be approached in the detection theory framework, common in other radar contexts, e.g. surveillance radars. Particularly, one should be interested to detect as many PS, or Distributed Scatterers (DS) introduced in the previous chapter, as possible for a given Probability of False Alarm P_{fa}.

Before addressing this point it is worth a further consideration on the assumed K scatterers model. The expression in (8.47) presumes that the backscattering coefficient of the generic k-th scatterer, assumed as a random variable, does not change with the observation, i.e. with the index running on vectors in (8.47). As previously explained in Section 8.1.1, this is tantamount to stating that the scatterers do not decorrelate over the different antennas: this is a reasonable assumption for an urban scenario or in the (not very frequent) case where data are assumed to be acquired at the same time. In the general case in which data are acquired over repeated passes, it becomes useful, at least for the data model, to account for the presence of scatterer decorrelation. Extending the modelling in the previous chapter for the single scatterer (7.11), this can be done as discussed in Section 8.1.1 by substituting the random variable γ_k with the random vector $\boldsymbol{\gamma}_k$ thus rewriting (8.47) as:

$$ y = \sum_{k=1}^{K} a(s_k, v_k) \odot \boldsymbol{\gamma}_k + w \tag{8.48} $$

Note that, strictly speaking the term "Tomography" refers to the class of algorithms that exploit both the amplitude and phase information in the data to reconstruct either in 3D the backscattering profile, by using the vertical aperture synthesis principle, or to directly extract the heights of the scatterers. The case in which the scatterers are assumed to have also a deformation over the time and the data are acquired in a temporal interval is known in literature as Differential Tomographic Synthetic Aperture Radar (DTomoSAR), being the mixture of tomography for 3D mapping and Differential Interferometry for the determination of surface displacement, or 4D Tomography (4D-Tomo) emphasizing the addition of time to the 3D tomography. It is also worth to note that for $K = 1$ the model in (8.48) reduces to the model in (7.11), provided that data calibration for atmospheric compensation and nonlinear deformations have been carried out on the data so that the vector $\boldsymbol{a}$ can be structured, similarly to PSI, as a function of only two unknowns, the residual topography of the scatterer s and the linear deformation velocity v.

At this stage it becomes of interest the problem to detect the scatterers and to estimate its geophysical parameters, i.e. residual height and linear deformation. We address in the following the cases of single and multiple scatterers analyzed at full spatial resolution (i.e., with a single-look processing), and single scatterer analyzed at reduced spatial resolution (i.e., with a multilook processing).

8.3.1 Detection and scatterer parameter estimation of single scatterers at full resolution

The problem to discriminate the scatterers capable to retain coherence over the different acquisitions of the tomographic array from the noise becomes now of interest. This issue, pertinent also to the Persistent Scatterers discussed in the previous chapter for the multitemporal case, is here addressed in the detection framework typical of classical surveillance radar. We consider the data model in (8.47) specified for a single scatterer ($k = 1$) and address the problem to discriminate the two hypothesis of the following test:

$$\begin{aligned} \mathcal{H}_0: & \quad \boldsymbol{y} = \boldsymbol{w} \\ \mathcal{H}_1: & \quad \boldsymbol{y} = \gamma \boldsymbol{a}(s, v) + \boldsymbol{w} \end{aligned} \tag{8.49}$$

$\boldsymbol{w}$ is assumed to be a zero mean, circularly complex Gaussian random vector, with $E[\boldsymbol{w}\boldsymbol{w}^H] = \sigma_w^2 \mathbf{I}$. Being in absence of decorrelation, the complex variable γ can be treated as a deterministic unknown or a random variable. Although leading to the same detection rule [111], for sake of simplicity here we assume the deterministic unknown model. The scatterer parameters (s, v) are also typically unknown.

The optimal test according to the Neyman-Pearson criterion, which maximizes the detection probability (P_d) for a given false alarm probability (P_{fa}),[3] is:

$$\mathcal{L}(\boldsymbol{y}) = \frac{f_{\boldsymbol{y}}\left(\boldsymbol{y}; \sigma_w^2, \gamma, s, v; \mathcal{H}_1\right)}{f_{\boldsymbol{y}}\left(\boldsymbol{y}; \sigma_w^2; \mathcal{H}_0\right)} \overset{\mathcal{H}_1}{\underset{\mathcal{H}_0}{\gtrless}} T \qquad (8.50)$$

where $\mathcal{L}(\boldsymbol{y})$ is the test statistic, T is the detection threshold, set to the value that guarantees the desired P_{fa}, and $f_{\boldsymbol{y}}\left(\boldsymbol{y}; \sigma_w^2, \mathcal{H}_0\right)$ and $f_{\boldsymbol{y}}\left(\boldsymbol{y}; \sigma_w^2, \gamma, z, \mathcal{H}_1\right)$ are the probability density functions (pdf) of the data under the $\mathcal{H}_0$ and $\mathcal{H}_1$ hypothesis, respectively:

$$f_{\boldsymbol{y}}\left(\boldsymbol{y}; \sigma_w^2; \mathcal{H}_0\right) = (\pi \sigma_w^2)^{-N} \exp\left(-\frac{\| \boldsymbol{y} \|^2}{\sigma_w^2}\right) \qquad (8.51)$$

$$f_{\boldsymbol{y}}\left(\boldsymbol{y}; \sigma_w^2, \gamma, s, v; \mathcal{H}_1\right) = (\pi \sigma_w^2)^{-N} \exp\left(-\frac{\| \boldsymbol{y} - \boldsymbol{a}\gamma \|^2}{\sigma_w^2}\right) \qquad (8.52)$$

The test in (8.50) can be manipulated according to the data models in (8.49) and the assumed noise statistic, leading to (see Appendix)

$$\Re(\gamma^* \boldsymbol{a}^H (s, v) \boldsymbol{y}) \overset{\mathcal{H}_1}{\underset{\mathcal{H}_0}{\gtrless}} T \qquad (8.53)$$

where $\Re$ indicates the real part operator. Expression (8.53) shows that the optimal test presumes the knowledge of γ, s, and v which, in real cases, are not known. We consider in the following the case in which the backscattering coefficient is, similarly to s and v, a deterministic unknown variable. The Generalized Likelihood Ratio Test (GLRT) provides a suboptimal test, in which all the unknowns are retrieved by the Maximum Likelihood Estimation (MLE). According to the GLRT criterion, the test statistic of the problem in

---

[3]With reference to a binary hypotheses test $(\mathcal{H}_0, \mathcal{H}_1)$, said $\mathcal{D}_i$ the event "decision for $\mathcal{H}_i$", $i = 0, 1$, the false alarm probability and the detection probability are defined as

$$P_{fa} = \Pr(\mathcal{D}_1 | \mathcal{H}_0)$$
$$P_d = \Pr(\mathcal{D}_1 | \mathcal{H}_1)$$

respectively, where $\Pr(\mathcal{A}|\mathcal{B})$ is the conditional probability of the event $\mathcal{A}$ given the event $\mathcal{B}$.

(8.49) is given by:

$$\mathcal{L}(y) = \frac{\max\limits_{\sigma_w^2, \gamma, s, v} f_y\left(y; \sigma_w^2, \gamma, s, v | \mathcal{H}_1\right)}{\max\limits_{\sigma_w^2} f_y\left(y; \sigma_w^2 | \mathcal{H}_0\right)} \tag{8.54}$$

According to [112] the solution of (8.54) can be found in a rather simple closed form (see Appendix):

$$\frac{|a^H(\hat{s}, \hat{v})y|}{\|a\|\|y\|} \overset{\mathcal{H}_1}{\underset{\mathcal{H}_0}{\gtrless}} T \tag{8.55}$$

where the threshold T is chosen, as usual, according to the desired P_{fa} and

$$(\hat{s}, \hat{v}) = \arg\max_{s, v} |a^H(s, v)y| \tag{8.56}$$

is the maximum likelihood estimate of residual topography and mean deformation velocity. Interestingly, under $\mathcal{H}_0$ the testing statistic compares the ratio between the maximum component of the noise in the steering directions and the noise norm to a threshold. It is thus intuitive that the P_{fa} becomes independent on the noise power (noise level) thus making the test a Constant False Alarm Rate (CFAR) test; this fact is mathematically demonstrated in [112]. The expression of the detector in (8.55) highlights that the GLRT test compares the maximum peak of the modulus of the reflectivity estimation achieved by beamforming at the numerator in (8.55), to a threshold after a proper normalization to the norms of the data and of the steering vector. The beamforming is in this case referred to the 4D (space-time) context and is simply achieved by extending the dependence of the steering vector also along the velocity direction, see (7.57). A resolution along the velocity can be thus introduced in analogy to (8.15):

$$\rho_v = \frac{\lambda}{2T_s}, \tag{8.57}$$

where T_s is the temporal span.

The GLRT is a suboptimal test, achieved by the optimal one after the substitution of the unknowns with their maximum likelihood estimates. It is shown to be asymptotically optimal (in terms of detection probability) with the number of acquisitions, i.e., for large datasize. Moreover, it can be shown that the statistical model assumed for the data leads the GLRT in (8.55) to be equivalent to the (suboptimal) Rao and Wald tests, see [112].

8.3.2 Detection and scatterer parameter estimation for multiple scatterers at full resolution

The case corresponding to single scatterers ($K = 1$) may be insufficient for the monitoring of buildings in urban areas due to the occurrence of the layover phenomenon. A higher order analysis can be carried out by extending the PS tomographic detection scheme discussed in the previous section. Below we limit the discussion to double scatterers being the extension to more scatterers straightforward. The problem can be cast in the following mathematical form

$$
\begin{aligned}
\mathcal{H}_0 : \quad & \boldsymbol{y} = \boldsymbol{w} \\
\mathcal{H}_1 : \quad & \boldsymbol{y} = \gamma_1 \boldsymbol{a}(s_1, v_1) + \boldsymbol{w} \\
\mathcal{H}_2 : \quad & \boldsymbol{y} = \gamma_1 \boldsymbol{a}(s_1, v_1) + \gamma_2 \boldsymbol{a}(s_2, v_2) + \boldsymbol{w}
\end{aligned}
\tag{8.58}
$$

and one wish to maximize the detection probability for a given false alarm probability. It should be noted that the overall detection probability also encompasses two factors, P_{d1} and P_{d2} corresponding to the decisions $\mathcal{D}_1 | \mathcal{H}_1$ and $\mathcal{D}_2 | \mathcal{H}_2$, respectively.[4] The false alarm probability is instead composed of three terms, corresponding to the decision $\mathcal{D}_1 | \mathcal{H}_0$, $\mathcal{D}_2 | \mathcal{H}_0$, and $\mathcal{D}_2 | \mathcal{H}_1$ [113]. Under the equiprobable hypothesis for $\mathcal{H}_0$, $\mathcal{H}_1$, and $\mathcal{H}_2$ the P_{fa} is defined as:

$$
P_{fa} = \frac{1}{2} [\Pr(\mathcal{D}_1 | \mathcal{H}_0) + \Pr(\mathcal{D}_2 | \mathcal{H}_0) + \Pr(\mathcal{D}_2 | \mathcal{H}_1)]
\tag{8.59}
$$

where $\Pr(\cdot)$ stands for probability.

A sequential scheme can be built based on the previously discussed single scatterer GLRT detector according to the following cancellation strategy [113]. Let us select the peak of the beamforming reconstruction of the backscattering profile in the s, v plane, i.e.:

$$
(\hat{s}_1, \hat{v}_1) = \arg\max_{s,v} |\boldsymbol{a}^H(s, v) \boldsymbol{y}|
\tag{8.60}
$$

The sequential scheme tests the hypothesis $\mathcal{H}_2$ according to the rule:

$$
\frac{\max\limits_{s,v} |\boldsymbol{a}^H(s, v) \boldsymbol{y}_c|}{\| \boldsymbol{a} \| \, \| \boldsymbol{y}_c \|} \overset{\mathcal{H}_2}{\underset{\overline{\mathcal{H}_2}}{\gtrless}} T_2
\tag{8.61}
$$

where $\overline{\mathcal{H}_2}$ is the hypothesis complementary to $\mathcal{H}_2$, and:

$$
\boldsymbol{y}_c = \mathbf{P}_{\boldsymbol{a}_1}^{\perp} \boldsymbol{y}
\tag{8.62}
$$

---

[4] $\mathcal{D}_i | \mathcal{H}_j$ is the decision for $\mathcal{H}_i$ when $\mathcal{H}_j$ is true.

is the cancelled data, i.e., the projection of the data $\boldsymbol{y}$ on the subspace orthogonal to the estimated steering direction $\boldsymbol{a}_1 = \boldsymbol{a}(\hat{s}_1, \hat{v}_1)$, achieved by means of the projection matrix:

$$\mathbf{P}_{\boldsymbol{a}_1}^{\perp} = \mathbf{I}_N - \frac{\boldsymbol{a}_1 \boldsymbol{a}_1^H}{\boldsymbol{a}_1^H \boldsymbol{a}_1} \tag{8.63}$$

If $\mathcal{H}_2$ is chosen, the two PS are selected with parameters $(\hat{s}_1, \hat{v}_1)$ in (8.60) and

$$(\hat{s}_2, \hat{v}_2) = \arg\max_{s,v} |\boldsymbol{a}^H(s, v) \boldsymbol{y}_c| \tag{8.64}$$

that is the bin (direction) that maximize the statistic in (8.61). Alternatively, if $\overline{\mathcal{H}}_2$ is chosen, the single scatterer GLRT test in the selected maximum is performed on the original data for the decision between $\mathcal{H}_1$ and $\mathcal{H}_0$:

$$\frac{|\boldsymbol{a}^H(\hat{s}_1, \hat{v}_1)\boldsymbol{y}|}{\|\boldsymbol{a}\| \|\boldsymbol{y}\|} \underset{\mathcal{H}_0}{\overset{\mathcal{H}_1}{\gtrless}} T_1 \tag{8.65}$$

It can be shown [113] that, if N is not small, for a given P_{fa} the best threshold pair (T_1, T_2) useful to maximize the scatterer detection probability provides $T_1 \simeq T_2$. It is worth noting that the above sequential scheme provides minor complexity in the increase of the scatterer order, because only one dimensional searches are involved in (8.60) and (8.61) or equivalently (8.64). It should be however note that the cancellation in (8.62) may interfere with the second scatterer, leading to a reduction of the energy coming from the second scatterer. This may lead to detection losses of the second scatterer, i.e. a reduction of P_{d2}.

An algorithm capable of improving the performances in the detection of double scatterers is referred to as support-based GLRT (sup-GLRT) and addressed in the following [114]. Similarly to the above discussed detector of multiple scatterers based on successive cancellation, the sup-GLRT exploit a sequential scheme. It tests the hypothesis of a given order of scatterers, say k_0 against the hypothesis of having more scatterers, i.e. k scatterers with $k \geqslant k_0$. For the double scatterer case it therefore performs a sequence of two tests, the first one involving the null hypothesis $\mathcal{H}_0$, i.e. $k_0 = 0$, and nonnull ($\overline{\mathcal{H}}_0$, i.e., $\mathcal{H}_1$ or $\mathcal{H}_2$ hypothesis:

$$\frac{\max\limits_{\sigma_w^2, \gamma_1 \ldots \gamma_k, s_1 \ldots s_k, v_1 \ldots v_k, k \in \{1,2\}} f_{\boldsymbol{y}}\left(\boldsymbol{y}; \sigma_w^2, \gamma_1 \ldots \gamma_k, s_1 \ldots s_k, v_1 \ldots v_k; \mathcal{H}_k\right)}{\max\limits_{\sigma_w^2} f_{\boldsymbol{y}}\left(\boldsymbol{y}; \sigma_w^2, ..., \mathcal{H}_0\right)} \underset{\mathcal{H}_0}{\overset{\overline{\mathcal{H}}_0}{\gtrless}} T_0 \tag{8.66}$$

and, if this test decides for $\overline{\mathcal{H}}_0$, the second one testing $\mathcal{H}_1$ against $\mathcal{H}_2$:

$$\frac{\max\limits_{\sigma_w^2,\gamma_1,\gamma_2,s_1,s_2,v_1,v_2} f_{\boldsymbol{y}}\left(\boldsymbol{y};\sigma_w^2,\gamma_1,\gamma_2,s_1,s_2,v_1,v_2;\mathcal{H}_2\right)}{\max\limits_{\sigma_w^2,\gamma_1,s_1,v_1} f_{\boldsymbol{y}}\left(\boldsymbol{y};\sigma_w^2,\gamma_1,s_1,v_1;\mathcal{H}_1,\right)} \overset{\mathcal{H}_2}{\underset{\mathcal{H}_1}{\gtrless}} T_1 \qquad (8.67)$$

Being the maximization with $k = 2$ a maximization with parameters including those of $k = 1$, the maximum with respect to $k = 1$ and $k = 2$ is simply $k = 2$ thus leading to:

$$\frac{\max\limits_{\sigma_w^2,\gamma_1,\gamma_2,s_1,s_2,v_1,v_2,} f_{\boldsymbol{y}}\left(\boldsymbol{y};\sigma_w^2,\gamma_1,\gamma_2,s_1,s_2,v_1,v_2;\mathcal{H}_2\right)}{\max\limits_{\sigma_w^2} f_{\boldsymbol{y}}\left(\boldsymbol{y};\sigma_w^2,...,\mathcal{H}_0\right)} \overset{\overline{\mathcal{H}}_0}{\underset{\mathcal{H}_0}{\gtrless}} T_0 \qquad (8.68)$$

The solution to (8.68) is simply provided by [114]:

$$\frac{\min\limits_{s_1,s_2,v_1,v_2} \boldsymbol{y}^H \mathbf{P}_{\mathbf{A}}^{\perp} \boldsymbol{y}}{\boldsymbol{y}^H \boldsymbol{y}} \overset{\mathcal{H}_0}{\underset{\overline{\mathcal{H}}_0}{\gtrless}} T_0 \qquad (8.69)$$

where, $\mathbf{A} = [\boldsymbol{a}_1(s_1,v_1),\boldsymbol{a}_2(s_2,v_2)]$ and $\mathbf{P}_{\mathbf{A}}^{\perp}$ is the operator projecting a generic vector in the subspace orthogonal to the space spanned by the vectors $\boldsymbol{a}_1$ and $\boldsymbol{a}_2$, that is $\mathbf{P}_{\mathbf{A}}^{\perp} = \mathbf{I} - \mathbf{A}(\mathbf{A}^H\mathbf{A})^{-1}\mathbf{A}^H$. The test in (8.69) can be equivalently cast as:

$$\frac{\max\limits_{s_1,s_2,v_1,v_2} \boldsymbol{y}^H \mathbf{P}_{\mathbf{A}} \boldsymbol{y}}{\boldsymbol{y}^H \boldsymbol{y}} \overset{\overline{\mathcal{H}}_0}{\underset{\mathcal{H}_0}{\gtrless}} T_0 \qquad (8.70)$$

where $\mathbf{P}_{\mathbf{A}}$ is the operator projecting in the subspace spanned by the vectors $\boldsymbol{a}_1$ and $\boldsymbol{a}_2$, that is $\mathbf{P}_{\mathbf{A}} = \mathbf{A}(\mathbf{A}^H\mathbf{A})^{-1}\mathbf{A}^H$. If the $\mathcal{H}_0$ hypothesis is selected, then the subsequent test compares the $\mathcal{H}_1$ hypothesis to the $\mathcal{H}_2$ one with the following test:

$$\frac{\min\limits_{s_1,s_2,v_1,v_2} \boldsymbol{y}^H \mathbf{P}_{\mathbf{A}}^{\perp} \boldsymbol{y}}{\min\limits_{s_1,v_1} \boldsymbol{y}^H \mathbf{P}_{\boldsymbol{a}_1}^{\perp} \boldsymbol{y}} \overset{\mathcal{H}_1}{\underset{\mathcal{H}_2}{\gtrless}} T_1 \qquad (8.71)$$

The joint maximization over the parameters space of both the steering vectors, significantly increases the computational complexity with respect to the detection scheme based on the cancellation. However this complexity increase is compensated by an increase in the detection performance, which is notable, especially for superresolving scatterers in elevation. Superresolution means the capability to discriminate scatterers below the Rayleigh resolution, which can be low (numerically high) in the presence of a small baseline span as in the case of the TerraSAR-X orbits.

8.3.3 Multilook detection of weak persistent scatterers

The detection performance of PS depends on the scattering power, specifically on the Signal to Noise Ratio, that for the generic k-th scatterer can be defined as:

$$SNR_k = \frac{|\gamma_k|^2}{\sigma_w^2} \tag{8.72}$$

Weak scatterers, typically located in rural areas and characterized by a distributed scattering, have generally low detection performances. As already discussed in Section 7.3 a possibility to improve the performances of the interferometric monitoring techniques is to resort to multilook approaches, see SqueeSAR and CAESAR. In this section we specifically focus on the use of multilook at the PS/DS detection stage. We limit to the case of single scatterers, which is typically the case of rural areas. The decision problem can be thus specified as:

$$\begin{aligned} \mathcal{H}_0: \quad & \mathbf{y}_l = \mathbf{w}_l \\ \mathcal{H}_1: \quad & \mathbf{y}_l = \gamma_l \mathbf{a}(s, v) + \mathbf{w}_l \end{aligned} \tag{8.73}$$

with $l = 1, \ldots, L$ being L the number of available looks. We assume in this case that the backscattering coefficients γ_l are independent and identically distributed (i.i.d.) random variables, with zero mean and variance σ_γ^2. The test provided by following the GLRT principle for the multilook case (MGLRT) is [111]:

$$\frac{\max\limits_{s,v} \sum_{l=1}^{L} |\mathbf{y}_l^H \mathbf{a}(s, v)|^2}{\| \mathbf{a} \|^2 \sum_{l=1}^{L} \| \mathbf{y}_l \|^2} \underset{\mathcal{H}_0}{\overset{\mathcal{H}_1}{\gtrless}} T \tag{8.74}$$

which can be cast as:

$$\frac{\max\limits_{s,v} \mathbf{a}^H(s, v)\hat{\mathbf{C}}_y \mathbf{a}(s, v)}{\| \mathbf{a} \|^2 \, \mathrm{tr}(\hat{\mathbf{C}}_y)} = \frac{\max\limits_{s,v} \mathrm{tr}[\mathbf{A}^H(s, v)\hat{\mathbf{C}}_y]}{\mathrm{tr}(\mathbf{A})\mathrm{tr}(\hat{\mathbf{C}}_y)} \underset{\mathcal{H}_0}{\overset{\mathcal{H}_1}{\gtrless}} T \tag{8.75}$$

where $\mathrm{tr}(\cdot)$ is the trace operator and $\mathbf{A}(s, v) = \mathbf{a}(s, v)\mathbf{a}^H(s, v)$ is the steering matrix associated with the steering vectors. Interpreting the trace operator as a scalar product simply states that, but for a normalization, the test compares the maximum scalar product between the steering matrix and the sample correlation matrix (i.e. the peak of the multilook Beamforming) to a threshold; the latter set according to the desired P_{fa}. In [111] it is shown that the use of multilook allows significantly improving (shift toward lower SNR) the detection curves with respect to the single-look GLRT in (8.54). This shift is provided by a decrease of the threshold for a given false alarm probability. MGLRT has been shown to achieve better performances with respect to other multilook detectors. However it has the drawback that the false alarm curves strongly depend on the number of averaged looks thus making the implementation more problematic especially in a multiresolution framework. Multiresolution detection aims at using a variable number of looks depending on the observed scene (urban or rural). One might use a mild, or ideally no multilook, over the built scenario (characterized by the presence of strong scatterers) to preserve the spatial resolution and exploit at the same time a larger number of looks in the areas characterized by distributed scattering. This strategy can provide a dramatic improvement in the density of monitored pixels.

The CAESAR [115] based detector, which is provided by substituting the sample correlation matrix in (8.75) with the one corresponding to its dominant component:

$$\frac{\max\limits_{s,v} \boldsymbol{a}^H(s,v)\hat{\mathbf{C}}_1 \boldsymbol{a}(s,v)}{\| \boldsymbol{a} \|^2} = \frac{\max\limits_{s,v} \text{tr}[\mathbf{A}^H(s,v)\hat{\mathbf{C}}_1]}{\| \boldsymbol{a} \|^2 \lambda_1} \overset{\mathcal{H}_1}{\underset{\mathcal{H}_0}{\gtrless}} T \qquad (8.76)$$

has been shown to achieve significant improvements in the detection capabilities with respect to the single look but sharing with the latter the same false alarm curve (P_{fa} vs T) independently of the number of averaged looks [115]. The detection performances in terms of P_d for a given P_{fa} are slightly lower than the MGLRT but with the advantage of being of easy implementation thanks to a setting of the threshold independent on the pixel in the multiresolution context [115].

8.3.4 Extension to multi-dimensional imaging

There might be cases in which the parametrization of the steering vector with respect to the mean deformation velocity (v) and elevation (s) might be insufficient. This is for instance the case of metallic structure observed with systems operating at high frequency (typically X-Band as for the case of COSMO-SkyMed and TerraSAR-X) [116]. In order to avoid detection losses, as well as incorrect parameter estimations it turns useful in this situation to model the steering vector as a function of additional parameters. Letting T_n be the temperature the n-th acquisition (typically it is used the average daily temperature) one can model the elements of the steering vector as:

$$\{a\}_n = e^{-j\frac{4\pi}{\lambda}(vt_n+\zeta T_n)+jc_{sn}s} \tag{8.77}$$

where ζ, measured in cm/K or cm/°C, accounts for the additional thermal deformation component with respect to (7.57) and all other symbols are specified as in the previous chapter; $\{\cdot\}_n$ extracts as usual the n-th element of a vector. The thermal coefficient represents the increase of the range corresponding to a change of one degree

Eq. (8.77) can be written in a compact and more general form:

$$\{a\}_n = e^{-j2\pi\xi_n p} \tag{8.78}$$

where p collects the unknowns parameters, in the above example v, ζ, and s whereas ξ_n collects the known "frequencies" associated in our example with the time, temperature and baseline values corresponding to the n-th acquisition so that the test can be also written in a compact form that allows accounting different parameter options. For instance the single look single scatterer detection problem can be formulated as:

$$\begin{aligned}\mathcal{H}_0: \quad & y = w \\ \mathcal{H}_1: \quad & y = \gamma a(p) + w\end{aligned} \tag{8.79}$$

Fig. 8.10 provides an example of reconstruction of the thermal coefficient of a structure in Las Vegas achieved by processing very high resolution (spotlight mode) SAR data characterized by a ground resolution of about 1 m. It should be noted that the increase of dimensionality of the problem with the introduction of additional parameters allows on one hand to follow unmodeled target phase, "signatures" which might be present on the phase response of the scatterer after the calibration for atmosphere and nonlinear small (rough) scale (i.e. low resolution) deformation component. On the other hand the dimensionality increase could lead to a phase overfitting. It can be in fact shown that, for fixed system parameters configuration (i.e. number of acquisitions, baselines, etc.), the introduction of the thermal coefficient leads to an increase of the false alarm probability for a fixed testing threshold [117], which translates to a reduction of the density of detected thermally insensitive scatterers.

Another problem is related to the possible correlation between the phase components, which may even lead to ambiguities. For

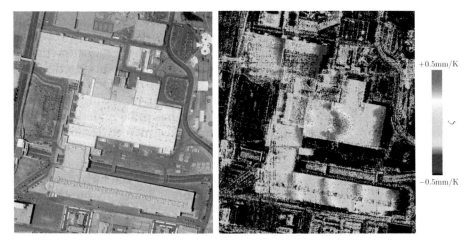

Figure 8.10. Comparison of the optical image relevant to the Las Vegas (USA) Convention Centre and the corresponding map of the thermal coefficient estimated from a dataset of very high resolution TerraSAR-X dataset acquired in Spotlight mode. Original SLC dataset provided by German Aerospace Center (DLR).

instance, if the temporal interval is limited to a couple of seasons with irregular time sampling, the time and the temperature vectors might be correlated thus leading generally to an intrinsic quasiambiguity between the estimated mean deformation velocity and thermal coefficient. Similar reasoning can be applied to small datasets in which the spatial baselines and the time are aligned in the baseline domain plane, even though this condition is less frequent. The problem has been addressed in [117].

8.4 TomoSAR applications to volume scattering scenarios

TomoSAR techniques can be usefully exploited for characterizing and monitoring forested areas, and more in particular their vertical structure that typically represents the arrangement of trees, trunks, and crown.

Forest structure reconstruction is important for forest monitoring, since it represents a significant indicator of productivity and biomass level. Moreover, the structure variation can reveal the dynamic of the forest, like growth, regeneration, decay, and the temporal evolution of the above ground biomass distribution.

Today, TomoSAR and airborne LiDAR are the two technologies that allow the measurement of 3D forest structure. LiDAR measures were first used to evaluate forest height and biomass. More recently, TomoSAR systems have been frequently used to obtain a continuous forest mapping. They benefit of the remote sensing systems advantages, such as the wide area coverage and the short revisiting time, which make them very attractive and highly competitive respect to in situ measurement systems. The latter are generally more precise, but allow only local measurements with very sparse coverage and involve typically high operative costs.

An important aspect that has to be underlined, is that for the 3D reconstruction of the forest structure, TomoSAR systems must have under foliage penetration capabilities. Then, low frequency SAR systems have to be used (typically P-band or L-band systems). Moreover, polarization diversity can enhance focusing ability and facilitate the characterization of different scattering components, which are related to geometrical structure, orientation and shape of the scatterers contributing to the electromagnetic response.

L-band SAR satellites include JERS-1 (1992–1998), ALOS (2006–2011), ALOS-2 (2014–present), and SAOCOM-1 (2017–present), which have a revisiting time in the most favorable case of the order of a few weeks. Such revisiting times do not allow monitoring with high temporal resolution.

Anyway, with the launch of ALOS-4, NISAR, and Tandem-L a revisiting time with L-band SAR systems of the order of few days will be obtained. The P-band frequency is less common in spaceborne SAR systems respect to L-band, while has superior abilities on penetration through the vegetation canopy and soil. Currently there are no P-band SAR satellite operating from the space. The upcoming P-band satellite BIOMASS of the European Space Agency (ESA) is expected to operate at a frequency of 435 MHz (69 cm wavelength) and 25 m spatial resolution.

Other operating low frequency airborne or UAV-borne sensors are the P-band airborne system SETHI by ONERA, and UAVSAR L-band system by Jet propulsion Laboratory (JPL)/NASA.

8.4.1 Single polarization multibaseline TomoSAR data model for vegetated areas

When the observed scene is a forested area, the relation between the SAR multibaseline data and the unknown elevation profile of the complex reflectivity $\gamma(s)$ is still given by (8.1). Anyway, in this case, $\gamma(s)$ is better described as a stochastic process, due to the high variability of the numerous scatterers located within each SAR resolution cell. For this reason, it follows that for the analysis of vegetated areas the interest is in the evaluation of the backscattered power as a function of the elevation, defined as:

$$p(s) = E\left[|\gamma(s)|^2\right] \tag{8.80}$$

where $E[\cdot]$ denotes the ensemble average.

The auto-correlation function of the multibaseline data is given by:

$$R_y(s'_m, s'_n) = E\left[y(s'_m)y^*(s'_n)\right] \qquad (8.81)$$

Referring to the discretized model (8.7) and (8.8) in presence of additive noise

$$y = A\gamma + w \qquad (8.82)$$

and assuming w to be a zero mean uncorrelated vector, and mutually uncorrelated with the reflectivity profile, the correlation matrix of the data vector can be written as:

$$\mathbf{R}_y = \mathbf{A}\mathbf{R}_\gamma \mathbf{A}^H + \sigma_w^2 \mathbf{I} \qquad (8.83)$$

where $\mathbf{I}$ is the identity matrix.

Assuming that the reflectivity profile is uncorrelated along the elevation s, i.e. that $R_\gamma(s_1, s_2) = \delta(s_1 - s_2)E\left[|\gamma(s_1)|^2\right]$, the correlation matrix of γ becomes:

$$\mathbf{R}_\gamma = E[\text{diag}\left(|\gamma_1|^2, \ldots, |\gamma_M|^2\right)], \qquad (8.84)$$

where γ_m is the $m - th$ element of γ, that is the reflectivity at the $m - th$ elevation bin.

The vertical distribution of scatterers given by the vector γ can be estimated exploiting (8.83), so that information on the vertical structure of the vegetation layer can be extracted. In (8.83), the data correlation matrix $\mathbf{R}_y$ is not known a priori and, as already specified, must be estimated from L realizations of the data vector (usually denoted as *looks*). Differently from the previous sections, for notation needs in the polarimetric case, the look dependence will be indicated between curl brackets. Therefore the following expression:

$$\hat{\mathbf{R}}_y = \frac{1}{L}\sum_{l=1}^{L} y(l) \, y^H(l) \qquad (8.85)$$

where $y(l)$ are the available looks, will be used hereafter in place of (8.20).

The power spectrum profile along the elevation can be estimated by using TomoSAR processing techniques using a stochastic model for the unknown reflectivity, such as classical beamforming, Capon beamforming, and MUSIC described in Sections 8.2.1, 8.2.3, and 8.2.4.

8.4.2 Multipolarization multibaseline TomoSAR data model for vegetated areas

The vegetation classification can be made by exploiting polarimetric signature, since the polarization information contained in the waves backscattered from the scene is related to the geometrical structure and orientation of the observed objects.

Most scattering models from vegetation include a mixture of contributions from the ground and canopy. The former is highly concentrated around a single value in the elevation direction (it can be ideally approximated by a Dirac delta function in a position corresponding to the elevation value of the ground), whereas the latter spreads throughout a larger range of elevations. The use of model based estimators for reconstructing forest structure would require the a priori knowledge of many parameters, such as the tree species and moisture, so that model-free approaches are usually preferred.

When N_P polarimetric channels are used, the signal model is modified by introducing the multibaseline polarimetric steering vectors and the polarimetric reflectivity vectors. Then, the spectral estimation techniques can be extended to the polarimetric case.

The number of polarization channels assumes the values $N_P = 2$ for dual-polarization configurations, $N_P = 3$ for fully polarimetric data with equal cross-polarizations and $N_P = 4$ for quad-polarization setups.

Using the fully polarimetric SAR configuration, the polarimetric information about backscattering from targets for each pixel can be represented in a 2×2 scattering matrix **S**. It is expressed in the backscatter alignment (BSA) convention in the linear horizontal (h) and linear vertical (v) polarization basis as:

$$\mathbf{S} = \begin{bmatrix} S_{hh} & S_{hv} \\ S_{vh} & S_{vv} \end{bmatrix} \tag{8.86}$$

where the subscripts of the elements identify the polarization channel on transmit and receive, respectively. The matrix's diagonal elements represent the copolarized scattering components, while the off-diagonal terms represent the cross-polarized ones. In the monostatic backscattering case, the reciprocity theorem constrains the scattering matrix to be symmetric, e.g. $S_{hv} = S_{vh}$.

In polarimetric SAR, **S** mainly is vectorized with a known target scattering vector, such as lexicographic vector:

$$\boldsymbol{\omega_3} = [S_{hh} \quad \sqrt{2}S_{hv} \quad S_{vv}]^T, \tag{8.87}$$

or Pauli vector:

$$\boldsymbol{f}_P = \frac{1}{\sqrt{2}}[S_{hh} + S_{vv} \quad S_{hh} - S_{vv} \quad 2S_{hv}]^T. \tag{8.88}$$

Note that Lexicographic and Pauli vectors can be obtained from the 2×2 scattering matrix $\mathbf{S}$ through the following relations:

$$\boldsymbol{\omega_3} = \frac{1}{2}\text{trace}\,(\mathbf{S}\boldsymbol{\Psi}_L) \tag{8.89}$$

$$\boldsymbol{f}_P = \frac{1}{2}\text{trace}\,(\mathbf{S}\boldsymbol{\Psi}_p) \tag{8.90}$$

where $\boldsymbol{\Psi}_L$ and $\boldsymbol{\Psi}_p$ are the orthogonal basis of Lexicographic and Pauli matrices:

$$\boldsymbol{\Psi}_L = \left(2\begin{bmatrix} 1 & 0 \\ 0 & 0 \end{bmatrix}, \; 2\sqrt{2}\begin{bmatrix} 0 & 1 \\ 0 & 0 \end{bmatrix}, \; 2\begin{bmatrix} 0 & 0 \\ 0 & 1 \end{bmatrix} \right), \tag{8.91}$$

$$\boldsymbol{\Psi}_p = \left(\sqrt{2}\begin{bmatrix} 1 & 0 \\ 0 & 1 \end{bmatrix}, \; \sqrt{2}\begin{bmatrix} 1 & 0 \\ 0 & -1 \end{bmatrix}, \; \sqrt{2}\begin{bmatrix} 0 & 1 \\ 1 & 0 \end{bmatrix} \right). \tag{8.92}$$

For the case of dual polarization SAR systems the Lexicographic vector is represented by:

$$\boldsymbol{\omega_2} = [S_{hh} \quad S_{vv}]^T. \tag{8.93}$$

For the monostatic case, the 3×3 Hermitian polarimetric correlation matrix $\mathbf{C}$ is [118] [15]:

$$\mathbf{C} = E\left(\boldsymbol{\omega_3}\boldsymbol{\omega}_3^H\right) = E\left(\begin{bmatrix} |S_{hh}|^2 & \sqrt{2}S_{hh}S_{hv}^* & S_{hh}S_{vv}^* \\ \sqrt{2}S_{hv}S_{hh}^* & 2|S_{hv}|^2 & \sqrt{2}S_{hv}S_{vv}^* \\ S_{vv}S_{hh}^* & \sqrt{2}S_{vv}S_{hv}^* & |S_{vv}|^2 \end{bmatrix} \right), \tag{8.94}$$

while the 3×3 Hermitian coherency matrix $\mathbf{T}$ is [118]:

$$\mathbf{T} = E\left(\boldsymbol{f}_P\boldsymbol{f}_P^H\right). \tag{8.95}$$

When considering multibaseline (MB) polarimetric TomoSAR configurations, the data vector can be formed by stacking in a single vector $\mathbf{y}_P$ the polarimetric responses, so that, referring to the fully polarimetric case, we have:

$$\boldsymbol{y}_P = [\boldsymbol{y}_1^T \quad \boldsymbol{y}_2^T \quad \boldsymbol{y}_3^T]^T \in \mathbb{C}^{3N \times 1}, \tag{8.96}$$

where $\boldsymbol{y}_i \in \mathbb{C}^{N \times 1}$ represents the MB data stack vector in a specific polarization channel i, with $i = 1, 2, 3$, satisfying the tomoSAR equation (8.9):

$$y_i = \mathbf{A}\boldsymbol{\gamma}_i + \boldsymbol{w}_i = \sum_{m=1}^{M} \boldsymbol{a}(s_m)\gamma_i(s_m) + \boldsymbol{w}_i \in \mathbb{C}^{N \times 1}, \qquad (8.97)$$

where $\mathbf{A}$ is the steering matrix defined in (8.8), and $\boldsymbol{\gamma}_i$ is the reflectivity in the polarimetric channel i. Here we refer to the reflectivity values instead of the scattering coefficients in (8.86): the reader can refer to (3.18) and (3.19) for the different meaning.

It should be noted that the vector $\boldsymbol{y}_P$, defined in (8.96), either can be generated in Lexicographic or Pauli representation. Analogously, the polarimetric reflectivity vector can be defined:

$$\boldsymbol{\gamma}_P = [\boldsymbol{\gamma}_1^T \ \boldsymbol{\gamma}_2^T \ \boldsymbol{\gamma}_3^T]^T \in \mathbb{C}^{3M \times 1}, \qquad (8.98)$$

where $\boldsymbol{\gamma}_1$, $\boldsymbol{\gamma}_2$ and $\boldsymbol{\gamma}_3$ are defined as:

$$\begin{aligned}
\boldsymbol{\gamma}_1 &= [\gamma_{hh}(s_1), \gamma_{hh}(s_2), \quad \ldots \quad , \gamma_{hh}(s_M)]^T \\
\boldsymbol{\gamma}_2 &= \sqrt{2}[\gamma_{hv}(s_1), \gamma_{hv}(s_2), \quad \ldots \quad , \gamma_{hv}(s_M)]^T \\
\boldsymbol{\gamma}_3 &= [\gamma_{vv}(s_1), \gamma_{vv}(s_2), \quad \ldots \quad , \gamma_{vv}(s_M)]^T
\end{aligned} \qquad (8.99)$$

in Lexicographic basis, and

$$\begin{aligned}
\boldsymbol{\gamma}_1 &= \frac{1}{\sqrt{2}}[\gamma_{hh}(s_1) + \gamma_{vv}(s_1), \gamma_{hh}(s_2) + \gamma_{vv}(s_2), \ \ldots \ \gamma_{hh}(s_M) + \gamma_{vv}(s_M)]^T \\
\boldsymbol{\gamma}_2 &= \frac{1}{\sqrt{2}}[\gamma_{hh}(s_1) - \gamma_{vv}(s_1), \gamma_{hh}(s_2) - \gamma_{vv}(s_2), \ \ldots \ \gamma_{hh}(s_M) - \gamma_{vv}(s_M)]^T \\
\boldsymbol{\gamma}_3 &= \sqrt{2}[\gamma_{hv}(s_1), \gamma_{hv}(s_2), \ \ldots \ \gamma_{hv}(s_M)]^T
\end{aligned}$$

$$(8.100)$$

in Pauli basis.

We can define the polarimetric span at the elevation s_m as:

$$\tau^2(s_m) = |\gamma_{hh}(s_m)|^2 + |\gamma_{vv}(s_m)|^2 + 2|\gamma_{hv}(s_m)|^2 \qquad (8.101)$$

so that the components of the polarimetric reflectivity vector at the elevation s_m can be expressed as:

$$\begin{aligned}
\gamma_1(s_m) &= \tau(s_m)k_1(s_m) \\
\gamma_2(s_m) &= \tau(s_m)k_2(s_m) \\
\gamma_3(s_m) &= \tau(s_m)k_3(s_m)
\end{aligned} \qquad (8.102)$$

where the vector $\boldsymbol{k}(s_m) = [k_1(s_m), k_2(s_m), k_3(s_m)]^T$ is the unitary polarimetric target vector at the elevation s_m represented either in Pauli or Lexicographic basis, with $\boldsymbol{k}^H(s_m)\boldsymbol{k}(s_m) = 1$.

The unknown real-valued and positive parameter $\tau^2(s_m)$ is usually denoted as the polarimetric reflectivity of the m-th source and specifies its power, while the absolute values of the elements of the unknown polarimetric target vector $\boldsymbol{k}(s_m)$ indicate the relative intensities of the scattering source at the elevation s_m in the different polarization channels.

When considering all the polarization channels, Eq. (8.97) can be written in terms of the polarimetric data $\boldsymbol{y}_P$ and reflectivity $\boldsymbol{\gamma}_P$ vectors, defined in (8.96) and (8.98), as:

$$
\boldsymbol{y}_P =
\begin{bmatrix}
\mathbf{A} & \mathbf{0}_{N\times M} & \mathbf{0}_{N\times M} \\
\mathbf{0}_{N\times M} & \mathbf{A} & \mathbf{0}_{N\times M} \\
\mathbf{0}_{N\times M} & \mathbf{0}_{N\times M} & \mathbf{A}
\end{bmatrix}
\boldsymbol{\gamma}_P + \boldsymbol{w}_P \quad \in \mathbb{C}^{3N\times 1} \qquad (8.103)
$$

where $\mathbf{0}_{N\times M}$ denotes an all zeros matrix of size $N \times M$ and $\boldsymbol{w}_P = [\boldsymbol{w}_1^T, \boldsymbol{w}_2^T, \boldsymbol{w}_3^T]^T$ is the polarimetric noise vector.

Eq. (8.103) can be recast in the form:

$$
\begin{aligned}
\boldsymbol{y}_P &= \sum_{m=1}^{M} \mathbf{A}_P(s_m)[\gamma_1(s_m) \quad \gamma_2(s_m) \quad \gamma_3(s_m)]^T + \boldsymbol{w}_P \\
&= \sum_{m=1}^{M} \mathbf{A}_P(s_m)\tau(s_m)\boldsymbol{k}(s_m) + \boldsymbol{w}_P
\end{aligned}
\qquad (8.104)
$$

with

$$
\mathbf{A}_P(s_m) =
\begin{bmatrix}
\boldsymbol{a}(s_m) & \mathbf{0}_{N\times 1} & \mathbf{0}_{N\times 1} \\
\mathbf{0}_{N\times 1} & \boldsymbol{a}(s_m) & \mathbf{0}_{N\times 1} \\
\mathbf{0}_{N\times 1} & \mathbf{0}_{N\times 1} & \boldsymbol{a}(s_m)
\end{bmatrix}
\quad \in \mathbb{C}^{3N\times 3}. \qquad (8.105)
$$

We note that (8.106) involves the square root of the polarimetric reflectivity $\tau(s_m)$ and the polarimetric target vector $\boldsymbol{k}(s_m)$, which are both unknown and have to be estimated from the polarimetric data.

Two different approaches can be followed for the formulation of the polarimetric tomographic imaging methods: the target vectors can be included in the definition of appropriate polarimetric steering vectors or can be associated to the polarimetric reflectivity pattern. In the following we first consider the former approach.

Setting $\boldsymbol{k}_m = \boldsymbol{k}(s_m)$, the polarimetric data model can be written as:

$$
\boldsymbol{y}_P = \sum_{m=1}^{M} \boldsymbol{b}(s_m, \boldsymbol{k}_m)\tau(s_m) + \boldsymbol{w}_P \quad \in \mathbb{C}^{3N\times 1} \qquad (8.106)
$$

with

$$b\left(s_m, k_m\right) = \mathbf{A}_P\left(s_m\right) k_m \qquad \in \mathbb{C}^{3N \times 1}. \qquad (8.107)$$

Then, the polarimetric steering vector of m-th source can be also written as:

$$b(s_m, k_m) = \begin{bmatrix} a(s_m)k_{1(s_m)} \\ a(s_m)k_{2(s_m)} \\ a(s_m)k_{3(s_m)} \end{bmatrix} = k_m \otimes a(s_m), \qquad (8.108)$$

where $\otimes$ is the Kronecker product. It is reminded that the unitary target vector k_m follows the same polarimetric basis used to express y_P.

The vector $\tau = [\tau(s_1), \quad \tau(s_2), \quad \cdots, \tau(s_M)] \in \mathbb{R}^{M \times 1}$ contains the realization of the real process describing the scattering amplitude. The complex additive noise $w_P \in \mathbb{C}^{3N \times 1}$ is assumed to follow a Gaussian distribution with zero mean and standard deviation σ_w. Letting $\mathbf{B}(s, \mathbf{K})$ be the matrix whose columns are the polarimetric steering vectors (8.107) corresponding to the M scatterers, i.e.

$$\mathbf{B}(s, \mathbf{K}) = \begin{bmatrix} a\left(s_1\right)k_1\left(s_1\right) & \dots a\left(s_m\right)k_1\left(s_m\right) & \dots a\left(s_M\right)k_1\left(s_M\right) \\ a\left(s_1\right)k_2\left(s_1\right) & \dots a\left(s_m\right)k_2\left(s_m\right) & \dots a\left(s_M\right)k_2\left(s_M\right) \\ a\left(s_1\right)k_3\left(s_1\right) & \dots a\left(s_m\right)k_3\left(s_m\right) & \dots a\left(s_M\right)k_3\left(s_M\right) \end{bmatrix}$$
$$\in \mathbb{C}^{3N \times M},$$

$$(8.109)$$

where $\mathbf{B}(s, \mathbf{K}) \in \mathbb{C}^{3N \times M}$ is the matrix whose columns are the polarimetric steering vectors (8.107), $s = [s_1, s_2, ..., s_M]^T$ and $\mathbf{K}$ is the matrix of size $3 \times M$, whose columns are the polarimetric target vectors at the elevations $s_1, s_2, ..., s_M$.

Eq. (8.106) can be written in the compact form:

$$y_P = \mathbf{B}(s, \mathbf{K})\tau + w_P \qquad (8.110)$$

Assuming τ be composed by incoherent components, i.e. $E\left[\tau(s_i)\tau(s_j)\right] = \overline{\tau^2}(s_i)\delta_{i,j}$ we have that the correlation matrix of y_P can be written simply as:

$$\mathbf{R}_P = E\left[y_P y_P^H\right] =$$
$$= \mathbf{B}(s, \mathbf{K})\mathrm{diag}\left(\left[\overline{\tau^2}(s_1), \overline{\tau^2}(s_2), \quad \cdots, \overline{\tau^2}(s_M)\right]\right)\mathbf{B}^H(s, \mathbf{K}) + \sigma_w^2\mathbf{I}$$

$$(8.111)$$

where $\mathbf{I}$ is the identity matrix of size $3N \times 3N$.

In case of concentrated scatterers, the data inversion can be performed starting from the vector y_P. For distributed scatterers, the target vectors have to be assumed to be random vectors, whose second order statistic can be represented by polarimetric correlation matrix (8.94), which can be estimated from the data.

8.4.3 Polarimetric unitary rank TomoSAR imaging

The extension to the polarimetric configuration not only allows to increase the number of available data, but also provides the extraction of the polarimetric target vectors enabling the investigation of the physical scattering mechanism.

Fully polarimetric unitary rank TomoSAR aims to estimate the reflectivity profile along the elevation direction and the polarimetric target vectors of the observed ground scene starting from the vector y_P. In this case, target vectors are assumed deterministic and the polarimetric correlation matrix is not considered.

Similarly to the case of single polarization, PolTomoSAR reconstruction can be performed in the context of spectral estimation. Classical and Capon beamforming can be easily extended to the polarimetric case, by substituting the polarimetric steering vectors (8.107) to the single polarization steering vectors, in the corresponding optimization problems.

Let us consider first the polarimetric classical beamforming optimization problem that can be written as [119]:

$$\boldsymbol{h}_{BF} = \min_{\boldsymbol{h}} \boldsymbol{h}^H \boldsymbol{h} \quad s.t. \quad \boldsymbol{h}^H \boldsymbol{b}(s_m, \boldsymbol{k}_m) = 1 \qquad (8.112)$$

Polarimetric BF optimization can be obtained by substituting the polarimetric steering vectors in (8.19), so that:

$$\hat{p}_{BF}^p(s_m) = \frac{\boldsymbol{b}^H(s_m, \boldsymbol{k}_m)\mathbf{R}_P \boldsymbol{b}(s_m, \boldsymbol{k}_m)}{N^2} \qquad (8.113)$$

that is depending on the polarization vector $\boldsymbol{k}_m$, which is unknown.

Since the objective of PolTomoSAR is the estimation of the largest of the local maxima of the reflectivity, we can maximize (8.113) respect to the polarization vector $\boldsymbol{k}_m$, obtaining:

$$\max_{||\boldsymbol{k}_m||^2=1} \frac{\boldsymbol{b}^H(s_m, \boldsymbol{k}_m)\mathbf{R}_P \boldsymbol{b}(s_m, \boldsymbol{k}_m)}{N^2}. \qquad (8.114)$$

Recalling that $\boldsymbol{b}(s_m, \boldsymbol{k}_m) = \mathbf{A}_P(s_m)\boldsymbol{k}_m$, with $\mathbf{A}_P$ given by (8.105), Eq. (8.114) becomes [119]:

$$\max_{||\mathbf{K}_m||^2=1} \frac{\boldsymbol{k}_m^H \mathbf{A}_P^H(s_m)\mathbf{R}_P \mathbf{A}_P(s_m)\boldsymbol{k}_m}{N^2} = \frac{\lambda_{Pmax}(s_m)}{N^2}, \qquad (8.115)$$

with $\lambda_{Pmax}(s_m)$ the maximum eigenvalue of the Hermitian matrix $\mathbf{A}_P^H(s_m)\mathbf{R}_P\mathbf{A}_P(s_m)$.

Then, the polarimetric beamforming estimator is given by:

$$P_{BF}^p(s_m) = \frac{\lambda_{Pmax}(s_m)}{N^2}. \qquad (8.116)$$

As far as Capon beamforming is concerned, the following optimization problem has to be solved:

$$h_{CP} = \min_{h} h^H \mathbf{R}_P h \quad s.t. \quad h^H b(s_m, k_m) = 1. \qquad (8.117)$$

Substituting the polarimetric steering vectors in the Capon reconstruction (8.34) and maximizing respect to the polarization vector k_m, polarimetric Capon reconstruction is obtained:

$$P_{CP}^p(s_m) = \frac{1}{\lambda_{Pmin}(s_m)} \qquad (8.118)$$

with $\lambda_{Pmin}(s_m)$ the minimum eigenvalue of the Hermitian matrix

$$\mathbf{A}_P^H(s_m)\mathbf{R}_P^{-1}\mathbf{A}_P(s_m).$$

It is worth noting that the physical behavior of the scattering source can be characterized using the unitary polarimetric scattering vectors $k_{BF}(s_m)$ and $k_{CP}(s_m)$, which for the cases of classical and Capon beamforming are given by:

$$k_{BF}(s_m) = u_{max}(s_m) \qquad (8.119)$$
$$k_{CP}(s_m) = u_{min}(s_m) \qquad (8.120)$$

where $u_{max}(s_m)$ and $u_{min}(s_m)$ denote the eigenvectors corresponding to the eigenvalues $\lambda_{max}(s_m)$ of the matrix

$$\mathbf{A}_P^H(s_m)\mathbf{R}_P\mathbf{A}_P(s_m)$$

and $\lambda_{min}(s_m)$ of the matrix

$$\mathbf{A}_P^H(s_m)\mathbf{R}_P^{-1}\mathbf{A}_P(s_m),$$

respectively.

Classical polarimetric spectral estimators are based on an unitary rank extension of the single polarization estimators to the polarimetric case, and provide the reflectivities and the target vectors $k_1, k_2, ..., k_M$, associated to M elevations, that optimize the criterion (8.112) or (8.117). In such an approach, polarimetric diversity is exploited to enhance elevation profile reconstruction and the scatterers discrimination.

8.4.4 Polarimetric full rank TomoSAR imaging

A full-rank polarimetric approach may be used for the reconstruction of the second order polarimetric information, i.e. polarimetric coherence $\mathbf{T}$ and correlation $\mathbf{C}$ matrices.

In this case, the polarimetric coherence or correlation matrix (depending on the generation form of $\mathbf{y}_p$) of the scatterer at elevation s_m, which can be written as:

$$\mathbf{X}_p(s_m) = E\left[\tau^2(s_m)\mathbf{k}(s_m)\mathbf{k}^H(s_m)\right] \in \mathbb{C}^{3\times 3} \tag{8.121}$$

has to be determined for fully characterize the scatterers.

From the model given in (8.106) we can write

$$\mathbf{y}_P = \sum_{m=1}^{M} \mathbf{A}_P(s_m)\,\mathbf{k}(s_m)\tau(s_m) + \mathbf{w}_P \in \mathbb{C}^{3N\times 1} \tag{8.122}$$

The polarimetric data correlation matrix can be written as:

$$
\begin{aligned}
\mathbf{R}_P &= E\left[\mathbf{y}_p\mathbf{y}_p^H\right] \\
&= \sum_{m=1}^{M}\sum_{n=1}^{M} \mathbf{A}_P(s_m)\mathbf{A}_P^H(s_n)\,E\left[\tau(s_m)\tau(s_n)\mathbf{k}(s_m)\mathbf{k}^H(s_n)\right] + \sigma_w^2 \mathbf{I} \\
&= \sum_{m=1}^{M} \mathbf{A}_P(s_m)\mathbf{A}_P^H(s_m)\,E\left[\tau^2(s_m)\mathbf{k}(s_m)\mathbf{k}^H(s_m)\right] + \sigma_w^2 \mathbf{I}
\end{aligned}
$$
$$\tag{8.123}$$

where the assumption of mutual incoherence between the targets at different elevations and between the targets and the noise has been exploited.

Using the definitions (8.105) and (8.121), we have:

$$\mathbf{R}_P = \sum_{m=1}^{M} \mathbf{X}_P(s_m) \otimes \boldsymbol{a}(s_m)\boldsymbol{a}^H(s_m) \tag{8.124}$$

The positive semidefinite matrices $\mathbf{X}_P$ can be reconstructed using the following beamforming and Capon estimators:

$$
\begin{aligned}
\mathbf{X}_{p_{BF}}(s_m) &= \mathbf{A}_P^H(s_m)\hat{\mathbf{R}}_P\mathbf{A}_P(s_m) \\
\mathbf{X}_{p_{CP}}(s_m) &= (\mathbf{A}_P^H(s_m)\hat{\mathbf{R}}_P^{-1}\mathbf{A}_P(s_m))^{-1},
\end{aligned}
\tag{8.125}
$$

where

$$\hat{\mathbf{R}}_P = \frac{1}{L}\sum_{l=1}^{L} \mathbf{y}_{pl}\mathbf{y}_{pl}^H, \tag{8.126}$$

is the polarimetric data sample correlation matrix, and L is the number of looks on which the averaging operation is performed.

8.4.5 Polarimetric backscattering separation

The polarimetric multibaseline TomoSAR technique presented in Section 8.4.3 allows the distinction and accurate positioning of the phase centers of the canopy scattering and ground scattering. However, due to the complexity of the forest structure and to the low resolution of the TomoSAR algorithms, there is no guarantee that reconstructed profile along the vertical direction exhibits two clear and separate peaks, so that often it is difficult to locate the phase centers of different scattering mechanisms. Some studies have shown that the reconstruction of the correlation matrices of different scattering mechanisms can lead to better results in determining the location of the corresponding scattering centers [120] [121].

The starting point consists in expressing the polarimetric TomoSAR signal as the superposition of the contributions of different scattering mechanisms, following a rationale in some sense similar to the PCA decomposition of the single polarization data discussed in Chapter 7. These algorithms can be applied in fully polarimetric configurations to provide an effective scattering mechanisms separation [118]. The effectiveness of the decomposition has been also proved in polarimetric interferometric SAR (PolInSAR), that is by using a single baseline with polarimetric channels [122].

Different decomposition approaches have been presented in literature [120,123,124]. Among them the most used one is the Sum of Kronecker Product (SKP) [120], hereafter briefly described. This method assumes that the backscattering from forested areas is the sum of the contributions of N_{SM} distinct Scattering Mechanisms (SMs), such as ground, canopy, ground-trunk scattering, or other. Each component of the polarimetric data vector (8.96) can, then, be expressed as:

$$y_i(s_n') = \sum_{q=1}^{N_{SM}} y_i^{(q)}(s_n') \tag{8.127}$$

with $i = 1, 2, 3$ denoting the polarization channel.

In (8.127), $y_i^{(q)}(s_n')$ represents the contribution of the q-th SM to the n-th track with baseline s_n', in the polarimetric channel i.

Three different hypotheses can be made:

A1) Different SMs are statistically uncorrelated:

$$E\left[y_i^{(q)}(s_m')y_j^{(h)*}(s_n')\right] = 0 \tag{8.128}$$

$\forall m, n, i, j$ if $q \neq h$.

A2) The interferometric coherence values of each SM, which can be written as:

$$\chi^{(q)}(s'_m, s'_n) = \frac{E\left(y_i^{(q)}(s'_m)y_j^{(q)*}(s'_n)\right)}{\sqrt{E\left(|y_i^{(q)}(s'_m)|^2\right) E\left(|y_j^{(q)*}(s'_n)|^2\right)}}, \qquad (8.129)$$

are invariant with respect to the choice of the polarimetric channels i and j. This is equivalent to assume that the electromagnetic properties of each SM (f.i. subsurface penetration, volume extinction, etc.) undergo negligible variations with respect to polarization.

A3) The polarimetric signature of each SM is invariant respect to the acquisition time and baseline:

$$E\left[y_i^{(q)}(s'_m)y_j^{(q)*}(s'_n)\right] = c_{ij}^{(q)}\chi^{(q)}(s'_m, s'_n), \qquad (8.130)$$

where $c_{ij}^{(q)}$ is the polarimetric correlation of the q-th SM in polarizations i and j,

$$c_{ij}^{(q)} = E\left[y_i^{(q)}(s'_n)y_j^{(q)*}(s'_n)\right] \qquad (8.131)$$

and $\chi^{(q)}(s'_m, s'_n)$ is the interferometric coherence of the m-th and n-th acquisitions for the q-th SM defined in (8.129).
This assumption implies that events altering the polarimetric signature (f.i. floods, fires, frosts) are assumed to not occur during the acquisition time interval.

In the above mentioned assumptions, we have:

$$
\begin{aligned}
E\left[y_i(s'_m)y_j^*(s'_n)\right] &= E\left[\sum_{q=1}^{N_{SM}} y_i^{(q)}(s'_m) \sum_{h=1}^{N_{SM}} y_j^{(h)*}(s'_n)\right] = \\
&= \sum_{q=1}^{N_{SM}}\sum_{h=1}^{N_{SM}} E\left[y_i^{(q)}(s'_m)y_j^{(h)*}(s'_n)\right] \\
&= \sum_{q=1}^{N_{SM}} E\left[y_i^{(q)}(s'_m)y_j^{(q)*}(s'_n)\right] \\
&= \sum_{q=1}^{N_{SM}} c_{ij}^{(q)} R^{(q)}(s'_m, s'_n).
\end{aligned}
\qquad (8.132)
$$

Then, the correlation matrix of the polarimetric data defined in (8.96) can be written as:

$$\mathbf{R}_P = E\left[\mathbf{y}_P \mathbf{y}_P^H\right] = \sum_{q=1}^{N_{SM}} \mathbf{C}^{(q)} \otimes \mathbf{R}^{(q)} \tag{8.133}$$

where the matrices $\mathbf{C}^{(q)} \in \mathbb{C}^{3\times3}$ and $\mathbf{R}^{(q)} \in \mathbb{C}^{N\times N}$ account for the correlation among different polarizations and baselines, and are denoted as the polarimetric signature and structure matrices of the k-th SM, respectively. The generic elements of these matrices are given by:

$$\begin{aligned}
\left\{\mathbf{C}^{(q)}\right\}_{ij} &= c_{ij}^{(q)} \\
\left\{\mathbf{R}^{(q)}\right\}_{mn} &= \chi^{(q)}(s_m', s_n').
\end{aligned} \tag{8.134}$$

The key of the SKP decomposition is based on the result that the matrix $\mathbf{R}_P$ can be also expressed by exploiting the SVD decomposition [125]. More specifically the $3N \times 3N$ matrix $\mathbf{R}_p$ in (8.133) is reshaped in a $9 \times N^2$ matrix by considering the column stackings (vectorizations) of the matrices $\mathbf{C}^{(q)}$ and $\mathbf{R}^{(q)}$ defining the following 9 and N^2 length vectors:

$$\begin{aligned}
\mathbf{c}^{(q)} &= \left[c_{11}^{(q)}, c_{21}^{(q)}, c_{31}^{(q)}, \ldots, c_{33}^{(q)}\right]^T \\
\mathbf{r}^{(q)} &= \left[r_{11}^{(q)}, \ldots, r_{N1}^{(q)}, r_{12}^{(q)}, \ldots, r_{N2}^{(q)}, \ldots, r_{NN}^{(q)}\right]^T
\end{aligned} \tag{8.135}$$

thus leading to:

$$\mathbf{W}_P = \sum_{q=1}^{N} \sigma_q \mathbf{c}^{(q)} [\mathbf{r}^{(q)}]^T \tag{8.136}$$

It is important to note that, thanks also to the linearity of the re-shaping operator, there is a element to element correspondence between the elements of $\mathbf{R}_p$ and $\mathbf{W}_p$ and the reshaping operation is invertible: in other words we can go back from $\mathbf{W}_p$ to $\mathbf{R}_p$. We can consider at this point the SVD decomposition of (8.136) thus leading to:

$$\mathbf{W}_P = \sum_{q=1}^{\min\{9,N^2\}} \sigma_q \mathbf{u}_q \mathbf{v}_q^H \tag{8.137}$$

$$\mathbf{R}_P = \sum_{q=1}^{N} \lambda_q \tilde{\mathbf{C}}^{(q)} \otimes \tilde{\mathbf{R}}^{(q)} \tag{8.138}$$

where $\lambda_q \geqslant 0$ and the sets of matrices $\tilde{\mathbf{C}}^{(q)}$ and $\tilde{\mathbf{R}}^{(q)}$ are obtained by reshaping the singular vectors $\boldsymbol{u}^{(q)}$ and $\boldsymbol{v}^{(q)}$ of the matrix $\mathbf{R}_P$ into 3×3 and $N \times N$ matrices, respectively.

If only N_{SM} dominant scattering mechanisms are present, the number of λ_k that are not negligible is equal to N_{SM}, regardless of the spatial structure of each SM (i.e. point-like, superficial, volumetric) and, by virtue of the Eckart-Young-Mirsky theorem, $\mathbf{R}_P$ can be written as:

$$\mathbf{R}_P \approx \sum_{q=1}^{N_{SM}} \lambda_q \tilde{\mathbf{C}}^{(q)} \otimes \tilde{\mathbf{R}}^{(q)} \qquad (8.139)$$

which is of the same form of (8.133), so that the matrices $\tilde{\mathbf{C}}^{(q)}$ and $\tilde{\mathbf{R}}^{(q)}$ can be easily related to the matrices $\mathbf{C}^{(q)}$ and $\mathbf{R}^{(q)}$. In particular it can be shown that [120]:

$$\mathbf{C}^{(q)} = \sum_{k=1}^{N_{SM}} \left\{ (\boldsymbol{\alpha}_0^{-1})^T \right\}_{kq} \tilde{\mathbf{C}}^{(k)} \qquad \mathbf{R}^{(q)} = \sum_{k=1}^{N_{SM}} \{\boldsymbol{\alpha}_0\}_{kq} \tilde{\mathbf{R}}^{(k)} \qquad (8.140)$$

where $\boldsymbol{\alpha}_0$ is a nonsingular, real valued, $N_{SM} \times N_{SM}$ matrix, that has to be determined.

Assuming $N_{SM} = 2$, i.e. considering the two SMs related to the ground reflection and the volume scattering from canopy, the polarimetric MB correlation matrix $\mathbf{R}_p$ can be expressed by the sum of two Kronecker products. It can be shown that the matrices $\tilde{\mathbf{C}}^{(q)}$ and $\tilde{\mathbf{R}}^{(q)}$ are related to the polarimetric correlation matrices $(\mathbf{C}^{(1)}, \mathbf{C}^{(2)})$ and interferometric structure matrices $(\mathbf{R}^{(1)}, \mathbf{R}^{(2)})$ of the ground and above ground volumetric scattering from canopy via a linear invertible transformation as [120]:

$$\mathbf{R}^{(1)}(\alpha, \beta) = \alpha \tilde{\mathbf{R}}^{(1)} + (1 - \alpha) \tilde{\mathbf{R}}^{(2)}$$
$$\mathbf{R}^{(2)}(\alpha, \beta) = \beta \tilde{\mathbf{R}}^{(1)} + (1 - \beta) \tilde{\mathbf{R}}^{(2)}$$
$$\mathbf{C}^{(1)}(\alpha, \beta) = \frac{1}{\alpha - \beta} ((1 - \beta)\tilde{\mathbf{C}}^{(1)} - \beta \tilde{\mathbf{C}}^{(2)}) \qquad (8.141)$$
$$\mathbf{C}^{(2)}(\alpha, \beta) = \frac{1}{\alpha - \beta} (-(1 - \alpha)\tilde{\mathbf{R}}^{(1)} + \alpha \tilde{\mathbf{R}}^{(2)})$$

where α and β are two real parameters that are selected by requiring the semipositive definitiveness condition for the matrices $\mathbf{C}^{(1)}(\alpha, \beta), \mathbf{C}^{(2)}(\alpha, \beta)$ and $\mathbf{R}^{(1)}(\alpha, \beta), \mathbf{R}^{(2)}(\alpha, \beta)$ determined by means of (8.141). This solution can be found by maximizing the diversity of the two coherence matrices $\mathbf{R}^{(1)}(\alpha, \beta)$ and $\mathbf{R}^{(2)}(\alpha, \beta)$ [120]. The solution for the parameters α and β can be selected by taking the center of the region where $\mathbf{C}^{(1)}(\alpha, \beta)$, $\mathbf{C}^{(2)}(\alpha, \beta)$, $\mathbf{R}^{(1)}(\alpha, \beta)$, $\mathbf{R}^{(2)}(\alpha, \beta)$ are semidefinite positive.

The polarimetric matrices of $\mathbf{C}^{(1)}$ and $\mathbf{C}^{(2)}$ represent information of backscattering mechanisms of the ground and volume, respectively, while matrices $\mathbf{R}^{(1)}$ and $\mathbf{R}^{(2)}$ contain interferometric information that can be employed for height mapping and reconstruction of vertical structure.

Appendices

A.1 Fourier Transform and properties

The Fourier Transform (FT) is a mathematical transformation that breaks a function into an alternative representation characterized by the sine and cosine functions at varying frequencies. The Fourier Transform shows that any function can be reformulated as the sum of sine waves with variable frequencies, thus allowing representing a signal from its original, space or time domain, into the dual frequency domain. The Fourier Transform originates from the theory of Fourier series which states that any real deterministic periodic function $x(t)$ with period T can be expressed as the discrete superposition of a set of complex sine wave functions:

$$x(t) = \sum_{k=-\infty}^{+\infty} X_k \cdot e^{j2\pi k \frac{t}{T}} \tag{A.1}$$

with

$$X_k = \frac{1}{T} \int_{-\frac{T}{2}}^{+\frac{T}{2}} x(t) \cdot e^{-j2\pi k \frac{t}{T}} \, dt \tag{A.2}$$

The Fourier series therefore relates the periodic signal $x(t)$ to a frequency representation with a sequence of equally spaced harmonics at frequencies $f_k = k/T$, whose amplitude is related to X_k. Specifically, for $k = 0$, the coefficient X_0 represents the average value of $x(t)$. Eq. (A.2) is an *analysis* equation that allows to determine the content in terms of harmonics, i.e. to analyze the signal. Conversely, Eq. (A.1) is a *synthesis* equation that, having known the amplitudes and phases of the various harmonics (i.e. known the coefficients X_k) allows to reconstruct, i.e. synthesize, the signal starting from its own frequency components (harmonics). Eqs. (A.1) and (A.2) allow to establish a correspondence between the signal $x(t)$ and the sequence X_k constituted by the Fourier coefficients of the series. This suggests that knowledge of the signal $x(t)$ in its own domain is equivalent to the knowledge of the sequence of the Fourier coefficients X_k in the frequency domain.

The theory of Fourier series for periodic signals can be extended also to nonperiodic functions. To extend this representation, it is convenient to find a simple relationship between periodic and nonperiodic signals. Let us assume $x_p(t)$ being the periodic signal generated from the repetition, with period T of a generic nonperiodic signal $x(t)$:

$$x_p(t) = \sum_{k=-\infty}^{+\infty} x(t - kT) \tag{A.3}$$

The original signal $x(t)$ can be considered as a limit case of a periodic signal: starting from $x_p(t)$, the original signal $x(t)$ can be obtained by assuming $T \to \infty$, such that:

$$x(t) = \lim_{T \to \infty} x_p(t) \tag{A.4}$$

According to (A.1), the periodic signal $x_p(t)$ can be expressed in terms of Fourier series with coefficients X_k. As T increases, the sequence of harmonics becomes denser, as the frequencies f_k become closer, and the amplitudes decrease because of the factor $1/T$. By multiplying X_k by a factor T, at the limit of $T \to \infty$ the sequence of harmonics ends up with a continuous, not tending towards zero, spectrum. Thus the nonperiodic signal $x(t)$ is described in the frequency domain by the superposition of an infinity of sine waves. At the limit, the Fourier transform $X(f)$ of $x(t)$ is given by:

$$X(f) = \int_{-\infty}^{+\infty} x(t) \cdot e^{-j2\pi ft} \, dt \tag{A.5}$$

and equivalently:

$$x(t) = \int_{-\infty}^{+\infty} X(f) \cdot e^{+j2\pi ft} \, df \tag{A.6}$$

While in case of periodic signal, this latter is represented by sinusoidal components at frequencies multiples of a single fundamental (i.e. $f_k = k/T$), in the general case of the nonperiodic signal, Eq. (A.6) still allows representing the nonperiodic signal $x(t)$ as the superposition of sinusoidal components distributed over all the continuous frequencies f.

- The Fourier transform $X(f)$ of a real signal is, in general, a complex signal. Its modulus and its phase are referred to as the modulus spectrum and phase spectrum of the signal, which are, respectively, even and odd functions for real signals: This symmetry can be summarized as:

$$X(f) = X^*(-f) \tag{A.7}$$

- The Fourier Transform is a linear operator, then for any complex constant a and b it results:

$$x(t) = ax_1(t) + bx_2(t) \rightarrow X(f) = aX_1(f) + bX_2(f) \tag{A.8}$$

- Based on the similarity between the relations (A.5) and (A.6) a duality property can be also derived:

$$\begin{aligned} if \quad & x(t) \rightarrow X(f) \\ then \quad & X(t) \rightarrow x(-f) \end{aligned} \tag{A.9}$$

- The scale expansion in one domain of a factor a reflects into a compression in the dual domain of the same amount:

$$x(at) \rightarrow \frac{1}{|a|} X\left(\frac{f}{a}\right) \tag{A.10}$$

- A translation in one domain corresponds to a multiplication by a phase ramp in the dual domain and reciprocally:

$$\begin{aligned} x(t - t_0) &\rightarrow X(f) \cdot e^{-j2\pi f t_0} \\ X(t) \cdot e^{j2\pi f_d t} &\rightarrow X(f - f_d) \end{aligned} \tag{A.11}$$

- The modulation property extends the temporal/frequency shifting: if a signal $x(t)$ is multiplied by $\cos(2\pi f_0 t)$, then its frequency spectrum gets translated up and down in frequency by f_0:

$$x(t) \cdot \cos(2\pi f_0 t) \rightarrow \frac{X(f - f_0) + X(f + f_0)}{2} \tag{A.12}$$

- The multiplication in one domain reflects into a convolution in the dual domain, and reciprocally:

$$\begin{aligned} x(t) \cdot y(t) &\rightarrow X(f) \otimes Y(f) \\ x(t) \otimes y(t) &\rightarrow X(f) \cdot Y(f) \end{aligned} \tag{A.13}$$

- The differentiation of a function in time domain is equivalent to the multiplication of its Fourier transform by a factor $j2\pi f$ in the frequency domain, and equivalently the integration in time domain corresponds to the division of its Fourier transform by a factor $j2\pi f$:

$$\frac{dx(t)}{dt} \rightarrow j2\pi f \cdot X(f)$$

$$\int_{-\infty}^{t} x(\alpha)\, d\alpha \rightarrow \frac{X(f)}{j2\pi f} \tag{A.14}$$

A.2 Baseband signal

The Low Pass Equivalent (LPE) representation of the signal can be considered as an extension of the phasor representation widely used in physics and engineering to simplify the operation of the analysis of systems in the pure sinusoidal regime. Given a sinusoidal tone:

$$x(t) = A\cos(\omega_0 t + \varphi) = A\cos(2\pi f_0 t + \varphi) \tag{A.15}$$

where f_0 is the frequency ($\omega_0 = 2\pi f_0$ the angular frequency) and φ the phase (at $t = 0$), the phasor representation is a bijective (given ω_0) association that relates the signal in (A.15) to the complex number

$$\tilde{x}(t) = x = Ae^{j\varphi} \tag{A.16}$$

The phasor association entails the following sequence of two transformations:

$$x(t) \rightarrow \mathring{x}(t) = A\cos(2\pi f_0 t + \varphi) + jA\sin(2\pi f_0 t + \varphi) = x(t) + j\hat{x}(t) \tag{A.17}$$

and

$$\mathring{x}(t) = Ae^{j2\pi f_0 t + j\varphi} \rightarrow \tilde{x} = \mathring{x}(t)e^{-j2\pi f_0 t} = Ae^{j\varphi} \tag{A.18}$$

The result is that the oscillation of $x(t)$, or equivalently the rotation in the complex plane of $\mathring{x}(t)$, at frequency f_0, i.e. with period $1/f_0$, is "frozen" by the substitution performed in (A.18), thus providing a signal $\mathring{x}$, that is called phasor, which is a vector fixed in the complex plane with amplitude A and phase (anomaly) φ. In the context of periodic signals the phase quantifies the amount of time elapsed from the starting of the period. This concept can be extended to signals, which are not rigorously periodic, but characterized by variations of the envelope and of the period that are "slow" with respect to the carrier, that is with respect to $1/f_0$. The phase is rigorously defined in a pure periodic signal- with reference to these signals, the LPE representation is an extension of the concept of phasor. More specifically, let us consider the signal:

$$x(t) = a(t)\cos(2\pi f_0 t + \varphi(t)) \qquad (A.19)$$

with envelope $a(t)$ and phase deviation $\varphi(t)$. The LPE representation is a sequence of two transformations in which the first transformation associates to $x(t)$ the analytic signal:

$$x(t) \to \mathring{x}(t) = a(t)\cos(2\pi f_0 t + \varphi(t)) + ja(t)\sin(2\pi f_0 t + \varphi(t)) =$$
$$= \mathring{x}(t) = a(t)e^{j2\pi f_0 t + j\varphi(t)} = x(t) + j\hat{x}(t)$$
$$(A.20)$$

and the second is a time-variant transformation that associate to $\mathring{x}(t)$ the LPE signal

$$\mathring{x}(t) \to \tilde{x}(t) = \mathring{x}(t)e^{-j2\pi f_0 t} = a(t)e^{j\varphi(t)} \qquad (A.21)$$
$$= x_c(t) + jx_s(t)$$

where $x_c(t)$ and $x_s(t)$ are referred to as in-phase and in-quadrature components. Comparing (A.17) and (A.18) with (A.20) and (A.21) one can easily recognize that, differently from the phasor, after the freezing of the high frequency oscillations (in radar systems f_0 is typically in the VHF/UHF or microwave band), the LPE signal $\tilde{x}(t)$ is still time varying. Such variations occur, however, generally on scales much larger than $1/f_0$, i.e., at base band. For this reason the LPE representation is also referred to as baseband equivalent representation of the signal $x(t)$.

The signal $\hat{x}(t)$ is called Hilbert transform of $x(t)$ and can be shown to be the transformation of $x(t)$ achieved by filtering $x(t)$ with a LTI filter having a pulse response function equal to $1/(\pi t)$ [26]. The transformation from $x(t)$ to $\mathring{x}(t)$ can be derived with reference to the sinusoidal case, i.e. (A.17), which is nothing else than a multiplication, in the spectral domain, of the Fourier Transform (FT) of $x(t)$ by the function $2U(f)$ (frequency response), where $U(\cdot)$ is the Heaviside (i.e. the unit step) function. The following FT relation:

$$\delta(t) + \frac{j}{\pi t} \longleftrightarrow 2U(f)$$

shows that the Hilbert transform of $x(t)$ is achieved by filtering $x(t)$ with a LTI filter having a pulse response function equal to $1/(\pi t)$ [26]. Furthermore, the relation between the power spectral densities of $x(t)$ and $\mathring{x}(t)$ is provided by the square modulus of the frequency response of the filter $2U(f)$, i.e. $4U(f)$. Considering the fact that $U(f)$ cancels the negative frequency, the net power gain is a factor of 2. This shows that the analytic signal has twice the power of the starting signal, in accordance with the fact that a pure imaginary component has been added to $x(t)$ to provide $\mathring{x}(t)$. It can be shown that $x(t)$ and its Hilbert transform $\hat{x}(t)$ are orthogonal and equi-power [26].

The modulus of the analytic signal in (A.20), which is equal to the modulus of the LPE in (A.21):

$$|\mathring{x}(t)| = |\tilde{x}(t)| = \sqrt{x^2(t) + \hat{x}^2(t)} = \sqrt{x_c^2(t) + x_s^2(t)} = a(t) \qquad \text{(A.22)}$$

is the envelope of the original signal $x(t)$: for this reason the LPE signal is also referred to as complex envelope. The phase of the analytic signal

$$\psi(t) = \angle \mathring{x}(t) = 2\pi f_0 t + \varphi(t) \qquad \text{(A.23)}$$

which is the argument of the original is referred to as instantaneous phase; $\varphi(t)$ is the instantaneous phase deviation. Its derivative, scaled by 2π, is commonly referred to as instantaneous frequency:

$$f_I(t) = \frac{1}{2\pi} \frac{d}{dt} \psi(t) = f_0 + \frac{1}{2\pi} \frac{d}{dt} \varphi(t) \qquad \text{(A.24)}$$

Moreover, the scaled derivative of the phase of the LPE, i.e. of the instantaneous phase deviation:

$$\Delta f_I(t) = \frac{1}{2\pi} \frac{d}{dt} \angle \tilde{x}(t) = \frac{1}{2\pi} \frac{d}{dt} \varphi(t) \qquad \text{(A.25)}$$

is generally called instantaneous frequency deviation.

Some remarks are now in order. First of all, we have so far analyzed the sequence of associating a LPE signal to a band-pass signal. The reverse operations are simply carried out as follows:

$$\tilde{x}(t) \rightarrow \mathring{x}(t) = \tilde{x}(t) e^{j2\pi f_0 t} \qquad \text{(A.26)}$$

and

$$\mathring{x}(t) \rightarrow x(t) = \Re\left\{\mathring{x}(t)\right\} = \Re\left\{\tilde{x}(t) e^{j2\pi f_0 t}\right\}$$
$$\hat{x}(t) = \Im\left\{\mathring{x}(t)\right\} = \Im\left\{\tilde{x}(t) e^{j2\pi f_0 t}\right\} \qquad \text{(A.27)}$$

From (A.27) it can be extrapolated the expression relating the original band-pass signal (and its Hilbert transform) to the in-phase and in-quadrature components:

$$x(t) = x_c(t) \cos(2\pi f_0 t) - x_s(t) \sin(2\pi f_0 t)$$
$$\hat{x}(t) = x_c(t) \sin(2\pi f_0 t) + x_s(t) \cos(2\pi f_0 t) \qquad \text{(A.28)}$$

above is referred to as Rice representation. A second remark concerns the effect of a delay on a band-pass and the equivalence with the delay of the associated LPE. This is the case of a delay corresponding to a radar echo. Let us consider the signal in and let us suppose that we observe a delayed version of a quantity $\tau = 2r/c$:

$$s(t) = x(t - \tau) = a(t - \tau)\cos(2\pi f_0 t + \varphi(t - \tau) - 2\pi f_0 \tau) \quad \text{(A.29)}$$

The LPE associated to $s(t)$ is therefore:

$$\tilde{s}(t) = a(t - \tau)e^{j\varphi(t-\tau) - j2\pi f_0 \tau} = \tilde{x}(t - \tau)e^{-j2\pi f_0 \tau} \neq \tilde{x}(t - \tau) \quad \text{(A.30)}$$

Above equation shows that, given a band-pass signal $x(t)$, the LPE associated to the shifted version $x(t - \tau)$ is different by the shifted version of the LPE associated to the original (nonshifted) band-pass signal. In other words when shifting LPE by a quantity $\tau = 2r/c$ a phase factor corresponding to the shift of the carrier, which is suppressed in the LPE representation, given by:

$$e^{-j2\pi f_0 \tau} = e^{-j\frac{4\pi}{\lambda}r}$$

must be added, i.e. multiplied as complex number. As seen in the rest of the book, this factor plays a fundamental role in the implementing imaging radars and interferometric applications.

A final remark concerns the notation. In the rest of the book we will generally refer to LPE representation. Therefore in order to avoid a weighting of the notation, the $\tilde{}$ will be suppressed. In doing this a pedix BP will be added to the original band-pass signal to distinguish it from the LPE.

A.3 Factorization property

We provide the evaluation of the integral

$$Q(t, f) = \int_{-\infty}^{+\infty} \Pi\left(\frac{\xi}{T}\right) \Pi\left(\frac{\xi - t}{T}\right) e^{j2\pi f \xi} d\xi \quad \text{(A.31)}$$

which represents the basis for several functions involved in radar processing, as f.i. the ACF of the chirp and the time-frequency ACF (ambiguity function) of the transmitted pulse.

To this aim, we exploit the factorization property of the function:

$$q(t, \xi) = \Pi\left(\frac{\xi}{T}\right) \Pi\left(\frac{\xi - t}{T}\right) \quad \text{(A.32)}$$

which can be easily understood by referring to Fig. A.1: the overlap of the two rectangular windows is equal to zero for values of the variable t outside the interval $[-T, T]$, and is the rectangular pulse centered in $t/2$ with extension $T - |t|$ inside the same interval. Accordingly, the function $q(t, \xi)$ can be rewritten as:

$$q(t, \xi) = \Pi\left(\frac{t}{2T}\right) \Pi\left(\frac{\xi - t/2}{T - |t|}\right) \qquad (A.33)$$

where the first rectangular window is independent of the integration variable ξ.

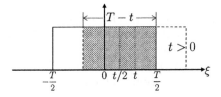

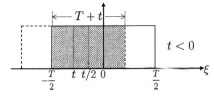

Figure A.1. Rectangular pulses overlapping for (top) $t > 0$ and (bottom) $t < 0$.

The function $Q(t, f)$ in (A.31) is thus given by:

$$Q(t, f) = \Pi\left(\frac{t}{2T}\right) \int_{-\infty}^{+\infty} \Pi\left(\frac{\xi - t/2}{T - |t|}\right) e^{j2\pi f \xi} d\xi \qquad (A.34)$$

where the integral is the Fourier transform (from the ξ domain to the f domain) of the ξ dependent rectangular window provided by the factorization property:

$$Q(t, f) = \Pi\left(\frac{t}{2T}\right) \mathcal{F}_\xi\left[\Pi\left(\frac{\xi - t/2}{T - |t|}\right)\right] \qquad (A.35)$$

A.4 Stationary phase approximation

Stationary Phase Approximation (SPA) asserts that for a real and continuous function $m(t)$, and a real function $\varphi(t)$ such that

$$\varphi'(t_0) = 0 \qquad \varphi''(t_0) \neq 0$$

at only a single point $t = t_0$, the approximation

$$\int_{-\infty}^{\infty} m(t)e^{jk\varphi(t)}dt \simeq \sqrt{\frac{2\pi j}{k\varphi''(t_0)}} m(t_0)e^{jk\varphi(t_0)} \qquad (A.36)$$

holds for sufficiently large k. Such a result is based on the extension of the Riemann-Lebesgue lemma, asserting that the integration over any interval that does not include the point t_0 tends to zero as $k \to \infty$, along with the representation of the delta function as the following limit:

$$\delta(t) = \lim_{\beta \to \infty} \sqrt{\frac{\beta}{j\pi}} e^{j(\beta t)^2} \qquad (A.37)$$

Indeed, according to the quoted lemma, the major contribution to the integral in (A.36) comes from a small interval around the stationary point t_0. In this interval

$$\varphi(t) \simeq \varphi(t_0) + \frac{\varphi''(t_0)}{2}(t - t_0)^2$$

Hence

$$\int_{-\infty}^{\infty} m(t)e^{jk\varphi(t)}dt \simeq e^{jk\varphi(t_0)} \int_{-\infty}^{\infty} m(t)e^{jk\frac{\varphi''(t_0)}{2}(t-t_0)^2}dt \qquad (A.38)$$

and (A.36) follows by exploiting the limit in (A.37) with $\beta = k\varphi''(t_0)/2$.

Some considerations on the practical applicability of the SPA are now in order. The function to be integrated is phase and amplitude modulated by the signals $\varphi(t)$ and $m(t)$, respectively. Its instantaneous frequency is:

$$f(t) = \frac{k}{2\pi}\varphi'(t) \qquad (A.39)$$

which increases with k and is zero in the phase stationary point t_0. When the oscillation rate is high enough with respect to the amplitude variations, the integral of the positive half waves compensates quite well the integral of the negative half waves, except for a very small interval around t_0. In the limiting case of $k \to \infty$ there is perfect compensation of the two contributions, except at stationary phase point, regardless of how fast the amplitude variations are. However, when k is a finite number, the higher the oscillation rate with respect the amplitude variations, the better the approximation given by the SPA. Oscillation rate and amplitude variations are measured by the bandwidths of the signals $\exp[jk\varphi(t)] \Pi(t/T)$

and $m(t)$, respectively, T being the duration of $m(t)$. Therefore, in order to profitable apply the SPA, the ratio between these two bandwidths has to be high enough: this is the practical condition to be verified.

For a chirp, the bandwidths of the phase modulated and amplitude pulses involved in the Fourier transform are αT and $1/T$, respectively. Accordingly, the SPA provides a good approximation of the chirp Fourier transform if the condition $\alpha T^2 \gg 1$ is satisfied. Such a condition, for a long pulse, can be achieved by properly tuning the chirp rate α, which acts as the k parameter: it increases the chirp bandwidth for a fixed amplitude bandwidth.

The SPA is a very useful tool for the evaluation of the Fourier Transform of phase modulated pulses as:

$$g(t) = m(t)e^{j2\pi \int_0^t F(\xi)d\xi} \tag{A.40}$$

where $F(t)$ is the instantaneous frequency. Indeed, its FT can be written as

$$G(f) = \int_{-\infty}^{\infty} m(t)e^{j\varphi(t,f)}dt \tag{A.41}$$

where

$$\varphi(t, f) = 2\pi \int_0^t F(\xi)d\xi - 2\pi f t \tag{A.42}$$

is, for each spectral frequency f, the phase of the complex function of t to be integrated by using the SPA. The FT approximation is thus:

$$G(f) \approx m(t_0)e^{j\varphi(t_0,f)} = g(t_0)e^{-j2\pi f t_0} \tag{A.43}$$

where, for each f, the phase stationary point t_0 is solution of the equation

$$f = F(t_0) \tag{A.44}$$

Eq. (A.43) highlights that, differently from the usual interpretation of the FT, in this case each harmonic does not depend on the whole signal $g(t)$ to be transformed, but only on its value on t_0. The latter generates the harmonic at the spectral frequency f given, according to (A.44), by the instantaneous frequency $F(t_0)$. In other words, each time instant t where the instantaneous frequency $F(t)$ of $g(t)$ is defined is a stationary point for the phase $\varphi(t, f)$ in (A.42) and generates the harmonic $G(f)$ at spectral frequency $f = F(t)$. Finally, from Eq. (A.42), it is worth noting that the maximum frequency deviation of the pulse to be integrated for achieving the FT in (A.41) is equal to the bandwidth (according to the Carson rule) of the pulse to be transformed. As a consequence, in this case, the greater the bandwidth of the phase modulated pulse in (A.40) with respect to the bandwidth of its amplitude signal, the better the FT approximation in (A.43) provided by the SPA.

A.5 Eigenvalue invariance under the Hadamard multiplication by a phase only dyadic product

Let $\mathbf{A}$ be an $N \times N$ hermitian matrix and $\mathbf{U}$, $\mathbf{\Lambda}$ be its $N \times N$ eigen-decomposition pair:

$$\mathbf{A} = \mathbf{U}\mathbf{\Lambda}\mathbf{U}^H = \sum_{i=1}^{N} \lambda_i \boldsymbol{u}_i \boldsymbol{u}_i^H \qquad (A.45)$$

We let $\boldsymbol{b}$ be a phase only N-element vector and consider the matrix $\mathbf{C}$ achieved by a Hadamard, i.e. element-by-element, product of $\mathbf{A}$ by the dyadic product $\boldsymbol{b}\boldsymbol{b}^H$, that is:

$$\mathbf{B} = \boldsymbol{b}\boldsymbol{b}^H \odot \mathbf{A} \qquad (A.46)$$

Combining the two equations we have:

$$\mathbf{B} = \boldsymbol{b}\boldsymbol{b}^H \odot \mathbf{U}\mathbf{\Lambda}\mathbf{U}^H = \sum_{i=1}^{N} \lambda_i \boldsymbol{b}\boldsymbol{b}^H \odot \boldsymbol{u}_i \boldsymbol{u}_i^H \qquad (A.47)$$

which can be further expressed as:

$$\mathbf{B} = \sum_{i}^{N} \lambda_i (\boldsymbol{b} \odot \boldsymbol{u}_i)(\boldsymbol{b}^H \odot \boldsymbol{u}_i^H) \qquad (A.48)$$

It is sufficient to demonstrate that $\boldsymbol{b} \odot \boldsymbol{u}_i$ form an orthonormal basis, that is:

$$(\boldsymbol{b}^H \odot \boldsymbol{u}_i^H)(\boldsymbol{b} \odot \boldsymbol{u}_j) = \boldsymbol{u}_i^H \boldsymbol{u}_j \qquad (A.49)$$

If the elements of $\boldsymbol{b}$ are scaled in amplitude by a given factor, the above reasoning can be still applied provided that the eigenvalues are scaled by the square of the amplitude scale factor.

A.6 Basic electromagnetic principles

Waves are the solution of the Maxwell equations in the absence of sources. Under this assumption, for a homogeneous media, it can be shown that the electric and magnetic fields are solutions of the following equation;

$$\nabla^2 \mathbf{F}(P, t) - \varepsilon\mu \frac{\partial^2}{\partial t^2}\mathbf{F}(P, t) = 0 \tag{A.50}$$

where F is field (electric or magnetic) and $P \equiv (x, y, z)$ is the generic point of the space, ε and μ are the (electric) permittivity and the (magnetic) permeability.

Essentially, referring to a scalar context, each field is a solution of the Helmotz wave equation:

$$\frac{\partial^2}{\partial z^2}f(z, t) - \frac{1}{v^2}\frac{\partial^2}{\partial t^2}f(z, t) = 0 \tag{A.51}$$

where v is the velocity and z is the propagation direction, that is:

$$f(z, t) = c_1 f_+(z - vt) + c_2 f_-(z + vt) \tag{A.52}$$

where f_+ is the forward solution, f_- is the backward solution and c_1 and c_2 are two real constants. The assumption on evanescent solution at $z \to \infty$ implies $f_- = 0$

In the sinusoidal regime, the monocromatic solution is provided by:

$$f(z, t) = A\cos(\omega_0 t - \beta_0 z) \tag{A.53}$$

that is a sinusoidal signal with temporal period $T = 2\pi/\omega_0$ and a phase related to the z with a spatial period (i.e. a wavelength) equal to $\lambda_0 = 2\pi/\beta_0$.

Eq. (A.53) represents a sinusoidal wave propagating with a velocity:

$$v = \frac{\omega}{\beta_0} = \frac{2\pi f_0}{\beta_0} = \lambda_0 f_0 \tag{A.54}$$

Having introduced the expression of a monocromatic wave propagating along z, we can write the general solution of a wave propagating along z as synthesized by its spectrum $A(\omega)$ as follows:

$$f(z, t) = \int_{-\infty}^{\infty} A(\omega)e^{-j(\omega t - \beta(\omega)z)}d\omega \tag{A.55}$$

where the expression highlights that in general:

$$\beta = \beta(\omega) \tag{A.56}$$

because each sinusoidal wave can propagate at a different velocity $v = \omega/\beta(\omega)$. It is interesting to note that the case corresponding to

β scaling with the frequency (i.e. proportional to ω):

$$\beta = \frac{\omega}{v} \tag{A.57}$$

corresponds to a propagation in a distortionless media. In such a case the synthesis equation simplifies in (A.55):

$$f(z, t) = \int_{-\infty}^{\infty} A(\omega) e^{-j\omega(t-z/v)} d\omega = f(t - v/z) \tag{A.58}$$

that is a propagation without distortion along z of the signal.

$$f(t) = \int_{-\infty}^{\infty} A(\omega) e^{-j\omega t} d\omega \tag{A.59}$$

Coming back to the vector case described in (A.50) and referring without loss of generality to the case of propagation along the z axis, we look, in the sinusoidal regime, for waves that are elementary solution characterized by the fact that the electric and magnetic field are constant across the plane orthogonal to the propagation direction, that is "plane waves". Mathematically we assume:

$$\frac{\partial}{\partial x} \mathbf{E}(x, y, z) = 0$$
$$\frac{\partial}{\partial y} \mathbf{E}(x, y, z) = 0$$
$$\frac{\partial}{\partial x} \mathbf{H}(x, y, z) = 0 \tag{A.60}$$
$$\frac{\partial}{\partial y} \mathbf{H}(x, y, z) = 0$$

From the Maxwell equations corresponding to the fields:

$$\nabla \times \mathbf{E} = -j\omega_0 \mu \mathbf{H}$$
$$\nabla \times \mathbf{H} = j\omega_0 \varepsilon \mathbf{E} \tag{A.61}$$

we have:

$$-j\omega_0 \mu H_z = \frac{\partial}{\partial x} E_y - \frac{\partial}{\partial y} E_x = 0$$
$$j\omega_0 \varepsilon E_z = \frac{\partial}{\partial x} H_y - \frac{\partial}{\partial y} H_x = 0 \tag{A.62}$$

that is $E_z = H_z = 0$: the field is therefore in this case transverse to the propagation direction. The following two independent systems of equation can be derived:

$$\frac{dE_x}{dz} = -j\omega_0\mu H_y$$

$$\frac{dH_y}{dz} = -j\omega_0\varepsilon E_x$$

(A.63)

$$\frac{dE_y}{dz} = j\omega_0\mu H_x$$

$$\frac{dH_x}{dz} = j\omega_0\varepsilon E_y$$

(A.64)

which states that, in the cross plane to the propagation direction, the electric field along one direction is related to the magnetic field component in the orthogonal direction. In other word if the wave has a zero transverse component of the electric field, say E_x then the orthogonal component of the magnetic field, i.e. H_y is also zero. In other words the electric and magnetic field, which are transverse to the propagation direction, are also orthogonal each other. Letting $k_z = \omega_0\sqrt{\varepsilon\mu}$, by derivation we get:

$$\frac{d^2 E_x}{dz^2} = -k_z^2 E_x$$

$$\frac{d^2 H_y}{dz^2} = -k_z^2 H_y$$

(A.65)

and

$$\frac{d^2 E_y}{dz^2} = -k_z^2 E_y$$

$$\frac{d^2 H_x}{dz^2} = -k_z^2 H_x$$

(A.66)

The solutions are given by:

$$E_x(z) = E_x^+ e^{-jk_z z} + E_x^- e^{+jk_z z}$$

$$H_y(z) = \frac{E_x^+}{\zeta} e^{-jk_z z} - \frac{E_x^+}{\zeta} e^{+jk_z z}$$

(A.67)

and

$$E_y(z) = E_y^+ e^{-jk_z z} + E_y^- e^{+jk_z z}$$

$$H_x(z) = \frac{E_y^+}{\zeta} e^{-jk_z z} - \frac{E_y^+}{\zeta} e^{+jk_z z}$$

(A.68)

with $\zeta = \sqrt{\varepsilon/\mu}$ being the impedance of the mean where the wave propagates and apexes indicating the forward and backward solution. The solutions in (A.67) and (A.68) provide the expression of the field components in the phasor domain. The expression in the time domain can be achieved by multiplying for $\exp(j\omega t)$ and taking the real part. For instance, considering the x forward component of the electric field we have:

$$e_x(t, z) = |E_x^+| \cos(\omega t - k_z z + \angle E_x^+) \qquad \text{(A.69)}$$

All above equations can be generalized to the propagation in a given direction $\hat{u}$ with a propagation vector $k \equiv (k_x, k_y, k_z)$. For instance, referring again to the x forward component of the electric field and letting $\boldsymbol{u} = u\hat{u}$ we have:

$$E_y(z) = E_y^+ e^{-jk\cdot u} \qquad \text{(A.70)}$$

that is

$$e_x(t, z) = |E_x^+| \cos(\omega t - k \cdot r + \angle E_x^+) \qquad \text{(A.71)}$$

in the time domain. Above expression considers identify plane waves propagating along k (the vector wavenumber), which are constant in the plane orthogonal to k identified by the $k \cdot u = 0$.

A.7 Elementary antennas

Radar applications usually require directive antennas radiating power in a narrow beam, which can be used either as a transmitting or a receiving device.

Let's start by briefly examining some systems that are used as antennas, which irradiate electromagnetic energy. The simplest is certainly constituted by the dipole antenna, consisting of a thin wire that is interrupted at the point where the excitation is applied, represented in Fig. A.2.

Usually, the source is positioned at the intermediate point of the wire, resulting in a symmetrical dipole. The typical feature of a dipole is that it is very short compared to the wavelength that characterizes the signal to be transmitted or received. This case is referred to as Hertzian dipole. Other widely used dipoles are the so-called resonant ones, as for instance the one where its length is equal to half a wavelength ($h = 2l = \lambda/2$).

To evaluate the field radiated by a dipole antenna, we consider the reference system shown in Fig. A.3, where a dipole oriented along the z axis is positioned, so short that the current I_0 flowing inside it can be considered uniform along z. Both a Cartesian $(\hat{x}, \hat{y}, \hat{z})$ and a spherical reference system $(\hat{r}, \hat{\phi}, \hat{\theta})$ are considered in Fig. A.3.

Figure A.2. Dipole antenna.

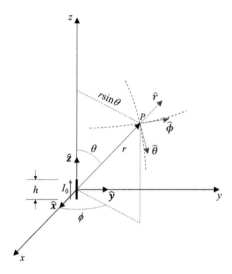

Figure A.3. Dipole antenna geometry.

The electromagnetic field radiated by this antenna at a point P at a large distance from the antenna is given by:

$$\boldsymbol{H}(r,\theta) = H_\phi(r,\theta)\hat{\boldsymbol{\phi}} = \frac{jkI_0 h}{4\pi r}\sin\theta e^{-jkr}\hat{\boldsymbol{\phi}}$$

$$\boldsymbol{E}(r,\theta) = E_\theta(r,\theta)\hat{\boldsymbol{\theta}} = \frac{j\omega\mu I_0 h}{4\pi r}\sin\theta e^{-jkr}\hat{\boldsymbol{\theta}} = \eta H_\phi(r,\theta)\hat{\boldsymbol{\theta}}$$

(A.72)

where $\boldsymbol{E}$ is the electric field, $\boldsymbol{H}$ is the magnetic field, $\omega = 2\pi f$, μ is the magnetic permittivity, $k = 2\pi/\lambda$, and $\eta = \sqrt{\mu\epsilon}$ is the impedance of the medium ($\sqrt{\mu_0\epsilon_0}$ in the free space case).

At a large distance from the antenna, the electric field and the magnetic field radiated by the dipole are orthogonal to each other, and both are orthogonal to the propagation direction of the electromagnetic field. Basically, both of them are tangent to a spherical surface which has the dipole as its center, as shown in Fig. A.3. Moreover, still at large distance from the dipole antenna, the curvature of the spherical surface can be considered very large, so that the surface can be locally well approximated by a planar surface.

At a large distance from the dipole antenna, the electromagnetic field does not depend on the angle ϕ, as expected given the symmetry of the dipole with respect to ϕ, and decades as $1/r$ with distance. Its dependence on elevation angle θ follows a sinusoidal behavior, with its maximum on the equatorial plane ($\theta = \pi/2$), and null of radiation in the direction of the two poles (for $\theta = 0$ and $\theta = \pi$).

Consider now the linear antenna, whose length is $h = 2l$ shown in Fig. A.4. Let us now consider a wire-like antenna in which a current distribution $I(z)$ flows. Considering the linearity of Maxwell's equations reported above, we can apply the superposition principle, and calculate the total field radiated by this antenna by superimposing the effects of the irradiation of the various elements in position z, of length $dz \ll \lambda$, each of which can be assimilated to a dipole antenna on which a current $I(z)$ flows.

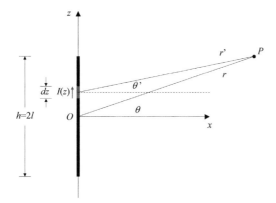

Figure A.4. Wire antenna geometry.

Limiting our analysis to the electric field, and considering only the component along $\hat{\theta}$, the electric field irradiated at a great distance by the antenna element at position z, when a current $I(z)$ flows inside it, is:

$$dE_\theta(r',\theta') = \frac{j\omega\mu I(z)dz}{4\pi r'}\sin\theta' e^{-jkr'} \tag{A.73}$$

The current value at position z can be obtained in the case of a thin antenna (length $\gg$ thickness), solving a Fredhold equation of 2nd kind:

$$I(z) = I_0 \frac{\sin k(l - |z|)}{\sin kl} \tag{A.74}$$

where I_0 value depends on the antenna dimension, antenna material, and input power. Now, substituting (A.74) in (A.73), and integrating (A.73) along the whole length of the antenna, we get the electric far-field irradiated by the wire-like antenna:

$$\boldsymbol{E}(r, \theta) = E_\theta(r, \theta)\hat{\boldsymbol{\theta}} = j\eta \frac{e^{-jkr}}{2\lambda r} I_0 \frac{2}{k \sin kl} \frac{\cos(kl \cos \theta) - \cos kl}{\sin \theta} \hat{\boldsymbol{\theta}} \tag{A.75}$$

The dependence of the electromagnetic field on the distance r is all concentrated in the term e^{-jkr}/r. This happens for every electromagnetic field radiated at a great distance by an arbitrary distribution of currents. For this reason, considering always only the electric field, this can be factored as the product between the term e^{-jkr}/r and a factor $F(\theta, \phi)$ that depends only on θ and ϕ (in this case, see the ϕ symmetry of the wire antenna, only from ϕ):

$$\boldsymbol{E}(r, \theta) = \frac{e^{-jkr}}{r} \boldsymbol{F}(\theta) \tag{A.76}$$

The function $F(\theta)$ is called radiation pattern of the antenna.

The behavior of the normalized radiation patterns of the wire antenna vs. the angle θ are plotted in Fig. A.5, for $2l = \lambda/2$, $2l = \lambda$ and $2l = 4/3\lambda$ with solid line, dotted line, and dashed line, respectively.

All plotted radiated electric fields exhibit the maximum value for $\theta = \pi$. It can be appreciated also that the larger the antenna dimension, the narrower the main radiation lobe is. As the length of the antenna increases over $4/3\lambda$, in addition to a further narrowing of the main lobe, there is also an increase in the number of lobes, and a shift of the maximum from the direction characterized by $\pi/2$.

From the point of view of the energy, the larger the electric field amplitude E in a given direction θ, the larger the power irradiated by the radar antenna in the same direction, that, as it is well known by the antenna e.m. theory, is related to the square of the electric field amplitude.

Now, we are ready to introduce the so-called antenna directivity, which is a measure of the ability of an antenna to radiate energy in a given direction. If P_T is the power transmitted (irradiated) by the radar antenna, the antenna directivity function is defined as the ratio between the power (spatial) density $S_i(r, \theta, \phi)$ irradiated

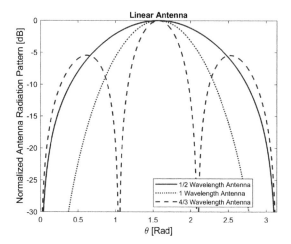

Figure A.5. Normalized antenna radiation pattern vs. the angle θ for $2l = \lambda/2$, $2l = \lambda$ and $2l = 4/3\lambda$ (solid line, dotted line, and dashed line, respectively).

by the radar antenna in a given direction (θ, ϕ), and the average irradiated power spatial density:

$$D(\theta, \phi) = \lim_{r \to \infty} \frac{S_i(r, \theta, \phi)}{\dfrac{P_T}{4\pi r^2}} \qquad \text{(A.77)}$$

The directivity function $D(\theta, \phi)$ can be also defined as the power density radiated by an antenna into a direction (θ, ϕ), compared to the power density radiated into that same direction by an isotropic radiator adopting the same power.

The maximum value assumed by the directivity function $D(\theta, \phi)$ in the direction identified by (θ_M, ϕ_M) is called the antenna directivity:

$$D_M = D(\theta_M, \phi_M) \qquad \text{(A.78)}$$

and is usually evaluated in dB. The larger the antenna directivity, the higher the capacity of the antenna to concentrate the radiated power in some directions.

In antenna theory it can also be introduced and used another parameter, the antenna gain. It differs from the antenna directivity because it takes into account also the electric loss inside the antenna. Neglecting the loss inside the antenna for sake of simplicity, antenna gain and directivity coincide.

Pass now to two dimensional radiating elements. A common type of two dimensional antennas is the aperture antenna, such as open-ended radiating waveguides, including horn-type antennas, slotted waveguides antennas, reflector and lens antennas. They find use in many applications, such as point-to-point radio communication links and radar systems.

The radiation field from aperture antennas is determined by the knowledge of the electromagnetic field distribution $\mathbf{E}_a$ and $\mathbf{H}_a$ over the antenna aperture, and can be computed with the help of the field equivalence principle, which states that the aperture fields may be replaced by equivalent electric and magnetic surface currents:

$$\mathbf{J}_s = \hat{n} \times \mathbf{H}_a$$
$$\mathbf{J}_m = -\hat{n} \times \mathbf{E}_a, \tag{A.79}$$

where $\hat{n}$ is a unit vector normal to the surface. The corresponding radiated fields can be calculated by computing the surface integral of the equivalent currents.

There are two alternative forms of the field equivalence principle, which may be used when only one of the aperture fields $\mathbf{E}_a$ or $\mathbf{H}_a$ is available. They are:

- perfect electric conductor:

$$\mathbf{J}_s = 0$$
$$\mathbf{J}_m = -2\left(\hat{n} \times \mathbf{E}_a\right), \tag{A.80}$$

- perfect magnetic conductor:

$$\mathbf{J}_s = 2\left(\hat{n} \times \mathbf{H}_a\right)$$
$$\mathbf{J}_m = 0. \tag{A.81}$$

The aperture surface A and the screen, shown in Fig. A.6, can be arbitrarily curved.

However, a common case is to assume that they are both flat. In this case, referring to the equivalence principle form for an aperture in a perfect electric conductor screen, radiated fields are given by:

$$\mathbf{E}_\theta(\theta, \phi) = j\beta \frac{e^{-j\beta r}}{2\pi \lambda r} \left[f_x \cos\phi + f_y \sin\phi\right]$$
$$\mathbf{E}_\phi(\theta, \phi) = j\beta \frac{e^{-j\beta r}}{2\pi \lambda r} \left[\cos\theta \left(f_y \cos\phi - f_x \sin\phi\right)\right], \tag{A.82}$$

where f_x and f_y are the components along the $\hat{x}$ and $\hat{y}$ axis of the vector:

$$\mathbf{f}(\theta, \phi) = \int_A \mathbf{E}_a(\mathbf{r}')e^{j\mathbf{k}\cdot\mathbf{r}'} dx' dy' = \int_A \mathbf{E}_a(x', y')e^{j\left(k_x x' + k_y y'\right)} dx' dy'. \tag{A.83}$$

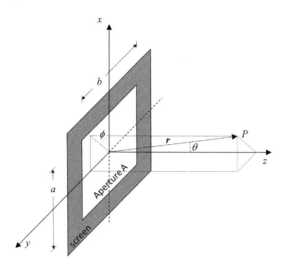

Figure A.6. Aperture antenna geometry.

We note that (A.83) assumes the form of 2-D Fourier transform.

In planar apertures, the fields $\mathbf{E}_a$ and $\mathbf{H}_a$ are tangential to the aperture xy-plane, and then can be expressed in terms of their cartesian components $\mathbf{E}_a = \hat{\mathbf{x}} E_{ax} + \hat{\mathbf{y}} E_{ay}$. In the uniform assumption, they are assumed to be constant over the aperture area, so that the radiated fields are depending on the following radiation integral:

$$\mathbf{f}(\theta, \phi) = \int_A \mathbf{E}_a(\mathbf{r}')e^{j\mathbf{k}\cdot\mathbf{r}'}dS' = \mathbf{E}_a \int_A e^{j\mathbf{k}\cdot\mathbf{r}'}dS' = Af(\theta, \phi)\mathbf{E}_a, \quad \text{(A.84)}$$

where the normalized scalar quantity $f(\theta, \phi)$ is defined as:

$$f(\theta, \phi) = \frac{1}{A} \int_A e^{j\mathbf{k}\cdot\mathbf{r}'}dS'. \quad \text{(A.85)}$$

The quantity $f(\theta, \phi)$ depends on the assumed geometry of the aperture and it, alone, determines the radiation pattern.

For a rectangular aperture of sides a and b, the surface integral (A.85) is separable in the x and y directions:

$$\begin{aligned} f(\theta, \phi) &= \frac{1}{ab} \int_{-a/2}^{a/2} \int_{-b/2}^{b/2} e^{j(k_x x' + k_y y')}dx'dy' \\ &= \frac{1}{a} \int_{-a/2}^{a/2} e^{jk_x x'}dx' \frac{1}{b} \int_{-b/2}^{b/2} e^{jk_y y'}dy', \end{aligned} \quad \text{(A.86)}$$

where we placed the origin of the r-integration in the middle of the aperture. Taking into account that $k_x = \beta \sin(\theta)\cos(\phi)$, $k_y =$

$\beta \sin(\theta) \sin(\phi)$, it is obtained:

$$f(\theta, \phi) = \text{sinc}(v_x)\text{sinc}(v_y) \qquad \text{(A.87)}$$

with:

$$\begin{aligned} v_x &= \frac{a}{\lambda} \sin\theta \cos\phi \\ v_y &= \frac{b}{\lambda} \sin\theta \sin\phi \end{aligned} \qquad \text{(A.88)}$$

Since $0° \leqslant \theta \leqslant 90°$ and $-180° \leqslant \phi \leqslant 180°$, the quantities v_x and v_y vary in the intervals $-a/\lambda \leqslant v_x \leqslant a/\lambda$ and $-b/\lambda \leqslant v_y \leqslant b/\lambda$. Moreover, the physically realizable values of v_x and v_y are those lying in the ellipse in the xy-plane:

$$\frac{v_x^2}{a^2} + \frac{v_y^2}{b^2} \leqslant \frac{1}{\lambda^2}. \qquad \text{(A.89)}$$

The values of v_x and v_y satisfying inequality (A.89) form a region referred to as the *visible region*. The 2D graph of (A.87) for $a = 8\lambda$ and $b = 4\lambda$, restricted to the visible region, is shown in Fig. A.7(a).

The pattern (A.87) simplifies along the two principal planes, the xz- and yz-planes, corresponding to $\phi = 0°$ and $\phi = 90°$. We have:

$$\begin{aligned} f(\theta, 0°) &= \text{sinc}\left(\frac{a}{\lambda} \sin\theta\right) \\ f(\theta, 90°) &= \text{sinc}\left(\frac{b}{\lambda} \sin\theta\right). \end{aligned} \qquad \text{(A.90)}$$

The patterns of a rectangular aperture with sides $a = 8\lambda$ and $b = 4\lambda$ in the two principal planes $\phi = 0°$ and $\phi = 90°$ are shown in Figs. A.7 (b) and (c).

We note that the 3 dB main lobe width of the patterns (A.90) in the two principal planes $\phi = 0°$ and $\phi = 90°$ are (in radians):

$$\begin{aligned} \Delta\theta_0 &= 0.886\lambda/a \\ \Delta\theta_{90} &= 0.886\lambda/b \end{aligned} \qquad \text{(A.91)}$$

where generally, as in this book, the coefficient is approximated to 1.

A.8 Antenna array

The radiation pattern of an antenna, such as wire or aperture one, in terms of the number of lobes, orientation of the main lobe, directivity, is specified by its size and current distribution (in the

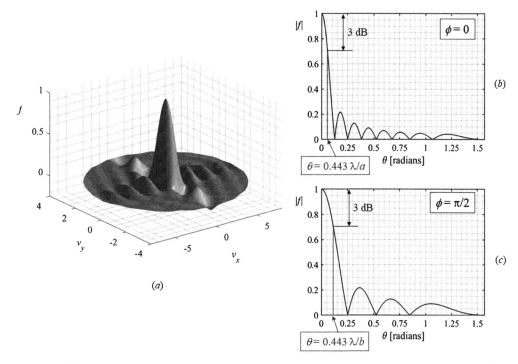

Figure A.7. Radiation pattern of a rectangular aperture with $a = 8\lambda$ and $b = 4\lambda$: (*a*) 2D representation; (*b*) principal plane $\phi = 0$; (*c*) principal plane $\phi = \pi/2$.

case of a wire antenna) or equivalent current distribution (in the case of an aperture antenna).

The angular width of the main lobe on a plane, which expresses the ability of an antenna to be directive, can be approximated by the ratio λ/L, where L is in this case the physical size of the antenna along that plane. Consequently, at a fixed frequency, small antennas cannot be very directive.

In many application cases, and among these remote sensing, a much greater directivity is required. The reason is twofold: on the one hand one wants to obtain much higher spatial resolutions, and on the other hand one wants to obtain a more favorable signal-to-noise ratio.

Using several antennas suitably spaced between them, in order to form what is called an array of antennas, in which it is possible to control the current (real or equivalent), it is possible to obtain much more directive radiation patterns. Hence, many small antennas can be arranged in space in various geometrical configurations to form array antennas, characterized by the highly directive radiation pattern of large antennas. In this case, the mechanical problems related to the realization of a unique large antenna are traded with the electrical problems connected to the feeding of many antenna elements. For purposes of analysis, arrays are usually assumed to consist of identical elements, although it is possible to create an array with different elements.

We therefore consider the system consisting of different antennas, shown in Fig. A.8 in the case of four antennas, which simultaneously radiate an electromagnetic field in space.

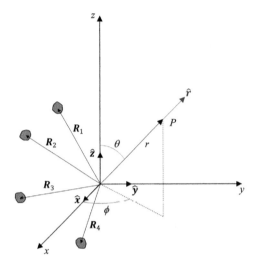

Figure A.8. Antenna array geometry.

The electric field E_A radiated by antenna array is the superposition of the electric fields E_n, $n = 1, \cdots, N$, radiated by the N different antennas:

$$E_A(r, \theta, \phi) = \sum_{n=1}^{N} E_n(r, \theta, \phi) = \sum_{n=1}^{N} F_n(\theta, \phi) \frac{e^{-jkr_n}}{r_n} \qquad (A.92)$$

where F_n, $n = 1, \cdots, N$, are the radiation patterns of the N antennas. At large distance from the antenna system, $r_n = r - R_n \cdot \hat{r}$. Consequently, (A.92) becomes:

$$E_A(r, \theta, \phi) = \frac{e^{-jkr}}{r} \sum_{n=1}^{N} F_n(\theta, \phi) e^{jkR_n \cdot \hat{r}} \qquad (A.93)$$

Arrays can be structured in many geometrical shapes. If we begin considering the case on N antennas equal from each other, all oriented in the same direction, so that they have the same radiation pattern, but for a complex valued term A_n related to the

N different current source, not depending from (θ, ϕ), (A.93) becomes:

$$E_A(r, \theta, \phi) = \frac{e^{-jkr}}{r} F(\theta, \phi) \sum_{n=1}^{N} A_n e^{jk R_n \cdot \hat{r}} \qquad \text{(A.94)}$$

The term:

$$f_A(\theta, \phi) = \sum_{n=1}^{N} A_n e^{jk R_n \cdot \hat{r}} \qquad \text{(A.95)}$$

is the so-called array factor. Consequently, the electric field radiated by the antenna array can be expressed as:

$$E_A(r, \theta, \phi) = \frac{e^{-jkr}}{r} F(\theta, \phi) f_A(\theta, \phi) = \frac{e^{-jkr}}{r} F_A(\theta, \phi) \qquad \text{(A.96)}$$

and its radiation pattern $F_A(\theta, \phi)$ can be expressed as the product of the radiation pattern $F(\theta, \phi)$ of the (all equal and equi-oriented) antennas used in the array, and the array factor $f_A(\theta, \phi)$.

Consider now the case of a linear antenna array, in which all the antennas are oriented in the same way, and their phase centers lie along a x axes. Suppose also that the antennas are equi-spaced with a distance d between two adjacent elements is, as shown in Fig. A.9.

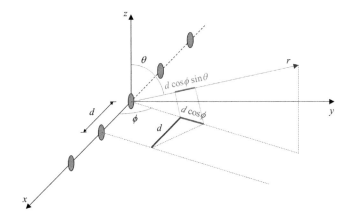

Figure A.9. Antenna linear array geometry.

In this case, the positions of the N antenna are $R_n = x_n \hat{x} = n d \hat{x}$, for $n = 0, 1, \cdots, N - 1$, and the array factor (A.95) becomes:

$$f_A(\theta, \phi) = \sum_{n=1}^{N} A_n e^{jknd \sin\theta \cos\phi} = \sum_{n=1}^{N} |A_n| e^{j(knd \sin\theta \cos\phi + \alpha_n)} \qquad \text{(A.97)}$$

as $\hat{x} \cdot \hat{r} = \sin\theta \cos\phi$, and where the factors $A_n = |A_n|e^{j\alpha_n}$ represent the different excitation conditions of the N antennas.

Now, exploiting the following identity:

$$\sum_{n=0}^{N-1} e^{jn\psi} = \frac{1 - e^{jN\psi}}{1 - e^{j\psi}} = e^{j(N-1)\frac{\psi}{2}} \frac{\sin\left(\frac{N\psi}{2}\right)}{\sin\left(\frac{\psi}{2}\right)} \qquad (A.98)$$

and assuming that the N antennas are excited with the current such that $A_n = ae^{jn\alpha}$, where a is a real value, the array factor (A.97) becomes:

$$f_A(\theta, \phi) = ae^{j(N-1)\frac{(kd\sin\theta\cos\phi+\alpha)}{2}} \frac{\sin\left(\dfrac{N(kd\sin\theta\cos\phi + \alpha)}{2}\right)}{\sin\left(\dfrac{kd\sin\theta\cos\phi + \alpha}{2}\right)}$$

$$|f_A(\theta, \phi)| = \frac{\left|\sin\left(\dfrac{N(kd\sin\theta\cos\phi + \alpha)}{2}\right)\right|}{\left|\sin\left(\dfrac{kd\sin\theta\cos\phi + \alpha}{2}\right)\right|} \qquad (A.99)$$

Such equation represents the array factor (and its module) of a so-called uniform linear array. Consider now the case of an array with all elements excited in phase ($\alpha = $ constant $= 0$), and assume that $\theta = \pi/2$. The array factors for four different configurations are shown in Fig. A.10.

The case of a spacing between antennas $d = \lambda/2$, and number of elements $N = 40$ is reported in the top left subfigure. The array factor on the plane $\theta = \pi/2$ exhibits two maxima for $\phi = \pm\pi/2$. For the symmetry of the array geometry, the same behavior occurs for all planes of the sheaf sharing the same common line (x axes) containing the antenna array. In this case, using as the element of the array an isotropic or an almost isotropic one, as the short antennas introduced in the previous appendix section, the radiation pattern of the entire array, given by the product between the radiation pattern of the antenna and the array factor, practically coincides with the array factor. The poor or null directivity characteristic of the single antenna is transformed into significant directivity, thanks to the array factor.

Increasing the spacing between adjacent antennas above $\lambda/2$, as for the other three cases presented in Fig. A.10 (top right: $d = \lambda$ and $N = 20$; bottom left: $d = 2\lambda$ and $N = 10$; bottom right: $d = 10\lambda$ and $N = 10$), we notice the appearance of other lobes, in addition to those for $\phi = \pm\pi/2$. The new lobes of the array factor corresponding to the new maxima are the so-called *grating lobes* of the

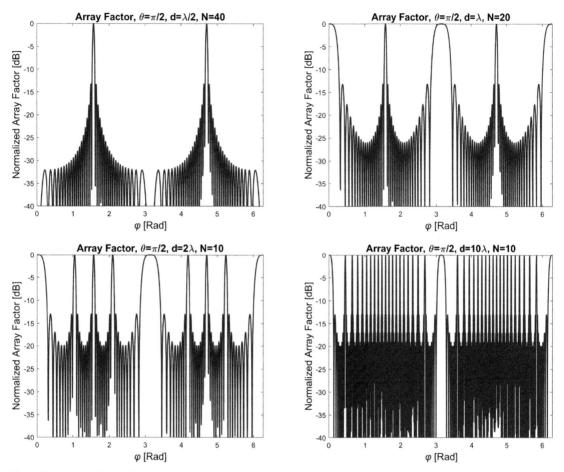

Figure A.10. Array factor for $\theta = \pi/2$, and for different antenna spacing d. Top left: $d = \lambda/2$ and $N = 40$; top right: $d = \lambda$ and $N = 20$; bottom left: $d = 2\lambda$ and $N = 10$; bottom right: $d = 10\lambda$ and $N = 10$.

array. If we use for these kinds of array (spacing larger than $\lambda/2$) a low directive antenna as array element, the radiation diagram of the resulting array (after the product between the radiation pattern of the antenna and the array factor) will exhibit radiation lobes comparable with the main lobe. In practice, the radiation pattern of the low directive antenna (slowly varying with respect to ϕ) is "sampled" by the array factor.

As it can be easily appreciated by Fig. A.10, the number of lobes of radiation pattern is related to the antenna spacing: the larger the spacing between antennas, the larger the number of lobes is. It has to be noted also that the width of the lobes (with the exception of the ones in $\phi = 0$, and π) is inversely proportional to the overall dimension of the array $L = N \times d$, according to the rule width $\cong$

L/λ. In the first three cases ($d = \lambda/2$, $N = 40$; $d = \lambda$, $N = 20$; and $d = 2\lambda$, $N = 10$), $L = 20\lambda$, and the lobes exhibit the same width, while in the fourth case ($d = 10\lambda/2$, $N = 10$), $L = 100\lambda$ and the lobes exhibit a much lower width.

Consider now the case of the synthetic antenna of a SAR system, and consider in particular the COSMO/SkyMed (CSK) case in the stripmap configuration. Its synthetic antenna is formed by the physical antenna, whose dimensions are $L_x = 5.7$m $\times$ $L_r = 1.4$m (along azimuth and cross azimuth directions, respectively), which moves along azimuth direction (x in Fig. A.9) with a velocity $v = 7.6$km/sec. The working frequency of CSK radar is $f = 9.6$GHz (X-band), so that the wavelength is $\lambda = 3.125$cm, and the Pulse Repetition Frequency (PRF) can vary in the interval (3000-4000)Hz. Assuming now a PRF value of 3800Hz, the resulting spacing between two adjacent positions in the synthetic array amounts to $d = 2$m, and taking into account the footprint dimension along azimuth (about 4.4km), the number of elements of the (synthetic) array is $N = 2200$. The spacing $d = 128(\lambda/2)$ between two adjacent antennas is very large in terms of wavelength, so that several grating lobes are present in the array factor of the synthetic antenna, as it is shown in top left plot of Fig. A.11. The radiation pattern of the real CSK antenna is shown in the bottom left plot Fig. A.11. The zoom of both previous plots are shown in the right part of Fig. A.11 (bottom left: zoom of the array factor of the synthetic array; bottom right: zoom of the radiation pattern of the real antenna).

The radiation pattern of the synthetic CSK aperture (array) is given by the product of the multiple grating lobes array factor represented in the plots on the top of Fig. A.11 and the radiation pattern of the physical CSK antenna reported in the plots on the bottom. It will exhibit a single lobe, as the radiation pattern of the physical antenna, as narrow as the grating lobe of the array factor aligned on $\phi = \pi/2$. Basically, a single very narrow lobe as would have resulted from a single 4.4km length physical antenna operating at 9.6GHz, despite the fact that the array spacing is enormously larger than $\lambda/2$ and the array factor has numerous grating lobes. In this case it is as if the radiation pattern of the physical antenna that "sample" the array factor.

Also for the case of synthetic array, as in the case of real array, the fields radiated from the different elements add constructively in some directions and destructively in others. This can lead, also in this case, to steering the main lobe of the radiation pattern toward any direction. While in the physical array this can be obtained by changing the phase of the excitation current at each array element, in the synthetic array the same behavior is obtained by changing the radiation pattern of the physical antenna. We can realize in this way a squinted version of a SAR system.

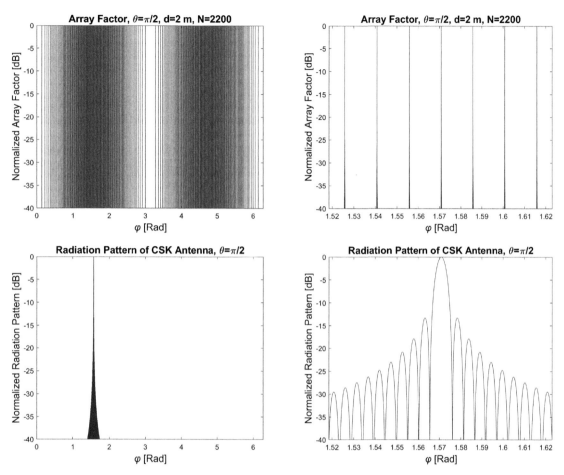

Figure A.11. Array factor of the synthetic array antenna and radiation pattern of the real antenna of COSMO/SkyMed satellites for $\theta = \pi/2$. Top left: array factor of the synthetic array; top right: radiation pattern of the real antenna; bottom left: zoom of the array factor of the synthetic array; bottom right: zoom of radiation pattern of real antenna.

A.9 Generalized likelihood ratio test detection of persistent scatterers

Under the given assumption on the noise statistic and data model, the pdf of the data under $\mathcal{H}_0$ and $\mathcal{H}_1$ is given by (8.51) and (8.52). The likelihood ratio is given by:

$$\mathcal{L}(\boldsymbol{y}) = \frac{f_{\boldsymbol{y}}\big(\boldsymbol{y}; \sigma_w^2, \gamma, s, v, \mathcal{H}_1\big)}{f_{\boldsymbol{y}}\big(\boldsymbol{y}; \sigma_w^2, \mathcal{H}_0\big)} = \exp\Big(-\frac{\|\,\boldsymbol{y} - \boldsymbol{a}\gamma\,\|^2}{\sigma_w^2} + \frac{\|\,\boldsymbol{y}\,\|^2}{\sigma_w^2}\Big)$$

$$\text{(A.100)}$$

which leads to:

$$\mathcal{L}(y) = \exp\left(-\frac{2\Re(y^H a\gamma) + \|a\|^2 |\gamma|^2}{\sigma_w^2}\right) \qquad \text{(A.101)}$$

which, after (monotonic) logarithmic transformation and absorption of the constant factors in the threshold leads to the test in (8.53).

The estimate of the unknown γ, obviously under $\mathcal{H}_1$, is the classical least square estimation given by solving the following equation:

$$\hat{\gamma} : \frac{\partial}{\partial\gamma} \ln f_y\left(y; \sigma_w^2, \gamma, s, v, \mathcal{H}_1\right) = -\frac{1}{\sigma_w^2}\frac{\partial}{\partial\gamma}(y - a\gamma)^H(y - a\gamma) = 0 \qquad \text{(A.102)}$$

which provides [126]:

$$y - a\hat{\gamma} = 0 \qquad \text{(A.103)}$$

that is:

$$\hat{\gamma} = \frac{a^H y}{\|a\|^2} = a^H y \qquad \text{(A.104)}$$

where the last equality is valid in the case of normalized steering vectors.

Moreover:

$$\hat{\sigma}_w^2|\mathcal{H}_0 : \frac{\partial}{\partial\gamma} \ln f_y\left(y; \sigma_w^2, \mathcal{H}_0\right) = 0 \qquad \text{(A.105)}$$

that is

$$-\frac{N}{\hat{\sigma}_w^2|\mathcal{H}_0} + \frac{\|y\|^2}{\hat{\sigma}_w^4|\mathcal{H}_0} = 0 \qquad \text{(A.106)}$$

which leads to:

$$\hat{\sigma}_w^2|\mathcal{H}_0 = \frac{\|y\|^2}{N} \qquad \text{(A.107)}$$

Similarly

$$\hat{\sigma}_w^2|\mathcal{H}_1 = \frac{\|y - a\hat{\gamma}\|^2}{N} = \frac{1}{N}\|y - \frac{aa^H}{a^H a}y\|^2 = \frac{1}{N}\|(I - \frac{aa^H}{a^H a})y\|^2 \qquad \text{(A.108)}$$

By defining the projector orthogonal to a, that is

$$P_a^\perp = I - \frac{aa^H}{a^H a} \qquad \text{(A.109)}$$

(A.108) can be written as

$$\hat{\sigma}_w^2 | \mathcal{H}_1 = \frac{1}{N} \parallel \mathbf{P}_a^\perp \boldsymbol{y} \parallel^2 \qquad (A.110)$$

Substitution of (A.104), (A.107) and (A.110) in (8.54) leads to

$$\mathcal{L}_{GLRT}(\boldsymbol{y}) = \frac{\max_{s,v} f_{\boldsymbol{y}}(\boldsymbol{y}; \hat{\sigma}_w^2 | \mathcal{H}_1, \hat{\gamma})}{f_{\boldsymbol{y}}(\boldsymbol{y}; \hat{\sigma}_w^2 | \mathcal{H}_1,)} \qquad (A.111)$$

that is:

$$\mathcal{L}_{GLRT}(\boldsymbol{y}) = \frac{\max_{s,v} \hat{\sigma}_w^2 | \mathcal{H}_1}{\hat{\sigma}_w^2 | \mathcal{H}_0} \qquad (A.112)$$

Accordingly,

$$\mathcal{L}_{GLRT}(\boldsymbol{y}) = \frac{\max_{s,v} \parallel \mathbf{P}_a^\perp \boldsymbol{y} \parallel^2}{\parallel \boldsymbol{y} \parallel^2} = \frac{\max_{s,v} \boldsymbol{y}^H \mathbf{P}_a^\perp \boldsymbol{y}}{\parallel \boldsymbol{y} \parallel^2} \qquad (A.113)$$

Substitution of the expression of the orthogonal projector in (A.109) finally leads to:

$$\mathcal{L}_{GLRT}(\boldsymbol{y}) = \max_{s,v} \frac{\left(\boldsymbol{y}^H \boldsymbol{y} - \frac{\boldsymbol{y}^H a a^H \boldsymbol{y}}{\|a\|^2} \right)}{\parallel \boldsymbol{y} \parallel^2} = \max_{s,v} \left(1 - \frac{|\boldsymbol{y}^H \boldsymbol{a}|^2}{\parallel \boldsymbol{a} \parallel^2 \parallel \boldsymbol{y} \parallel^2} \right) \quad (A.114)$$

and hence to test in (8.55).

List of symbols

Matrices (real and complex) are bold roman upper case.
Vectors (real and complex) are bold italic lower case.
Complex vectors representing electromagnetic fields are also bold roman upper case.
Points of the space of Euclidean geometry are bold italic upper case.

In the following list we do not include symbols denoting the estimate of true quantities, which are indicated with the addition of the "hat". For instance $\hat{x}$ indicates the estimates of x. We also omit to explicitly include symbols denoting the standard deviation of random variables, which are indicated with the introduction of the "σ". For instance σ_x indicates the standard deviation of x. Moreover, the introduction of the "primed" symbol generally indicates the variable corresponding to the sampled radar coordinate. For instance x' indicates the sampled coordinate x. Finally, different definitions of the same symbol is separated by semicolon.

(x_0, r_0)	azimuth and range coordinates of the scene center
α	chirp rate or incidence angle
α_b	baseline tilt angle
$\mathbf{A}$	steering matrix collecting the steering vectors corresponding to each tomographic bin along the height ($N \times M$)
$\boldsymbol{a}(z, v), \boldsymbol{a}(s, v)$	steering vector depending on the height/velocity and slant-height (or elevation)/velocity, respectively
$\boldsymbol{a}_m$	steering vector corresponding to the generic m-th tomographic bin ($N \times 1$)
$\boldsymbol{b}$	(Euclidean) geometric vector connecting two acquisitions in the plane orthogonal to the azimuth for a single interferometric pair; vector collecting all the orthogonal baseline components in the multiacquisition interferometric case
$\mathbf{C}_\gamma$	covariance matrix of the backscattering coefficient at the different antennas ($N \times N$)
$\mathbf{C}_y$	data covariance matrix ($N \times N$)
$\boldsymbol{d}$	deformation (time series) vector ($N \times 1$)
$\boldsymbol{d}_{NL}$	nonlinear component of $\boldsymbol{d}$ ($N \times 1$)
$\boldsymbol{d}_{F,NL}$	full resolution residual (w.r.t. the low resolution) nonlinear deformation vector ($N \times 1$)
$\boldsymbol{d}_L$	full resolution residual (w.r.t. the low resolution) deformation vector ($N \times 1$)
$\mathbf{E}$	electric field phasor (two element complex vector)
β	terrain slope for a planar topography; ratio between the radial velocity and the speed of light
γ	backscattering vector. The elements are collected at the antennas ($N \times 1$) or on the tomographic bins ($M \times 1$) depending on the context
$\boldsymbol{k}_s$	vector containing the slant height (elevation) wavenumbers ($N \times 1$)
$\boldsymbol{k}_z$	vector containing the height (vertical) wavenumbers ($N \times 1$)
$\mathbf{M}_\gamma$	matrix collecting the coherence values describing the correlation structure of the backscattering at the different antennas ($N \times N$)
$\mathbf{M}_y$	matrix collecting the coherence values describing the correlation structure of the data at the different antennas, accounting also for the additive noise, ($N \times N$)

$\mathbf{R}_y$	data correlation matrix ($N \times N$)
$\mathbf{S}$	scattering matrix (2x2)
t	temporal baselines (acquisition times) vector ($N \times 1$)
$\boldsymbol{\varphi}$	(unrestricted) phase vector ($N \times 1$)
$\boldsymbol{\varphi}_a$	Atmospheric Phase Delay (APD) vector ($N \times 1$)
$\boldsymbol{\varphi}_d$	phase vector associated with the deformation ($N \times 1$)
$\boldsymbol{\varphi}_w$	phase vector associated with the noise, including the phase of the scatterer ($N \times 1$)
$\boldsymbol{w}$	complex data noise vector ($N \times 1$)
$\boldsymbol{\xi}_n$	frequencies vector for the n-th antenna, (2×1) in the case of elevation and velocity
$\boldsymbol{y}$	complex data vector
χ	cross-correlation index between two focused and registered images
χ_p	ambiguity function for the pulse $p(t)$
d_p	degree of polarization
δr	range variation between the two acquisitions (range parallax)
δr_d	deformation (displacement) component of δr
δr_n	range variation at the n-th acquisition
δr_T	topographic component of δr
$\Delta \boldsymbol{\varphi}$	(measured) interferometric phase vector ($M \times 1$)
γ_{t_n}	scene reflectivity at the acquisition time t_n
$\gamma(s)$	backscattering along the elevation (slant height) in the 3D model
$\gamma(s, v)$	backscattering along the elevation(slant height)/velocity domain in the 4D model
$\gamma(x, r)$	radar reflectivity pattern of the scene along the azimuth and (slant) range
γ_m	m-th sample (on the tomographic domain) of the backscattering coefficient
$\gamma_{p,q}$	polarimetric radar reflectivity for the p-th polarization on transmit and q-th polarization on receive
$\hat{\boldsymbol{r}}$	line of sight (los) versor
L_A	real antenna azimuth length
λ	transmitted wavelength
$\mathcal{B}_N$	equivalent noise bandwidth
$\mathcal{E}_p$	pulse energy
$\mathcal{E}_T$	transmitted energy
$\mathcal{G}$	amplifier gain
$\mathcal{P}_p$	transmitted peak power
ϕ_{DC}	off-broadside (aspect) angle corresponding to the Doppler Centroid
ρ	coherence
ρ_A, ρ_a	azimuth resolution before and after compression, respectively
ρ_R, ρ_r	range resolution before and after compression, respectively
ρ_s	Rayleigh resolution in the slant height (elevation) direction
ρ_t	temporal coherence
ρ_v	Rayleigh resolution in the velocity direction
ρ_z	Rayleigh resolution in the height (vertical) direction
σ	Radar Cross Section (RCS)
σ_0	Normalized Radar Cross Section (NRCS)
τ	echo round trip (i.e. total) delay
τ_c	echo round trip (i.e. total) delay of the reference point, typically corresponding to the range beam center r_0
τ_{min}, τ_{max}	near and far range echo delay

τ_{SS}	echo round trip (i.e. total) delay corresponding to the Start-Stop approximation
θ	off nadir (look angle)
θ_0	off nadir (look angle) at the reference point, typically the scene mid range
φ	unrestricted interferometric phase
φ_a	atmospheric phase delay (APD) component of φ
φ_n	(unrestricted) phase corresponding to the n-th acquisition (n-th element of $\boldsymbol{\varphi}$)
φ_w	noise component associated with φ
$\varphi_{\gamma n}$	phase contribution of the scatterer at the n-th antenna
φ_{a_n}	atmospheric phase delay (APD) component at the n-th antenna (n-th element of $\boldsymbol{\varphi}_a$)
φ_{w_n}	phase noise component at the n-th antenna (n-th element of $\boldsymbol{\varphi}_n$)
$b_\parallel$	parallel baseline length
$b_\perp, b$	orthogonal baseline length
B_d	Doppler bandwidth [Hz]
b_n	orthogonal baseline length at the n-th acquisition (n-th element of $\boldsymbol{b}$)
B_r	range spatial bandwidth [m^{-1}]
B_t	transmitted bandwidth [Hz]
B_x	azimuth spatial bandwidth [m^{-1}]
b_{cr}	critical baseline
c	speed of the light
f_0	carrier frequency
f_d	Doppler frequency
f_r	range spatial frequency [m^{-1}]
f_x	azimuth spatial frequency [m^{-1}]
f_{DC}	Doppler Centroid
G_T, G_R	transmitting and receiving antenna gain
$h(x', r')$	azimuth and range postfocusing Point Spread Function (PSF)
L	number of looks
M	number of interferograms; number of bins in the tomographic domain
N	number of acquisitions
N_P	number of polarimetric channels ($N_P = 1$ for the single polarization case).
$p(\cdot)$	transmitted pulse
P_N	noise power
P_T, P_{BS}, P_R	transmitted, backscattered, and received power
$R(t)$	time dependent sensor-target distance
r, r'	target and sensor range coordinates, respectively
R_0	azimuth variant range of the scene center along the illumination interval (scene center range migration)
r_n	range at the n-th acquisition
s	slant height (elevation) target coordinate, absolute or residual (with respect to a Digital Elevation Model known a priori) depending on the context
S_i, S_r	incidence and received power density
S_T	span of the spatial baselines along the elevation (orthogonal baseline span)
S_{pq}	scattering coefficient for the p-th polarization on transmit and q-th polarization on receive
T	transmitted pulse duration; detection threshold

T_e	noise equivalent temperature
T_I	pulse repetition interval (PRI)
t_n	n-th time corresponding to the n-th acquisition; time of the transmission of the n-th pulse
T_p	transmitted burst duration
T_s	span of the temporal baselines
T_{int}	Integration time
v	deformation mean velocity
v_r	radial (i.e. los) velocity component
w_n	complex noise contribution at the n-th antenna (n-th element of w)
X	azimuth swath size
x, x'	target and sensor azimuth coordinates, respectively
x_{DC}	azimuth displacement of the beam corresponding to the Doppler Centroid
Y	(ground) range swath size
$y(\cdot)$	received (raw) signal; postfocused signal after registration and calibration
y_n	focused and registered complex data at the n-th antenna; focused, registered, and calibrated complex data at the n-th antenna
Y_r	slant range swath size
$z_{2\pi}$	ambiguity height
$\phi_A; \phi_a$	azimuth aperture before and after the focusing, respectively

Bibliography

[1] N. Levanon, Radar Principles, A Wiley-Interscience Publication, Wiley, 1988.

[2] C.F. Bohren, D.R. Huffma, Absorption and Scattering of Light by Small Particles, Wiley-Interscience, New York, 1983.

[3] A.K. Gabriel, R.M. Goldstein, H.A. Zebker, Mapping small elevation changes over large areas: differential radar interferometry, Journal of Geophysical Research. Solid Earth 94 (B7) (1989) 9183–9191.

[4] D. Massonnet, M. Rossi, C. Carmona, F. Adragna, G. Peltzer, K. Feigl, T. Rabaute, The displacement field of the landers earthquake mapped by radar interferometry, Nature 364 (1993) 138–142.

[5] M. Richards, W. Holm, J. Scheer, Principles of Modern Radar: Basic Principles, Electromagnetics and Radar, Institution of Engineering and Technology, 2010, [Online]. Available: https://books.google.it/books?id=AWh3vwEACAAJ.

[6] M. Richards, Fundamentals of Radar Signal Processing, second edition, McGraw-Hill Education, 2014, [Online]. Available: https://books.google.it/books?id=qGZCAQAACAAJ.

[7] J.G. Proakis, M. Salehi, N. Zhou, X. Li, Communication Systems Engineering, vol. 2, Prentice Hall, New Jersey, 1994.

[8] A. Papoulis, Systems and Transforms with Applications in Optics, McGraw-Hill, New York, 1968.

[9] A. Meta, P. Hoogeboom, L.P. Ligthart, Signal processing for FMCW SAR, IEEE Transactions on Geoscience and Remote Sensing 45 (11) (2007) 3519–3532.

[10] D. Wehner, High-Resolution Radar, Artech House, 1995.

[11] L. Tsang, J.A. Kong, R.T. Shin, Theory of Microwave Remote Sensing, Wiley-Interscience, 1985.

[12] J.A. Kong, Electromagnetic Wave Theory, EMW Publishing, Cambridge, MA, 2005.

[13] W.-M. Boerner, Basics of SAR Polarimetry I, 02 2007, p. 41.

[14] P. Imperatore, SAR imaging distortions induced by topography: a compact analytical formulation for radiometric calibration, Remote Sensing 13 (16) (2021), [Online]. Available: https://www.mdpi.com/2072-4292/13/16/3318.

[15] S. Cloude, E. Pottier, An entropy based classification scheme for land applications of polarimetric SAR, IEEE Transactions on Geoscience and Remote Sensing 35 (1) (1997) 68–78.

[16] Y. Cui, Y. Yamaguchi, J. Yang, H. Kobayashi, S.-E. Park, G. Singh, On complete model-based decomposition of polarimetric SAR coherency matrix data, IEEE Transactions on Geoscience and Remote Sensing 52 (4) (2014) 1991–2001.

[17] G. Singh, S. Mohanty, Y. Yamazaki, Y. Yamaguchi, Physical scattering interpretation of POLSAR coherency matrix by using compound scattering phenomenon, IEEE Transactions on Geoscience and Remote Sensing 58 (4) (2020) 2541–2556.

[18] A. Ishimaru, Wave Propagation and Scattering in Random Media, Academic Press, New York, 1978.

[19] P. Beckman, A. Spizzichino, The Scattering of Electromagnetic Waves from Rough Surfaces, Artech House Publishers, 1987.

[20] F. Ulaby, R. Moore, A.K. Fung, Microwave Remote Sensing: Active and Passive. Radar Remote Sensing and Surface Scattering and Emission Theory, vol. 2, Artech House, 1985.

[21] A.K. Fung, Microwave Scattering and Emission Models and Their Applications, Artech House, 1994.

[22] G. Franceschetti, M. Migliaccio, D. Riccio, G. Schirinzi, SARAS: a synthetic aperture radar (SAR) raw signal simulator, IEEE Transactions on Geoscience and Remote Sensing 30 (1) (1992) 110–123.

[23] S.O. Rice, Reflection of electromagnetic waves by slightly rough surfaces, in: M. Kline (Ed.), The Theory of Electromagnetic Waves, Wiley, New York, 1951.

[24] C. Elachi, J. van Zyl, Introduction to the Physics and Techniques of Remote Sensing, Wiley, 2006.

[25] D.X. Yue, F. Xu, A.C. Frery, Y.Q. Jin, Synthetic aperture radar image statistical modeling: part one-single-pixel statistical models, IEEE Geoscience and Remote Sensing Magazine 9 (1) (2021) 82–114.

[26] A. Papoulis, S.U. Pillai, Probability, Random Variables, and Stochastic Processes, Tata McGraw-Hill Education, 2002.

[27] O. Frey, C.L. Werner, R. Coscione, Car-borne and UAV-borne mobile mapping of surface displacements with a compact repeat-pass interferometric SAR system at l-band, in: IGARSS 2019 - 2019 IEEE International Geoscience and Remote Sensing Symposium, 2019, pp. 274–277.

[28] A. Moreira, P. Prats-Iraola, M. Younis, G. Krieger, I. Hajnsek, K.P. Papathanassiou, A tutorial on synthetic aperture radar, IEEE Geoscience and Remote Sensing Magazine 1 (1) (2013) 6–43.

[29] M. Crosetto, O. Monserrat, M. Cuevas-González, N. Devanthéry, B. Crippa, Persistent scatterer interferometry: a review, in: Theme Issue 'State-of-the-art in photogrammetry, remote sensing and spatial information science', ISPRS Journal of Photogrammetry and Remote Sensing 115 (2016) 78–89, [Online]. Available: https://www.sciencedirect.com/science/article/pii/S0924271615002415.

[30] C.A. Wiley, Synthetic aperture radars, IEEE Transactions on Aerospace and Electronic Systems 3 (1985) 440–443.

[31] A. Love, In memory of Carl A. Wiley, IEEE Antennas and Propagation Society Newsletter 27 (3) (1985) 17–18.

[32] M. Soumekh, Reconnaissance with ultra wideband UHF synthetic aperture radar, IEEE Signal Processing Magazine 12 (4) (1995) 21–40.

[33] N. Gebert, G. Krieger, A. Moreira, Digital beamforming on receive: techniques and optimization strategies for high-resolution wide-swath SAR imaging, IEEE Transactions on Aerospace and Electronic Systems 45 (2) (2009) 564–592.

[34] G. Krieger, MIMO-SAR: opportunities and pitfalls, IEEE Transactions on Geoscience and Remote Sensing 52 (5) (2014) 2628–2645.

[35] G. Fornaro, E. Sansosti, R. Lanari, M. Tesauro, Role of processing geometry in SAR raw data focusing, IEEE Transactions on Aerospace and Electronic Systems 38 (2) (2002) 441–454.

[36] A. Moreira, J. Mittermayer, R. Scheiber, Extended chirp scaling algorithm for air- and spaceborne SAR data processing in stripmap and ScanSAR imaging modes, IEEE Transactions on Geoscience and Remote Sensing 34 (5) (1996) 1123–1136.

[37] P. Prats, R. Scheiber, J. Mittermayer, A. Meta, A. Moreira, Processing of sliding spotlight and tops SAR data using baseband azimuth scaling, IEEE Transactions on Geoscience and Remote Sensing 48 (2) (2010) 770–780.

[38] A. Gersho, Characterization of time-varying linear systems, Proceedings of the IEEE 51 (1) (1963) 238.

[39] L. Zadeh, Frequency analysis of variable networks, Proceedings of the IRE 38 (3) (1950) 291–299.

[40] W. Kozek, On the transfer function calculus for underspread LTV channels, IEEE Transactions on Signal Processing 45 (1) (1997) 219–223.

[41] R. Bamler, A comparison of range-Doppler and wavenumber domain SAR focusing algorithms, IEEE Transactions on Geoscience and Remote Sensing 30 (4) (1992) 706–713.

[42] R.H. Stolt, Migration by Fourier transform, Geophysics 43 (1) (1978).

[43] G. Franceschetti, R. Lanari, V. Pascazio, G. Schirinzi, WASAR: a wide-angle SAR processor, IEE Proceedings. Part F. Radar and Signal Processing 139 (2) (1992) 107–114.

[44] F. Rocca, Synthetic Aperture Radar: a New Application for Wave Equation Techniques, Stanford Exploration Project Report, vol. SER-56, 1987, p. 167.

[45] R.K. Raney, H. Runge, R. Bamler, I.G. Cumming, F.H. Wong, Precision SAR processing using chirp scaling, IEEE Transactions on Geoscience and Remote Sensing 32 (4) (1994) 786–799.

[46] G. Franceschetti, R. Lanari, Synthetic Aperture Radar Processing, CRC Press, 1999.

[47] C. Pickover, The Math Book: From Pythagoras to the 57th Dimension, 250 Milestones in the History of Mathematics, Sterling Milestones Series, Sterling, 2009, [Online]. Available: https://books.google.it/books?id=JrslMKTgSZwC.

[48] C.G. Brown, K. Sarabandi, L.E. Pierce, Validation of the Shuttle Radar Topography Mission height data, IEEE Transactions on Geoscience and Remote Sensing 43 (8) (Aug. 2005) 1707–1715.

[49] B. Wessel, M. Huber, C. Wohlfart, U. Marschalk, D. Kosmann, A. Roth, Accuracy assessment of the global tandem-x digital elevation model with GPS data, ISPRS Journal of Photogrammetry and Remote Sensing 139 (2018) 171–182, [Online]. Available: https://www.sciencedirect.com/science/article/pii/S0924271618300522.

[50] M. Costantini, A novel phase unwrapping method based on network programming, IEEE Transactions on Geoscience and Remote Sensing 36 (3) (1998) 813–821.

[51] D.C. Ghiglia, M.D. Pritt, Two-Dimensional Phase Unwrapping, Theory, Algorithms, and Software, Wiley-Interscience, New York (USA), 1998.

[52] C.W. Chen, H.A. Zebker, Two-dimensional phase unwrapping with use of statistical models for cost functions in nonlinear optimization, Journal of the Optical Society of America. A 18 (2) (2001) 338–351.

[53] G. Fornaro, V. Pascazio, Chapter 20 - SAR interferometry and tomography: theory and applications, in: N.D. Sidiropoulos, F. Gini, R. Chellappa, S. Theodoridis (Eds.), Academic Press Library in Signal Processing: Volume 2, in: Academic Press Library in Signal Processing, Elsevier, 2014, pp. 1043–1117.

[54] G. Franceschetti, G. Fornaro, Chapter 4 - synthetic aperture radar interferometry, in: G. Franceschetti, R. Lanari (Eds.), Synthetic Aperture Radar Processing, CRC Press, Boca Raton, 1999, pp. 167–223.

[55] R.F. Hanssen, Radar Interferometry - Data Interpretation and Error Analysis, Kluwer Academic Publishers, 2001.

[56] B. Puysségur, R. Michel, J.-P. Avouac, Tropospheric phase delay in interferometric synthetic aperture radar estimated from meteorological model and multispectral imagery, Journal of Geophysical Research. Solid Earth 112 (B5) (2007).

[57] M.-P. Doin, C. Lasserre, G. Peltzer, O. Cavalié, C. Doubre, Corrections of stratified tropospheric delays in SAR interferometry: validation with global atmospheric models, Journal of Applied Geophysics 69 (1) (2009) 35–50.

[58] P. Mateus, G. Nico, R. Tome, J. Catalao, P.M.A. Miranda, Experimental study on the atmospheric delay based on GPS, SAR interferometry, and numerical weather model data, IEEE Transactions on Geoscience and Remote Sensing 51 (1) (2013) 6–11.

[59] G. Fornaro, N. D'Agostino, R. Giuliani, C. Noviello, D. Reale, S. Verde, Assimilation of GPS-derived atmospheric propagation delay in DInSAR data processing, IEEE Journal of Selected Topics in Applied Earth Observations and Remote Sensing 8 (2) (2015) 784–799.

[60] F. Chaabane, A. Avallone, F. Tupin, P. Briole, H. Maitre, A multitemporal method for correction of tropospheric effects in differential SAR interferometry: application to the gulf of Corinth earthquake, IEEE Transactions on Geoscience and Remote Sensing 45 (6) (2007) 1605–1615.

[61] S. Frasier, A. Camps, Dual-beam interferometry for ocean surface current vector mapping, IEEE Transactions on Geoscience and Remote Sensing 39 (2) (2001) 401–414.

[62] A. Budillon, C.H. Gierull, V. Pascazio, G. Schirinzi, Along-track interferometric SAR systems for ground-moving target indication: achievements, potentials, and outlook, IEEE Geoscience and Remote Sensing Magazine 8 (2) (2020) 46–63.

[63] W.B.J. Davenport, W.L. Root, An Introduction to the Theory of Random Signals and Noise, IEEE Press, New York, 1987.

[64] F. Gatelli, A.M.M. Guarnieri, F. Parizzi, P. Pasquali, C. Prati, F. Rocca, The wavenumber shift in SAR interferometry, IEEE Transactions on Geoscience and Remote Sensing 32 (4) (1994) 855–865.

[65] A.M. Guarnieri, F. Rocca, Combination of low- and high-resolution SAR images for differential interferometry, IEEE Transactions on Geoscience and Remote Sensing 37 (4) (1999) 2035–2049.

[66] G.W. Davidson, R. Bamler, Multiresolution phase unwrapping for SAR interferometry, IEEE Transactions on Geoscience and Remote Sensing 37 (1) (1999) 163–174.

[67] G. Fornaro, A.M. Guarnieri, Minimum mean square error space-varying filtering of interferometric SAR data, IEEE Transactions on Geoscience and Remote Sensing 40 (1) (2002) 11–21.

[68] R. Bamler, P. Hartl, Synthetic aperture radar interferometry, Inverse Problems 14 (4) (1998) R1.

[69] J.-S. Lee, K. Hoppel, S. Mango, A. Miller, Intensity and phase statistics of multilook polarimetric and interferometric SAR imagery, IEEE Transactions on Geoscience and Remote Sensing 32 (5) (1994) 1017–1028.

[70] J.-S. Lee, K. Papathanassiou, T. Ainsworth, M. Grunes, A. Reigber, A new technique for noise filtering of SAR interferometric phase images, IEEE Transactions on Geoscience and Remote Sensing 36 (5) (1998) 1456–1465.

[71] C.-A. Deledalle, L. Denis, G. Poggi, F. Tupin, L. Verdoliva, Exploiting patch similarity for SAR image processing: the nonlocal paradigm, IEEE Signal Processing Magazine 31 (4) (2014) 69–78.

[72] C.-A. Deledalle, L. Denis, F. Tupin, A. Reigber, M. Jäger, NL-SAR: a unified nonlocal framework for resolution-preserving (pol) (in)SAR denoising, IEEE Transactions on Geoscience and Remote Sensing 53 (4) (2015) 2021–2038.

[73] E. Sansosti, P. Berardino, M. Manunta, F. Serafino, G. Fornaro, Geometrical SAR image registration, IEEE Transactions on Geoscience and Remote Sensing 44 (10) (2006) 2861–2870.

[74] R. Hanssen, R. Bamler, Evaluation of interpolation kernels for SAR interferometry, IEEE Transactions on Geoscience and Remote Sensing 37 (1) (1999) 318–321.

[75] K. Itoh, Analysis of the phase unwrapping algorithm, Applied Optics 21 (14) (1982) 2470.

[76] B. Delaunay, Sur la sphère vide, Bulletin de l'Académie des Sciences de l'URSS. Classe des sciences mathématiques et na 1934 (6) (1934) 793–800.

[77] G. Fornaro, G. Franceschetti, R. Lanari, E. Sansosti, M. Tesauro, Global and local phase-unwrapping techniques: a comparison, Journal of the Optical Society of America. A 14 (10) (Oct 1997) 2702–2708, [Online]. Available: https://opg.optica.org/josaa/abstract.cfm?URI=josaa-14-10-2702.

[78] T.J. Flynn, Two-dimensional phase unwrapping with minimum weighted discontinuity, Journal of

the Optical Society of America. A 14 (10) (Oct 1997) 2692–2701, [Online]. Available: https://opg.optica.org/josaa/abstract.cfm?URI=josaa-14-10-2692.

[79] D.C. Ghiglia, L.A. Romero, Minimum Lp-norm two-dimensional phase unwrapping, Journal of the Optical Society of America. A 13 (10) (Oct 1996) 1999–2013, [Online]. Available: https://opg.optica.org/josaa/abstract.cfm?URI=josaa-13-10-1999.

[80] G. Fornaro, A. Pauciullo, D. Reale, A null-space method for the phase unwrapping of multitemporal SAR interferometric stacks, IEEE Transactions on Geoscience and Remote Sensing 49 (6) (2011) 2323–2334.

[81] P. Berardino, G. Fornaro, R. Lanari, E. Sansosti, A new algorithm for surface deformation monitoring based on small baseline differential SAR interferograms, IEEE Transactions on Geoscience and Remote Sensing 40 (11) (2002) 2375–2383.

[82] S. Usai, A least squares database approach for SAR interferometric data, IEEE Transactions on Geoscience and Remote Sensing 41 (4) (2003) 753–760.

[83] T. Greville, Note on the generalized inverse of a matrix product, SIAM Review 8 (4) (1966) 518–521.

[84] A. Ferretti, C. Prati, F. Rocca, Permanent scatterers in SAR interferometry, IEEE Transactions on Geoscience and Remote Sensing 39 (1) (2001) 8–20.

[85] A. Ferretti, C. Prati, F. Rocca, Nonlinear subsidence rate estimation using permanent scatterers in differential SAR interferometry, IEEE Transactions on Geoscience and Remote Sensing 38 (5) (2000) 2202–2212.

[86] A. Ferretti, A. Fumagalli, F. Novali, C. Prati, F. Rocca, A. Rucci, A new algorithm for processing interferometric data-stacks: SqueeSAR, IEEE Transactions on Geoscience and Remote Sensing 49 (9) (2011) 3460–3470.

[87] A.M. Guarnieri, S. Tebaldini, On the exploitation of target statistics for SAR interferometry applications, IEEE Transactions on Geoscience and Remote Sensing 46 (11) (2008) 3436–3443.

[88] G. Fornaro, S. Verde, D. Reale, A. Pauciullo, Caesar: an approach based on covariance matrix decomposition to improve multibaseline–multitemporal interferometric SAR processing, IEEE Transactions on Geoscience and Remote Sensing 53 (4) (2015) 2050–2065.

[89] G. Fornaro, A. Pauciullo, D. Reale, S. Verde, Multilook SAR tomography for 3-d reconstruction and monitoring of single structures applied to COSMO-SKYMED data, IEEE Journal of Selected Topics in Applied Earth Observations and Remote Sensing 7 (7) (2014) 2776–2785.

[90] G.H. Golub, C.F. Van Loan, Matrix Computations, JHU Press, 2013.

[91] S. Verde, D. Reale, A. Pauciullo, G. Fornaro, Improved small baseline processing by means of CAESAR eigen-interferograms decomposition, ISPRS Journal of Photogrammetry and Remote Sensing 139 (2018) 1–13, [Online]. Available: https://www.sciencedirect.com/science/article/pii/S0924271618300546.

[92] A. Pepe, R. Lanari, On the extension of the minimum cost flow algorithm for phase unwrapping of multitemporal differential SAR interferograms, IEEE Transactions on Geoscience and Remote Sensing 44 (9) (2006) 2374–2383.

[93] G. Strang, Introduction to Linear Algebra, vol. 3, Wellesley-Cambridge Press, Wellesley, MA, 1993.

[94] F. Marvasti, Nonuniform Sampling: Theory and Practice, vol. 1, Springer, 01 2001.

[95] F. Meglio, G. Panariello, G. Schirinzi, Three dimensional SAR image focusing from non-uniform samples, in: 2007 IEEE International Geoscience and Remote Sensing Symposium, 2007, pp. 528–531.

[96] C. Rambour, A. Budillon, A.C. Johnsy, L. Denis, F. Tupin, G. Schirinzi, From interferometric to tomographic SAR: a review of synthetic aperture radar tomography-processing techniques for scatterer unmixing in urban areas, IEEE Geoscience and Remote Sensing Magazine 8 (2) (2020) 6–29.

[97] E. Sansosti, A simple and exact solution for the interferometric and stereo SAR geolocation problem, IEEE Transactions on Geoscience and Remote Sensing 42 (8) (2004) 1625–1634.

[98] J. Capon, High-resolution frequency-wavenumber spectrum analysis, Proceedings of the IEEE 57 (8) (1969) 1408–1418.

[99] S.S. Haykin, Adaptive Filter Theory, Prentice Hall Information and System Science Series, Prentice-Hall Inc., Upper Saddle River, NJ, 1991.

[100] F. Gini, F. Lombardini, M. Montanari, Layover solution in multibaseline SAR interferometry, IEEE Transactions on Aerospace and Electronic Systems 38 (4) (2002) 1344–1356.

[101] J. Li, P. Stoica, Z. Wang, On robust capon beamforming and diagonal loading, IEEE Transactions on Signal Processing 51 (7) (2003) 1702–1715.

[102] A. Budillon, A. Evangelista, G. Schirinzi, Three-dimensional SAR focusing from multipass signals using compressive sampling, IEEE

Transactions on Geoscience and Remote Sensing 49 (1) (2011) 488–499.

[103] E. Aguilera, M. Nannini, A. Reigber, Wavelet-based compressed sensing for SAR tomography of forested areas, IEEE Transactions on Geoscience and Remote Sensing 51 (12) (2013) 5283–5295.

[104] D. Donoho, Compressed sensing, IEEE Transactions on Information Theory 52 (4) (2006) 1289–1306.

[105] E.J. Candes, M.B. Wakin, An introduction to compressive sampling, IEEE Signal Processing Magazine 25 (2) (2008) 21–30.

[106] J.A. Tropp, S.J. Wright, Computational methods for sparse solution of linear inverse problems, Proceedings of the IEEE 98 (6) (2010) 948–958.

[107] S.S. Chen, D.L. Donoho, M.A. Saunders, Atomic decomposition by basis pursuit, SIAM Review 43 (1) (2001) 129–159, [Online]. Available: http://dblp.uni-trier.de/db/journals/siamrev/siamrev43.html#ChenDS01.

[108] E. Candes, T. Tao, The Dantzig selector: statistical estimation when p is much larger than n, The Annals of Statistics 35 (6) (2007) 2313–2351, https://doi.org/10.1214/009053606000001523, [Online].

[109] R. Tibshirani, Regression shrinkage and selection via the lasso, Journal of the Royal Statistical Society, Series B, Methodological 58 (1) (1996) 267–288.

[110] L.I. Rudin, S. Osher, E. Fatemi, Nonlinear total variation based noise removal algorithms, Physica D. Nonlinear Phenomena 60 (1) (1992) 259–268, [Online]. Available: https://www.sciencedirect.com/science/article/pii/016727889290242F.

[111] A. Pauciullo, D. Reale, W. Franzé, G. Fornaro, Multi-look in GLRT-based detection of single and double persistent scatterers, IEEE Transactions on Geoscience and Remote Sensing 56 (9) (2018) 5125–5137.

[112] A. De Maio, G. Fornaro, A. Pauciullo, Detection of single scatterers in multidimensional SAR imaging, IEEE Transactions on Geoscience and Remote Sensing 47 (7) (Jul. 2009) 2284–2297.

[113] A. Pauciullo, D. Reale, A. De Maio, G. Fornaro, Detection of double scatterers in SAR tomography, IEEE Transactions on Geoscience and Remote Sensing 50 (9) (2012) 3567–3586.

[114] A. Budillon, G. Schirinzi, GLRT based on support estimation for multiple scatterers detection in SAR tomography, IEEE Journal of Selected Topics in Applied Earth Observations and Remote Sensing 9 (3) (2016) 1086–1094.

[115] S. Verde, A. Pauciullo, D. Reale, G. Fornaro, Multiresolution detection of persistent scatterers: a performance comparison between multilook GLRT and CAESAR, IEEE Transactions on Geoscience and Remote Sensing 59 (4) (2021) 3088–3103.

[116] G. Fornaro, D. Reale, S. Verde, Bridge thermal dilation monitoring with millimeter sensitivity via multidimensional SAR imaging, IEEE Geoscience and Remote Sensing Letters 10 (4) (july 2013) 677–681.

[117] D. Reale, G. Fornaro, A. Pauciullo, Extension of 4-d SAR imaging to the monitoring of thermally dilating scatterers, IEEE Transactions on Geoscience and Remote Sensing 51 (12) (2013) 5296–5306.

[118] S.R. Cloude, E. Pottier, A review of target decomposition theorems in radar polarimetry, IEEE Transactions on Geoscience and Remote Sensing 34 (2) (1996) 498–518.

[119] S. Sauer, L. Ferro-Famil, A. Reigber, E. Pottier, Three-dimensional imaging and scattering mechanism estimation over urban scenes using dual-baseline polarimetric InSAR observations at l-band, IEEE Transactions on Geoscience and Remote Sensing 49 (11) (Nov 2011) 4616–4629.

[120] S. Tebaldini, Algebraic synthesis of forest scenarios from multibaseline PolInSAR data, IEEE Transactions on Geoscience and Remote Sensing 47 (12) (Dec 2009) 4132–4142.

[121] M. Pardini, K. Papathanassiou, On the estimation of ground and volume polarimetric covariances in forest scenarios with SAR tomography, IEEE Geoscience and Remote Sensing Letters 14 (10) (2017) 1860–1864.

[122] K.P. Papathanassiou, S.R. Cloude, Single-baseline polarimetric SAR interferometry, IEEE Transactions on Geoscience and Remote Sensing 39 (11) (Nov 2001) 2352–2363.

[123] M. Neumann, L. Ferro-Famil, A. Reigber, Estimation of forest structure, ground, and canopy layer characteristics from multibaseline polarimetric interferometric SAR data, IEEE Transactions on Geoscience and Remote Sensing 48 (3) (March 2010) 1086–1104.

[124] H. Aghababaee, M.R. Sahebi, Model-based target scattering decomposition of polarimetric SAR tomography, IEEE Transactions on Geoscience and Remote Sensing 56 (2) (Feb 2018) 972–983.

[125] C.F. Van Loan, N. Pitsianis, Approximation with Kronecker products, in: M.S. Moonen, G.H. Golub, B.L.R. De Moor (Eds.), Linear Algebra for Large Scale and Real-Time Applications, in: NATO ASI Series (Series E: Applied Sciences), vol. 232, 1993.

[126] H.L. Van Trees, Optimum Array Processing: Part IV of Detection, Estimation, and Modulation Theory, John Wiley & Sons, 2002.

Index